# Lecture Notes in Geoinformation and Cartography

## Publications of the International Cartographic Association (ICA)

More information about this subseries at http://www.springer.com/series/10036

Alexander James Kent · Soetkin Vervust ·
Imre Josef Demhardt · Nick Millea
Editors

# Mapping Empires: Colonial Cartographies of Land and Sea

7th International Symposium of the ICA
Commission on the History of Cartography,
2018

 Springer

*Editors*
Alexander James Kent
School of Human and Life Sciences
Canterbury Christ Church University
Canterbury, Kent, UK

Imre Josef Demhardt
Department of History
University of Texas at Arlington
Arlington, TX, USA

Soetkin Vervust
Department of Art Sciences
and Archaeology
Vrije Universiteit Brussel
Brussels, Belgium

Nick Millea
Bodleian Library
University of Oxford
Oxford, Oxfordshire, UK

ISSN 1863-2246          ISSN 1863-2351  (electronic)
Lecture Notes in Geoinformation and Cartography
ISSN 2195-1705          ISSN 2195-1713  (electronic)
Publications of the International Cartographic Association (ICA)
ISBN 978-3-030-23446-1          ISBN 978-3-030-23447-8  (eBook)
https://doi.org/10.1007/978-3-030-23447-8

This Springer imprint is published by the registered company Springer Nature Switzerland AG
The registered company address is: Gewerbestrasse 11, 6330 Cham, Switzerland

# Preface

This volume comprises a selection of research papers that were presented at the 7th International Symposium of the ICA Commission on the History of Cartography, which took place in Oxford, UK, from 13 to 15 September 2018. It is the fifth volume in a series of proceedings which has been made possible through the partnership between the International Cartographic Association (ICA) and Springer International Publishing.

The 7th International Symposium of the ICA Commission on the History of Cartography adopted the general theme of 'Mapping Empires: Colonial Cartographies of Land and Sea'. The event was jointly organized between the ICA Commission on the History of Cartography, the ICA Commission on Topographic Mapping and the Bodleian Libraries of the University of Oxford. All paper and poster sessions were held at the Bodleian's Weston Library that is situated in the heart of Oxford.

The chapters in this volume discuss the colonial mapping of various regions around the world (the Far East, the Middle East, India, Africa and the Americas), as influenced by cosmopolitan exploration and imperialistic activity during, but not limited to, the 'long nineteenth century' (mid-eighteenth to mid-twentieth centuries). They also provide a focus on some mapmakers of the period and on issues surrounding boundary mapping and toponymy (place names). The rise of European hegemony, which formed the overall historical and geopolitical context of the theme of the Symposium, coincided with a scientific turn that underpinned the evolution of topographic mapping and hydrographic charting. The colonial cartographies examined in the Symposium and discussed in this volume brought forth a rich legacy of mapping that continues to influence the aesthetics and authority of mapmaking today.

The Symposium could not have been possible without the generous support and assistance of staff at the Bodleian Libraries and at Canterbury Christ Church University. In particular, we extend our thanks to Richard Ovenden, Bodley's

Librarian; to Stuart Ackland, Debbie Hall, Peter Hawksworth and Peter Jolly of the Map Room; and to John Hills and Kathryn Roberts of the School of Human and Life Sciences at Canterbury Christ Church University.

Canterbury, UK                                               Alexander James Kent
Brussels, Belgium                                                  Soetkin Vervust
Arlington, TX, USA                                             Imre Josef Demhardt
Oxford, UK                                                          Nick Millea

# Contents

**The Far East**

**Sketching Layers in Japan: Mineral Wealth, Geo-bodies and Imperial Territory** . . . . . . . . . . . . . . . . . . . . . . . . . . . . . . . . . . . . . . . . . . . 3
Edward Boyle

**Putting America's First Empire on the Map: American Early Efforts to Map the Philippine Islands** . . . . . . . . . . . . . . . . . . . . . . . . . . . . . 23
Eric H. Losang

**The Exploration and Survey of the Outlying Islands of the Dutch East Indies** . . . . . . . . . . . . . . . . . . . . . . . . . . . . . . . . . . . . . . . . . . 37
Ferjan J. Ormeling

**A View from Inside: Chinese Mapping of the World Against the Backdrop of Colonial Experience** . . . . . . . . . . . . . . . . . . . . . . . . . . 61
Laura Pflug

**The Middle East and India**

**French Cartographic Services in the Levant: Putting Syria and Lebanon on the Map of the Empire** . . . . . . . . . . . . . . . . . . . . . . . 77
Louis Le Douarin

**Surveying Empires: Archaeologies of Colonial Cartography and the Great Trigonometrical Survey of India** . . . . . . . . . . . . . . . . . . 101
Keith D. Lilley

**War Cartography in the Survey of India, 1920–1946** . . . . . . . . . . . . . . 121
Oyndrila Sarkar

## Mapping the World

**Red Star to Red Lion: The Soviet Military Mapping of Oxford** ...... 143
John Davies and Alexander James Kent

**Maps Against Imperialism: Frank Horrabin and Alexander
Radó's Atlases in the Interwar Period** ........................... 159
Gilles Palsky

**Empire as Spectacle:** *Harmsworth's Atlas of the World and Pictorial
Gazetteer with an Atlas of the Great War* ........................ 177
Peter Vujakovic

## Mapping Boundaries

**Mapping Changes in Ottoman-Austrian Borders During
the Eighteenth Century** ........................................ 197
Uğur Kurtaran

**Lines on the Map: International Boundaries** ..................... 207
Rose Mitchell

## Toponyms

**German Names in the Kilimanjaro Region** ...................... 229
Wolfgang Crom

**The French Map of Beirut (1936)** ............................. 247
Jack Keilo

## Mapmakers

**Military or Missionary Map? The First Topographic Map of Northern
New Spain (1725–1729)** ........................................ 263
Mirela Altić

**'Dead on Arrival': The Unused Cartographic Legacy of Carl
Friedrich Reimer** .............................................. 287
Jeroen Bos

**Head-Hunters, Cannibals and Pirates: Surveying in the 1960s** ........ 309
Roy Wood

# Editors and Contributors

## About the Editors

**Alexander James Kent** is Reader in Cartography and Geographic Information Science at Canterbury Christ Church University (UK), where he lectures on cartographic history and design, GIS, remote sensing and on European and political geography. His research explores the relationship between maps and society, particularly the intercultural aspects of topographic mapping and the aesthetics of cartography. He is also Editor of *The Cartographic Journal*, Immediate Past President of the British Cartographic Society, Chair of the ICA Commission on Topographic Mapping and Chair of the ICA World Cartographic Forum.

**Soetkin Vervust** is Postdoctoral Research Fellow at the Vrije Universiteit Brussel (Belgium) and Newcastle University (UK). Her research interests lie in eighteenth- and nineteenth-century military cartography, the use of digital techniques for the study of old maps and their applicability to historical geography and landscape archaeology. She has served as Executive Secretary of the ICA Commission on the History of Cartography since 2015.

**Imre Josef Demhardt** is interested in post-enlightenment cartography, colonialism and regional studies with a focus on Central Europe, sub-Saharan Africa and North America. Besides numerous articles and several books on these subjects, he is involved as Co-Editor of Vol. 5 (Nineteenth Century) in the encyclopedic project on the *History of Cartography*. He holds the Garrett Chair in the History of Cartography at the University of Texas at Arlington and currently serves as Chair of the ICA Commission on the History of Cartography.

**Nick Millea** has been Map Librarian at the Bodleian Library, University of Oxford, since 1992. He served as Bibliographer for *Imago Mundi* (2005–2010 and 2012–2015). He is also Founding Member and Co-Convenor of The Oxford

Seminars in Cartography (TOSCA). Most recently, he has co-curated the *Talking Maps* exhibit at the Bodleian Library and has written the exhibit's complementary books: *Talking Maps* and *Fifty Maps and the Stories They Tell*.

## Contributors

**Mirela Altić** Institute of Social Sciences, Zagreb, Croatia

**Jeroen Bos** Leiden University Libraries, Leiden, The Netherlands

**Edward Boyle** Kyushu University, Fukuoka, Japan

**Wolfgang Crom** Staatsbibliothek zu Berlin, Berlin, Germany

**John Davies** Woodford Green, UK

**Jack Keilo** Université Paris-Sorbonne, Paris, France

**Alexander James Kent** Canterbury Christ Church University, Canterbury, UK

**Uğur Kurtaran** Karamanoğlu Mehmetbey University, Karaman, Turkey

**Louis Le Douarin** European University Institute, Florence, Italy

**Keith D. Lilley** Queen's University Belfast, Belfast, Northern Ireland, UK

**Eric H. Losang** Leibniz Institute for Regional Geography, Leipzig, Germany

**Rose Mitchell** The National Archives, Richmond, UK

**Ferjan J. Ormeling** University of Amsterdam, Amsterdam, The Netherlands

**Gilles Palsky** University of Paris1 Panthéon-Sorbonne, Paris, France

**Laura Pflug** Leibniz Institute for Regional Geography, Leipzig, Germany

**Oyndrila Sarkar** Department of History, Presidency University, Kolkata, India

**Peter Vujakovic** Canterbury Christ Church University, Canterbury, UK

**Roy Wood** Independent Researcher, Newbury, UK

# The Far East

# Sketching Layers in Japan: Mineral Wealth, Geo-bodies and Imperial Territory

Edward Boyle

**Abstract** In 1876, an American by the name of Benjamin Smith Lyman submitted to the Japanese government a geological map of 'Yesso', which had been compiled under his direction. This map displayed the assumed stratigraphy of Hokkaido, in northern Japan, and is considered the first modern geological map to be produced by an Asian state. This provided a new means of comprehending territory, at exactly the moment the land in question was being re-presented as Hokkaido. The strata exhumed in the course of mapping this land at depth were not limited to those under the Earth. The map was assembled atop a history of Japanese control over the region, one which accounted for the precocious presence of an earlier American survey, conducted under the previous Tokugawa government, which had sought to map mineral deposits in this land of Yesso. These in turn reflected a longer history of mineral extraction, present in the earliest accounts of *Ezo*, and the motivation for Japan to have long 'held the reins' over this amorphous region. The 1876 geological map is a striking example of colonial modernity, through which we are able to observe the institutional mimicry characteristic to, and increasingly emphasized in the study of, late-nineteenth century inter-imperial society. The presence of this map challenges us to recover the various strata atop of which this imperial sociability was able to flourish, and examine the role of the map in incorporating a modernizing Japan within a globally-comprehensible means of territorial authority and control.

## 1 Introduction

On 10 May 1876, an American geologist, Benjamin Smith Lyman, submitted to his employers a striking map he had made showing the mineralogical make-up of a defined portion of the Earth. The map was intended to provide a summary to the work of the Geological Survey of Hokkaido, which for the previous five years had

E. Boyle (✉)
Kyushu University, Fukuoka, Japan
e-mail: tedkboyle@gmail.com

© Springer Nature Switzerland AG 2020
A. J. Kent et al. (eds.), *Mapping Empires: Colonial Cartographies of Land and Sea*,
Lecture Notes in Geoinformation and Cartography,
https://doi.org/10.1007/978-3-030-23447-8_1

3

sought to acquire knowledge of this Japanese island. Entitled the 'Geological Sketch Map of the Island of Yesso', it represents the land which is today known as Hokkaido, the northernmost of the four main islands of Japan (Fig. 1). At the time Lyman was drafting his map, however, its status was not quite so secure: not even a decade had passed since the Japanese state had officially laid claim to the region and granted it the name by which it is now known. In August 1869, Japan's new Meiji government had proclaimed that this region, formerly referred to as Ezo, or barbarian land, would henceforth be known as Hokkaido, the 'North Sea Circuit'. This emphasized that the region was now as much a part of Japan as the other seven Imperial Circuits which had traditionally defined the geography of the Japanese state (Boyle 2016: 69–70).

Lyman's reference to the 'Island of Yesso' in the map's title implicitly draws a distinction between the land of Yesso, or Ezo, and the new administrative geography of Hokkaido that would overlay it in the service and interests of Japan. While the notion of Ezo's geography already existed, with the topography of the region having for several centuries been traced out upon layers and layers of paper, a new administrative entity of Hokkaido would be actively brought into existence by the state through the incorporation and exploitation of Hokkaido's territory as another part of Japan. A month prior to the change in the region's administrative designation, the Meiji government had established the *Kaitakushi*, the administrative body responsible for opening up this land and encouraging its utilization in the interests of the Japanese state.[1] It was to the *Kaitakushi* that Lyman submitted his map, and their employment of geologists like Lyman is indicative of the new state's determination to exploit this territory to the full, not only at its surface, but deep within the bowels of the Earth.

This geological sketch is representative of a new vision for the territory, one made possible by the global circulation of a form of geological knowledge able to be represented on the map. This was knowledge with obvious applicability for the ability of the Meiji state to extract resources from its territory. The 'Geological Sketch Map of the Island of Yesso' showed the state's ambition to exploit its territory to a greater depth than before, with the map crucial for connecting 'the territory with what comes with it' (Wood and Fels 1992: 195). The map offered the Meiji government a valuable means of representing its knowledge regarding the qualities of a portion of its territorial body. This chapter will excavate and bring to light the conditions under which this new means of mapping territory became possible for Japan.

---

[1]There are numerous translations, such as Development Commission, Colonial Office, or Colonization Board, reflecting the name's meaning of 'office for opening the land'. This chapter will continue to refer it as the *Kaitakushi*, which is becoming more common in English-language histories.

**Fig. 1** A Geological Sketch Map of the Island of Yesso, Japan. Drawn at a scale of 1:2,000,000, the map is a coloured lithograph of 1.3 feet long × 0.9 feet wide, or 39 cm × 27 cm. Folds to 21 cm × 14 cm. The map shows the main island of Hokkaido and its surrounding islands. Those off Hokkaido's eastern coast out to Edorop (Etorofu, Iturup in Russian) are today known in Japan as the Northern Territories and are currently disputed with Russia. The map is dense with text in English and Japanese. It lists all the members of the Geological Survey of Hokkaido, and includes geological survey notes, tables showing the amount of coal, and a legend. Published in Tokei (Tokyo) by the *Kaitakushi*. Hokkaido University Northern Studies Collection, map 973 (Courtesy of Hokkaido University Library)

## 2   Representing Authority

### 2.1   Ezo on Maps of Japan

The rhetorical power inherent in graphical representations of a state's territory was brilliantly captured in Thongchai Winichakul's notion of the 'geo-body'. This is the state's cartographic claim to sovereign territorial authority that emerges in association with, and response to, the geometrical surveying conducted by European explorers and empires, able to lay down linear boundaries on the map (Winichakul 1994). Prior to this, he argues, Asian states had a much more fluid conception of the boundaries of their polities. In Japan, however, perhaps due to the inherent insularity of island governance, there emerges a stable and replicable representation of the state's geo-body at a much earlier date, in which the national body, while not geometrically-determined, was defined by islands. By the late thirteenth century, we already see the development of a graphic representation of territorial authority through the so-called *gyoki* maps, which offered a map of the provinces and circuits that constituted the administrative ideal of Japan associated with the seventh-century state. This representation displayed the extent of the known world, and covered the main islands of Honshu, Shikoku and Kyushu, as well as a number of outlying islands (Unno 1994: 366–371).

This early modern Japanese geobody did not, however, extend to showing the island known as Hokkaido today, which existed out beyond the administrative reach of the Japanese state, and thus beyond the boundaries of the known world. This was despite the presence of an extensive trade between Japan and the mysterious world of its north, which was known as the 'thousand isles of Ezo', after the 'barbarian' population resident there. By the fifteenth century, Ezo was recognized as a foreign yet actual place, and thus open to political claims, with rulers in northern Japan seeking recognition as governors of this amorphous Ezo region (Howell 1994: 78). The value of such a claim was managing trade with this mysterious northern land, which had assumed almost legendary status by this period. This is shown in the reports of a Japanese from the southern island of Kyushu, who in 1541 told the Jesuits in Goa about the country of 'Gsoo' (Ezo) north of Japan, overflowing with gold and inhabited by huge bearded natives (Boscaro and Walter 1994: 84). Ezo was thus positioned in relation to Japan even prior to Europeans arriving in the country.

On one of the earliest European maps showing the geography of 'Iapam' in detail, a 1585 Portuguese copy of a Japanese *gyoki*-type map, the Japanese geobody is shown as consisting of three main islands. There is no representation of Ezo, but the map, which is orientated to the south, does note the presence of a 'yezoga xima' (*Ezo-ga-shima*, or thousand isles of Ezo) to the east of Japan (Cattaneo 2014, Plate 1). This lack of representation indicates the amorphous character of Ezo in Japan at the time. While the existence of Ezo was dimly perceived, its actual geography remained obscure at the outset of Japan's early modern era, as reflected in both European maps and the Japanese materials on which they were based.

Nevertheless, the connection of Ezo with both strangeness and immense wealth was to be an enduring one, and was sufficient to bring it onto both maps of Japan and Japan's map over the course of the seventeenth century.

## 2.2   Securing Precious Metals

A key reason why Ezo came onto Europe's map of Japan in the seventeenth century is because of its connection with mineral wealth. Together with the search to control the supply of other valuable commodities, like spice, it was a desire to seize control of gold, silver, and other minerals that drove the expansion of European empires (Frank 1998). Visions of gold, in particular, drove a competitive imperialism which gave rise to much European exploration, including of Japan itself, the fabulous wealth of which had been reported to Europe by Marco Polo and others. Consequently, early Western visitors to Japan believed that they had come across a country with significant and easily mined mineral resources. One Spanish trader writing in 1594 believed that Japan was rich in gold, silver, copper, and iron, that mines were found throughout the country, and that ore was extracted with ease (Wittner 2008).

The tales that circulated within and beyond Japan of Ezo's immense wealth led to interest in the region being aroused abroad. European representatives in Japan sought out Ezo in order to secure this famed supply of precious metals. In 1609, Sebastian Vizcaino, an ambassador sent by the viceroy of New Spain to thank the shogunate for aiding the shipwrecked governor of the Philippines, was also charged with finding the truth with regards Ezo's rich gold and silver reserves (Oka 2016: 23). The head of the English factory at Hirado, John Saris, was noting in 1613 that the inhabitants of Yedzo 'have much silver and sand-gold' (Purchas 1905: 488). A desire to attempt to locate this place is visible in the way in which maps of Japan made in Europe did come to feature an island with designations referring to Ezo ('Jezo', 'Jedso', 'Yesso', and so forth), with the earliest extant being the 1617 map of Christophorus Blancus, almost certainly based upon an original map by Ignacio Moreira (Schütte 1962). The famous VOC voyage under Maerten Gerritsz. Vries in 1643 was motivated by a desire to gain access to Ezo's mineral wealth (Hesselink 2002; for more on early European understandings of Ezo, see Boyle forthcoming).

Yet this concern with seeking out and securing sources of gold was not solely the preserve of European states. The arrival of Europeans to Japan coincided with an intense internal struggle for political supremacy, which fuelled the country's first mining boom (Sippel 2006). The massive expansion in output resulted from the desire of domestic political actors to both control mineral resources and foreign trade. The years prior to 1600 saw the opening of numerous new gold, silver, and copper mines, which continued into the early part of the Tokugawa era (1603–1868). The new rulers sought to control the flow of minerals into and out of the country, money supply, and production, with many of the mines, including the famous gold and silver mines on Sado Island, being managed directly by the Tokugawa state itself (Morris-Suzuki 1994: 43–49).

This was the political context within which the isles of Ezo began to be mapped, as part of a general concern of state entities in Japan with the acquisition and management of mineral wealth. The founder of the Tokugawa shogunate, Tokugawa Ieyasu, in 1604 allegedly suggested to Matsumae Yoshihiro, ruler of the Matsumae family ensconced at the southern tip of Ezo, that he should be placed in charge of managing gold mining in Ezo. This expanded the Matsumae's existing monopoly over shipping dues, conferred on the Matsumae family by Ieyasu's predecessor, Toyotomi Hideyoshi (Walker 2001: 36). Gold was therefore crucial to the expansion of Japan's political engagement with this mysterious land of Ezo.

While the story of Ieyasu's suggestion is first recorded in a Matsumae family history from the 1640s, and so may well be apocryphal, it does show the importance of mineral wealth for Ezo by this later date. The Spanish Jesuit Diego Carvalho, visiting Ezo in 1621, has left an account of the placer mining operations being conducted there at the time, through which the rights to utilize sections of the river to pan for gold were effectively rented from the Matsumae (Cooper 1965: 235–236). By 1635, a gold rush, which had begun in 1617 with the licensing of mining at two sites near to the Matsumae's base at the southern tip of the island, had expanded to include at least ten locations much further north, with miners from all over Japan involved in the mining and panning for gold (Kikuchi 2003: 234–238). Consequently, some of the earliest cartographic material that we have of Ezo itself appear to have been privately-made manuscripts illustrating the location at which resources were able to be found. Maps such as the *Matsumae Ezochi ezu* (Map of Matsumae and Ezo) offer a sprawling representation of the region, upon which mountains of gold marked in several places, and may have been made either by or for gold prospectors (Akitsuki 1999: 23, Fig. I-6). Their effect was to show the entire expanse of this mysterious northern land as the source of great wealth, with Ezo represented as a space of extraction, of precious metals, furs and skins, and goods from the continent.

It seems to have been the disruption caused by Japanese gold miners penetrating further and further into Ezo that resulted in the most significant threat to the Matsumae's claim to manage Japan's interests in the region. The Shakushain Disturbance of 1669 at least partly stems from the impact these miners had on Ainu society, due to the disruption of traditional Ainu fishing practices caused by placer mining, which necessitated the demining and diversion of rivers (Walker 2001: 82–84). Intra-Ainu disputes were likely sharpened by conflict over who possessed the right to rent out these sections of the river to Japanese miners, who formed a significant proportion of the Japanese victims of the conflict. A number of maps were produced in the conflict's aftermath, including that reproduced here (Fig. 2). From today's perspective, these show an exceedingly limited grasp of the region's geography, but provide an image of Ezo as a space of extraction, including of precious metals.

Although groups of miners rapidly resumed their travels to Ezo following the Disturbance, however, the place of gold in the region's economy declined rapidly. By the end of the eighteenth century the proceeds from 'the gold-mine taxes fell off so that the continually prospering Ezo trade became the main source of Matsumae

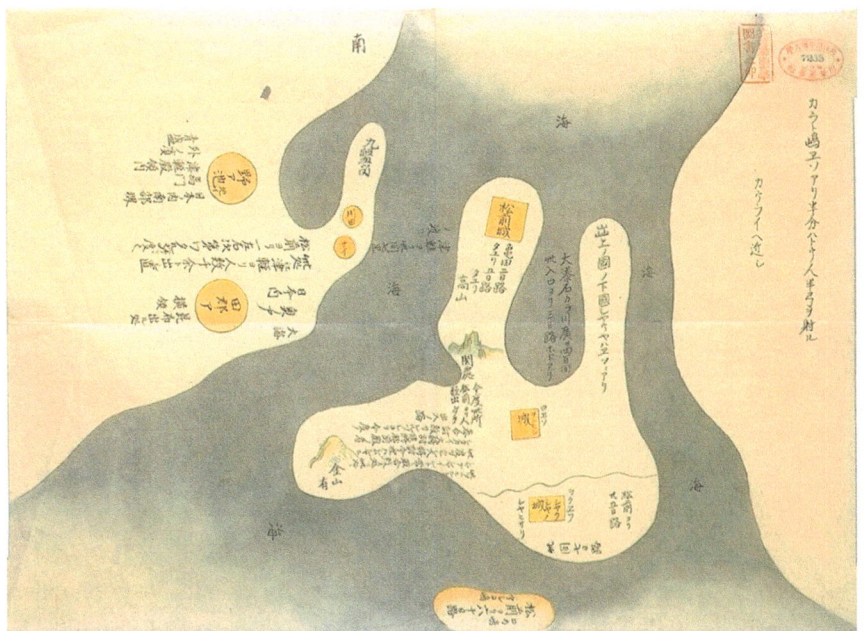

**Fig. 2** Ezo Matsumae Nihon no zu [Map of Japan, Matsumae and Ezo]. Anonymous, original c. 1670s. Reproduction of a manuscript map in the possession of Koreto Ashida. 50 cm × 40 cm. Hand-drawn and coloured. Oriented broadly to the West, although it states to the south, Matsumae Castle is in the yellow box at the map's centre. To the left is northern Honshu, to the right is *Karato (Karafuto)*, today's Sakhalin, shown here as part of the continent. The Kuril Islands, whose Ainu inhabitants came to trade fur with the Japanese at Matsumae, are amalgamated as a single 'Roka' (Rakko–Sea Otter) Island at the bottom of the map, sixty days travel from Matsumae. The map offers a schematic (and illusory) political outline of the main island of Ezo divided into three areas, Matsumae, Kuchi-Ezo (Inner Ezo) ruled by Onibishi, and Oku-Ezo (Anterior Ezo) ruled by Shakushain. It was conflict between these two Ainu chieftains which precipitated the Shakushain Disturbance. The 'mountains of gold' are located in Onibishi's territory of Kuchi-Ezo, in the south-east corner of the island. Hokkaido University Northern Studies Collection, map 871 (Courtesy of Hokkaido University Library)

income' (Takakura 1960: 26). Rather than roving Japanese miners, the Japanese presence in Ezo came to be characterized by 'trade fiefs' contracted out to merchant operators under the 'contract fishery system'. The role of the Matsumae continued to be the management of this system of extraction, and their own mapping of Ezo would reflect their claims to the entirety of this space (for an overview of this administrative cartography, see Boyle 2018). Due to the changing economic structure that underpinned the Matsumae's claims to political authority, the presence of gold largely disappears from maps of Ezo, even as Ezo itself comes to be increasingly represented as part of Japan's map.

In the nineteenth century, the representation of Ezo on maps of Japan indicated an expansion in the state's geo-body, and by the time of Benjamin Smith Lyman's appointment, the representation of the island as a 'large frying pan with a short crocked handle' (Lyman 1877: 2) had been traced out upon layers and layers of paper. For both Japanese and those further afield, the space of Ezo was partially defined by its association with the extraction of precious metals. As in Europe, state expansion and consolidation were associated with territorial control over areas of production, and indeed the wealth and health of the state was explicitly connected to the production and circulation of minerals within it. Governance of the earth's surface could not be separated from activity beneath the soil, as shown by the efforts to administer and control such spaces of extraction. Nevertheless, the representation of such spaces remained at the planar surface of the map itself, making a visual claim to the horizontal extent of the state's authority. Until the nineteenth century, the nation's geo-body was a distinctly flat one.

## 3  Embodying the Earth

### 3.1  The Development of Geological Mapping

The rise of geological surveying in the nineteenth century is a notable development in extending the territorial control of the state along a vertical, as well as horizontal, axis. This was obviously an outgrowth of the industrial revolution, literally fuelled by resources extracted from beneath the Earth's crust, as the world's powers began their transition to an 'inorganic' economy (Wrigley 2016). The desire of governments and industrialists to develop mineral resources encouraged a growing reliance upon those specialists able to decipher the world beneath their feet. In response emerged the science of geology, an empirical system of field observation that focussed its attention on the three-dimensional structure of the Earth in order to 'rationalize the search for minerals demanded by industry' (Stafford 1984: 6). The most important means of representing this structure was the geological map, which sought to locate and 'order' the positions of distinct rock formations on a flat topographical surface.

   The possibility of representing the positions of these formations across wide areas was demonstrated by William Smith's beautiful geological map of England and Wales of 1815, showing the stratigraphy, the order of distinct mineral strata, of this section of the earth (Winchester 2001). This style of mapping provided the means through which spatial relations in the world below were brought to order and represented in their proper positions, as cartography would the world above (Rudwick 1982). From its inception, the new geological mapping was driven by a belief in the possibility of its universal application. The concern with classifying the various strata being uncovered through the process of mapping led to geology becoming 'a science concerned with correlation of strata across space, leading [...] to attempts to develop an internationally accepted nomenclature of strata that could be applied globally' (Braun 2000: 22).

This cemented the importance of such stratigraphic representations to nineteenth century geology, which became a territorial science [with] its essence [...] embodied in maps' (Secord 1982). By 1835, the Geological Society of London had persuaded the British government to undertake the geological mapping of the entire country, with the Geological Survey of Great Britain emerging as a branch of the Ordnance Survey and utilizing the latter's material as base maps upon which to evaluate the nation's mineral wealth. A shared concern with territory and extraction meant that colonial geological mapping emerged essentially in tandem with its national equivalent, and thus came to be reflected in a parallel process of institutional development. Already in the year of its founding, the Geological Survey of Great Britain organized its first overseas expedition, to Mesopotamia, and conducted more than 40 overseas geological surveys sponsored by the imperial government prior to the death of its second Director-General in 1871 (Stafford 1984).

In the United States, meanwhile, the efforts of a number of individual American states in the 1820s to conduct geological surveys of their own territory would come to be developed as part of an expansionary national enterprise. The geological survey became another means by which a territory would come to be known, with the geologist expected to locate exploitable mineral resources as sources of wealth (Hendrickson 1961: 358). 'The scientific establishment [was] instrumental in emphasizing to the federal government the importance of the scientific information these expeditions would produce' as federally funded geological expeditions came to be attached to private enterprise's opening of the American West (Kues 2008: 103).

By the mid-nineteenth century, then, geological mapping had emerged as both a universally-applicable means of representing territorial order, and one that would evaluate the extractive potential of that territory. It was a form of knowledge that circulated worldwide, able to be applied in different parts of the world in developing knowledge about territory. It was for this purpose Benjamin Smith Lyman was appointed Chief Geologist and Mining Engineer for the Geological Survey of Hokkaido in 1873, which resulted in the presentation of such knowledge upon the 'Geological Sketch Map of the Island of Yesso' three years later. Lyman's work was part of a new state working to 'state' itself into existence upon layers and layers of paper. The land upon which it was seeking to do so was not, though, a blank canvas atop which a new geological map could be unrolled. While the designation Hokkaido was a mere five years old when Lyman was appointed, it referred to a region which had been constituted on the map through the extraction, of labour, of wealth, and of mineral resources. Although the latter were no longer generally shown on maps of Ezo, they remained present in mental maps of the region.

## 3.2 Sketching Layers

Initially absent from Japan's geo-body, by the mid-nineteenth century, Ezo had come to exist on Japan's map as an extractive space. The importance of minerals and precious metals in the constitution and condition of the state was an ongoing concern throughout the Tokugawa period, and the marginal and barbarous space of Ezo was understood as possessing the potential for further precious metal extraction. This provided a policy response able to be advocated in response to a variety of crises. This is visible in the proposals of successive administrators and intellectuals, who put forward gold-mining in Ezo as a policy response to the political problems of the day. In the *Hokkai Zuihitsu* of 1739, for example, Sakakura Genjiro argued for mining Ezo's mineral wealth as an alternative to a previous administration's devaluation of the currency (Stephan 1969: 31–33). Fifty years later, in the midst of increasing concern over the threat offered by Russia to Japan's position in Ezo, the Sendai physician Kudo Heisuke argued that 'Trade with Russia will be a good way to help foster the development of Ezo. If Ezo can be brought under Japanese control, all of Ezo's products including precious metals will be available to us' (Keene 1969: 108). These abstract proposals had occasionally been supplemented by more empirical investigation, such as in 1766 when a prospector from Edo, Yamashiroya Yasuemon, reportedly crossed over to Ezo to investigate these famous gold fields, and returned home disappointed. Despite this absence of actual gold production, Ezo was retained as a space of precious metal extraction on the mental maps of Japanese, as well as Europeans.

In the political confusion that gripped Japan in the aftermath of the arrival of Commodore Perry's 'Black Ships' and the Tokugawa Shogunate's reluctant agreement to 'open' the country, advocating for the exploitation of Ezo as a source of precious metal emerged once again as a policy option. This is shown in the Tokugawa government's engagement of a pair of American geologists in the early 1860s. The central government had assumed responsibility for most of Ezo from the Matsumae family in 1855, and administered the region from Hakodate, which was also declared a Treaty Port in 1858. The plan to hire the two Americans appears to have originated with Muragaki Norimasa, who in 1860 served as Vice-Ambassador on the Japanese Embassy to the United States. Upon his return to Japan in 1861, he was appointed as one of the two governors of Hakodate responsible for the administration of Ezo. On 23 April 1861, the American Consul, Townsend Harris, was instructed by Norimasa to write to C. Walcott Brooks, a commercial agent employed by the Japanese government. Brooks was commissioned to 'engage two gentlemen who are thorough Mineralogists and practical Mining Engineers to examine their mines of Gold, Silver, Copper and Lead, and give them instructions as to the best manner of working them […] they will be conveyed to Hakodada [Hakodate]' (Hasegawa 1986: 99). One of them later noted that 'the object of our engagement with the Government was the exploration of its lands on the island of Yesso, and the introduction, if found advisable, of foreign methods of mining and working metals' (Pumpelly 1869: 144).

The biographies of the two men selected for this task reveals how surveying the state's subsoil had become a global phenomenon by the mid-nineteenth century. The senior of the two, William Phipps Blake, had graduated from Yale in 1846, served as official government geologist on the Pacific Railroad Surveys in the American southwest from 1853–1855 (the surveys of Whipple and Williamson) and investigated mining resources in Carolina. Returning to California to seek the position of state geologist, he lost out to a rival, Josiah Whitney, and surveyed the California gold fields before agreeing to go to Japan (Testa 2002). A prolific author of over 200 articles, a review of his work in the year of his death noted that he made significant contributions to the geology of Arizona, California, Utah and Wisconsin, and of Mexico, Alaska and England, in addition to his work in Ezo (University of Arizona 1910). His junior colleague, Raphael Pumpelly was a fellow New Yorker who had travelled to Europe rather than attending Yale, graduating from the Freiberg University of Mining and Technology in 1859, before returning to the United States and being employed as a mining engineer in Arizona in 1860–1861. While his own notes suggest he was busier fighting off Apaches than collecting samples, he presented his studies at the California Academy of Sciences in August of 1861, prior to heading to Japan (Kues 2008). Following the conclusion of his work there, he conducted a survey of northern China and went on to lead a distinguished career, as a nominal Harvard Professor and leader of archaeological expeditions to Central Asia.

Blake and Pumpelly left San Francisco on 23 November 1861, landed in Yokohama on 21 February 1862, and in Hakodate on 5 May that year. They established the School of Mines and Applied Science in the town, which they struggled to equip in order to instruct their five students. Blake offered instruction in gold mining, metallurgy and chemistry, while Pumpelly taught geology, mining, blasting and surveying. Their employment may be seen as a perfect example of colonial modernity, with a series of practices unable to be carried out in the metropole, due to the opposition they would arouse, being displaced to marginal areas. It was from this marginal Ezo region that certain of these practices were transmitted back to the rest of the country. A notable example of this was Pumpelly's powder blasting, which he recorded '[...] was so successful that before I left Japan I was told that several princes had sent men to Yurup [in Ezo] to learn the new process' (Pumpelly 1869: 190). Here, we see the movement of a particular practice that has first occurred in Ezo to the rest of Japan, and consequently this marginal and barbarian space comes to serve as a contact zone (Pratt 1992) able to mediate between Japan and the world. In that respect, Ezo continued to serve as the margins of Japan.

However, the practices which the two men brought with them to Ezo did not only work to define it in relation to Japan, but the rest of the world, as well. This is shown in their geological investigations, which signified the incorporation of this region within a globalizing geological science, in which all of the Earth's subsurface would be correlated and represented upon paper. This was a form of knowledge with a global reach, through which Pumpelly's experience of Corsica and Vesuvius would 'enable him to distinguish the foundation rocks of the island, a sequence of metamorphosed sediments penetrated by granitic and basic eruptives, and the superjacent volcanic deposits of various kinds' (Willis 1925: 37). This ability to relate the subterranean territory of Ezo to elsewhere in the world enabled the two American geologists to seek to decipher its stratigraphy, and to communicate that knowledge through the sketching of geological maps.

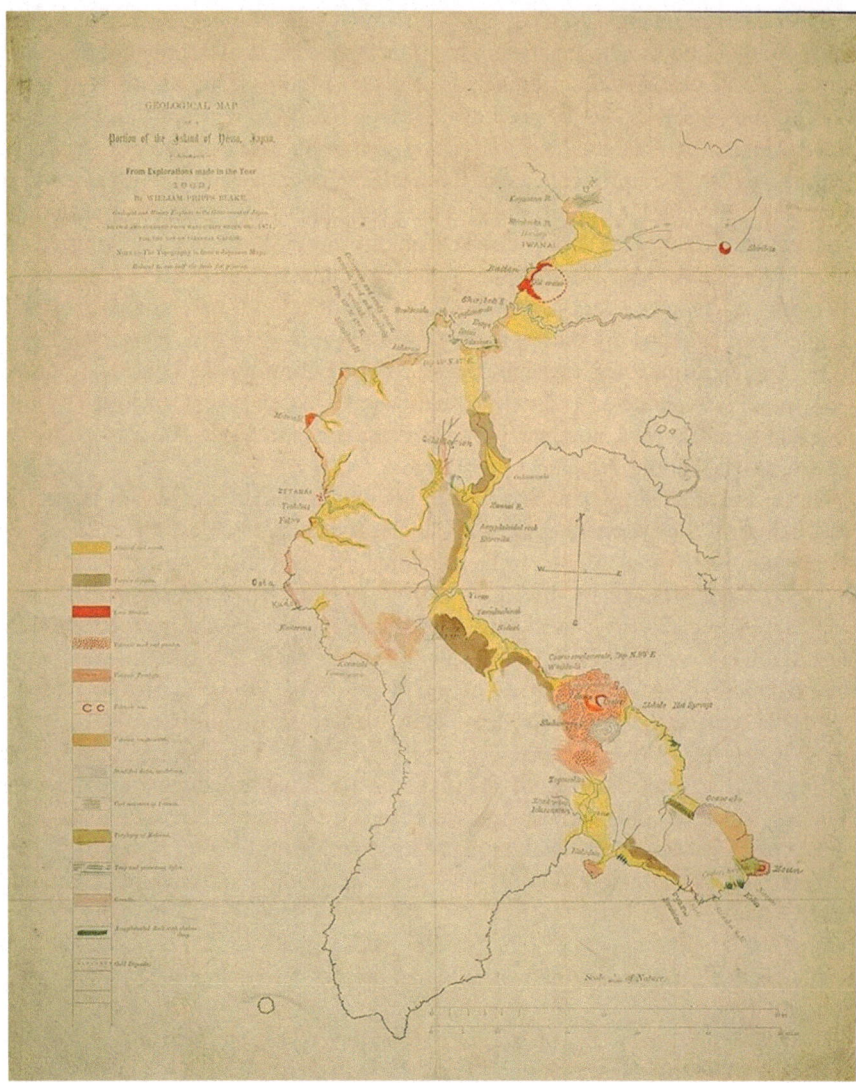

**Fig. 3** Geological Map of a Portion of the Island of Yesso, Japan by William Phipps Blake. Scale 1:460,000. 56 cm × 44 cm. Lithograph printed, hand-coloured. Made for the use of the *Kaitakushi*'s General Capron in 1871, this map accompanied the report on the 1862 survey, which he submitted to the *Kaitakushi* in that year. As with Pumpelly's map, it sketched the stratigraphy of the 'short crocked handle' of Hokkaido, the Oshima Peninsula, where Japan's control of Ezo was formalized and long-established. The presence of gold was noted near Kunni, reflecting his and Pumpelly's surveys of the old workings there. Hokkaido University Northern Studies Collection, map 895 (Courtesy of Hokkaido University Library)

The experiences of Blake and Pumpelly in Ezo were summarized on the maps that they produced. Each of Pumpelly's three books that mentioned his engagement in Yesso, which were published over more than thirty years, incorporated the same map of the Oshima Peninsula in southwest Hokkaido, entitled a 'Geological Route-Sketch, Southern Yesso, Japan' (Pumpelly 1866: 159–160, Plate 8; Pumpelly 1869: 143; Pumpelly 1918: 307). This was based on the three expeditions that himself and Blake were able to conduct over the course of the year they were employed. Blake, the senior of the two men, did not append a map to his only published report on the Yesso Expedition (Blake 1874), but was able to supply a 'Geological Map of a Portion of the Island of Yezo, Japan' when requested a decade later. The copy of this map made by Benjamin Smith Lyman still exists in the latter's papers in Amherst (Kim 2009: 49). In addition, Hokkaido University Library holds eight copies of Blake's map, 'drawn and colored from manuscript notes' by Benjamin Smith Lyman, and printed by the *Kaitakushi*. It is this map which is reproduced here (Fig. 3).

These maps made by Blake and Pumpelly share two notable characteristics; acknowledgements of their limitations, and that their representation is literally dependent upon topography 'from a Japanese map'. The two geologists both took pains to emphasize the preliminary nature of their findings, emphasizing that their maps were mere sketches. Their ability to produce such sketches, though, stemmed from the partial incorporation of Ezo into Japan's geobody, which furnished the two geologists with the cartographic material atop which Ezo's body was able to be fleshed out. 'The geographical basis of this map is taken mainly from an unpublished Japanese survey of Yesso, in the Imperial archives of the vice-royalty of Yesso' (Pumpelly 1866), access to which provided the two geologists with the means to construct this composite image of Ezo's stratigraphical layers. This process of constructing a picture of Ezo's territory at depth was one that would continue apace a decade later.

# 4 Finding Value

## 4.1 Institutional Depth

The *Kaitakushi* was the institutional vehicle through which the new Meiji government would officially incorporate Ezo into Japan, and through which Ezo's colonially-modern status would be formalized. Hokkaido would subsequently become Japan's 'internal colony' (Imanishi 2008), serving as an experimental station from which experiences and administrators would be subsequently exported to other Japanese colonies elsewhere in Asia. In that sense, as has been widely recognized in recent years, Hokkaido needs to be seen as part of the history of the Japanese empire (Oguma 1998). Hokkaido was also, however, imperial in another sense; with its territory being drawn into intra-imperial networks of production and circulation, of knowledge, personnel, and materials. The employment of foreigners by the new Meiji government in its first fifteen years is much commented upon, both at the time and today, with numerous British, French, German and American advisors attached to various ministries (perhaps 220 individuals in 1872, for example). The circulation of these men (and they were exclusively men) from other countries into Japanese government service was crucial to Japan's recognition as a 'sovereign' nation and later accession to the hallowed ranks of 'international society' (Gong 1984; Suzuki 2009).

While an earlier era of scholarship perceived development through a national lens, however, the fact that this society consisted of expansionist powers whose relations were characterized by intra-imperial cooperation as well as competition has increasingly come into focus over the past two decades. The role of such administrators and technicians in aiding the establishment of a modern, expansionist imperial state is visible in its most concentrated form in the activities of the *Kaitakushi* in Hokkaido. Its Commissioner, Kuroda Kiyotaka, decided that the experience of the American West offered the most obvious parallels for Japan's efforts to colonize this territory, and in January 1871 went to the United States in order to consult with President Ulysses Grant about hiring experts to assist in Japan's mission. At Grant's recommendation, the Commissioner of Agriculture, Horace Capron, was put in charge of this search, before ultimately taking the job himself. Capron initially travelled to Japan with three assistants, one of whom, Thomas Antisell, was responsible for geology.

Antisell was born in Dublin, trained as a surgeon there and in London, before studying chemistry in Paris and Berlin, returning to Dublin in 1844 and lecturing in botany whilst contributing to the geological mapping of Ireland. Having been forced to emigrate to the United States in 1848 for his role in publishing a Republican newspaper, Antisell opened a clinic in New York while lecturing on chemistry. He then served, like William Blake, as official geologist to a Pacific Railroad Survey, that of Parkes, which surveyed those parts of southern Arizona, New Mexico, and California lying along one of four proposed routes for the so-called Southern Pacific railroad. 'Antisell's report is the first detailed description and interpretation of the geology of southwestern and south-central New Mexico' (Kues 2008: 92), and was to be the only other geological survey with which he was involved. Antisell subsequently became the chief examiner of the US Patent Office, a lecturer in chemistry at Georgetown University, and, from 1866, chief chemist to the Department of Agriculture. It was through this connection that he was taken to Japan.

While Capron remained in Tokyo after arriving in August 1871, it was Antisell and Major Warfield, a civil engineer, who were initially sent to Hokkaido later that year, where they spent two months surveying the southwest of the island. In addition to reporting on soils, flora, topography, possibilities for water power and the climate, Antisell undertook what he termed a 'geological reconnaissance of Hokkaido'. It appears that Antisell's pessimism regarding Hokkaido's development prospects led to conflict between Capron and himself, and his removal as geologist. Relations between the two men deteriorated to the extent that Antisell's reports were excised from those that Capron submitted at the end of his term, in the *Reports and Official Letters to the Kaitakushi* (Capron 1875), where they were replaced by the 'Abstract of the Report of W. P. Blake'. By this time, Antisell had left the *Kaitakushi* and was working as a chemist for the Ministry of Finance, before leaving Japan in 1876. Although Antisell's contribution was largely excised from the official record, his surveys and maps did end up in the hands of his successor as geologist, Benjamin Smith Lyman. Lyman's papers at the University of Massachusetts in Amherst contain a copy of Antisell's own geological sketch map of the Oshima peninsula (Kim 2009: 50, Fig. 12), where their presence shows how successive rounds of geological investigation overlaid one another in the creation of Hokkaido's stratigraphical map.

## 4.2   The Value of Representation

Benjamin Smith Lyman arrived in Japan with an Assistant, Henry Smith Munroe, in January 1873. Lyman had been born in Massachusetts in 1835 and graduated from Harvard two decades later. He did some work for geological surveys being conducted by his wife's uncle, J. Peter Lesley, and went to study in Paris at Ecole Imperiale des Mines in 1859. He then, as Raphael Pumpelly had, attended the Freiberg University of Mining and Technology, the world's oldest mining school, in 1861–1862. Returning to the United States, he began working as a consulting mining engineer, and conducted surveys in Pennsylvania and Nova Scotia, as well as also experiencing the US frontier in Arizona and California. In 1869, he was hired by the British government to evaluate the oil lands of the Punjab. He returned to the United States in mid-1871, advocated for the importance of surveying in geological expeditions (Lyman 1873), and was hired by the Japanese government at the end of the following year.

Lyman conducted geological surveys of Yesso in 1873, 1874, and 1875. The 1873 expedition was undertaken with Munroe and six students, who split up and conducted geological surveys at a number of mountains and oil fields on the island's south-western peninsula. Lyman here was following Kuroda's initial instructions, in which the survey was to confine itself to the 'four southwestern provinces', or about one-third of the island (Lyman 1874: 3). In addition to this, Lyman also ventured forth to investigate locations like Shakotan, Jozankei, Noboribetsu and Esan. The following year saw the ten students split into two parties, with half surveying gold fields under Munroe, and the other half investigating the geology of what would become the coal fields of Horonai, Ishikari, and Sorachi. Lyman himself, with one translator and ten assistants, made a virtual circumnavigation of the island in the course of a 150-day long journey that lasted from May to October. This was his only opportunity to gather data for much of the territory he was charged with mapping. The 1875 survey ran for about 100 days between June and October, and was primarily for the close investigation of coal fields around Onuma and Sorachi (Imai 1965).

During the winter, the members of the survey spent their time at the *Kaitakushi*'s
school in Tokyo, and engaged in producing the results of this surveying work in
material form. As Lyman's final report made clear, the production of these materials
also involved the students and was considered part of their training. By the conclusion of the survey, the *Kaitakushi*'s press in Tokyo had produced over 900 pages
of reports, together with numerous small-scale maps of various coal fields 'well
lithographed and photographed', if rather crowded as a result of giving all the
information in both English and Japanese (W 1877: 523). The *Geological Sketch
Map of the Island of Yesso* was intended to provide a synoptic overview of the
entire survey, even though, as Lyman himself admitted, in many places 'it cannot
lay claim to much exactness' (Lyman 1877: 102).

The map was drawn on the basis of the special surveys conducted by Lyman and
his Assistant Geologists, together with 'hundreds of dips observed' outside of these.
It also, though, laid claim to the earlier geological investigations which had been
undertaken on the island, giving details of four separate routes of geological
expeditions (Fig. 4). In adding those of Blake and Pumpelly in 1862 to the surveys
of Antisell, Munroe, and Lyman himself, the map demonstrated how the institutional knowledge embodied on the map was more extensive than the institutional
setting within which the map was made. Although the survey of Blake and
Pumpelly preceded the establishment of the *Kaitakushi* by several years, the
globalizing geological milieu within which the two American geologists operated
meant that the knowledge they acquired was made available to and utilized by their
successors. Yet the continuity demanded for the map's completion was more
extensive than this, for as Lyman noted, the actual representation of territory was

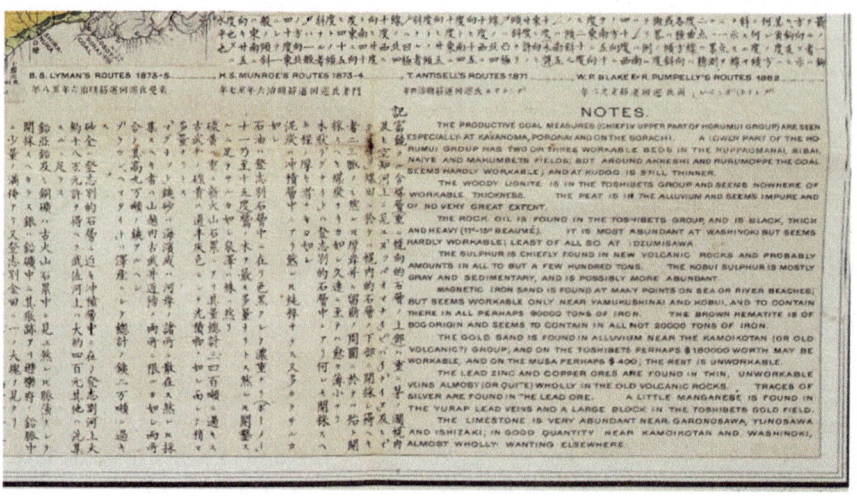

**Fig. 4** Detail from *A Geological Sketch Map of the Island of Yesso, Japan*, bottom-right of the
map. Shows the four separate surveying expeditions marked on the map, while below this are
'Notes on useful minerals present in Yesso', including those for 'gold sand'. Hokkaido University
Northern Studies Collection, map 871 (Courtesy of Hokkaido University Library)

'copied from Matsuura's large map of 1860' that in turn based its representation of the coastline on the maps associated with Ino Tadataka's survey of 'nearly a hundred years ago' (Lyman 1877: 101). In this respect, his map overlays the same terrain as that of Blake and Pumpelly, while also incorporating the latter's discoveries.

The finished geological map by Lyman and his Assistant Geologists therefore provides us with a layered picture of the land of Hokkaido, one that emphasizes its incorporation within a global practice of geological surveying. An Ezo region that had been slowly and unevenly incorporated into Japan's geo-body over several centuries was resurveyed and remapped as a geological space. This served to grant Hokkaido corporeal form, in which the territory would be made useful to the state through the 'working of Yesso minerals'. Most significantly, it served to engender knowledge of the vast coal reserves in Ezo, a matter of 'great national importance' that would assure 'future great wealth to the empire'. The maps very existence, indeed, would improve the government's 'credit' (Lyman 1877: 107), providing it with a future revenue source against which to make loans. Yet this new mineral wealth, with the discovery of vast fields of coal, was in contrast to that of gold, whose extraction had been central to the emergence of Ezo on the map. Both the map and Lyman's report were unsparing, noting of the gold sands on the island that, with two partial exceptions, 'the rest is unworkable'. Under Lyman's unsparing gaze, Ezo had moved from a rich repository of precious metals to one whose wealth was bound up in its coal beds.

Two centuries earlier, maps made in Japan had represented Ezo as a region in which gold 'came with' the territory, its presence serving to justify the determination of Japanese mapmakers to incorporate this barbarian space within the nation's geobody. The globalizing practice of geological mapping in the mid-nineteenth century provided another means through which Ezo's place within this Japanese geobody was able to be represented. The presence of these geologists in Japan points to the emergence of a strata of specialists, able to circulate freely within not only national contexts, but one broadened into an intra-imperial web of knowledge that extended to a marginal, remote Ezo region, in order to provide depth to Japan's knowledge of its own geo-body. Through its ability to make visible knowledge regarding Ezo's subsurface composition, the representation of stratigraphic layers on the 'Geological Sketch Map of the Island of Yesso' granted literal depth to Hokkaido's incorporation into Japan's geobody.

# 5 Conclusion

The geological map produced by Benjamin Smith Lyman in 1876 represented a new form of stratigraphical knowledge, whereby different subsoil mineral formations were compared and correlated globally, in order to attempt to capture the value of a territory through discerning 'what comes with it'. The Sketch Map which brought this new, stratified body of Hokkaido into existence was not only dependent upon the area's representation upon layers and layers of paper, but the ability to bring a particular form of knowledge to bear. It was this knowledge, enabling geological 'surveys and maps that will answer all practical mining requirements', that made it possible to draw the connection between the representation of territory on the surface and knowledge of its depths. This drew Hokkaido into a world of imperial competition that was dependent upon the control and exploitation of resources beneath the Earth's surface, while a state's control of territory was represented and demarcated across the flat planar surfaces of the map itself.

The map presaged the future value of the internal colony of Hokkaido to the Japanese state in important ways. The coal beds which Benjamin Smith Lyman saw layered into its territory would power the nation's inorganic economy for over a century, bringing fuel for the engine of progress with the territory represented on the map. These geological maps provided a different means of representing the wealth of Ezo to Japan, one which overwrote visions of gold with geological knowledge of this body of the Earth. In the event, Lyman's confident assertions regarding the 'unworkable' nature of Ezo's gold fields were to be undermined by a gold rush of the 1890 s and the 1915 discovery of the Konomai Mine, which remained in production until 1973. This did not, though, diminish the value of Lyman's Sketch Map for the state, which re-presented Ezo as an integral part of a modernizing Japan though the island's incorporation within a globally-applicable means of territorial representation.

**Acknowledgements** This work was supported by JSPS KAKENHI Grant Number JP 16K17071.

# References

Akitsuki T (1999) Nihon Hokuhen No Tanken to Chizu No Rekishi [A history of the exploration and cartography of the Northwest Pacific]. Hokkaido Daigaku Tosho Kankokai, Sapporo

Blake WP (1874) Notes on the geology of Island of Yesso, Japan. Geol Mag 2(1):464–465

Boscaro A, Walter L (1994) Ezo and Its Surroundings through the Eyes of European Cartographers. In: Walter L (ed) Japan: a cartographic vision: European printed maps from the early 16th to the 19th century. Prestel-Verlag, Munich

Boyle E (2016) Imperial practice and the making of modern Japan's territory: towards a reconsideration of Empire's boundaries. Geogr Rev Japan (Series B) 88(2):66–79

Boyle E (2018) The Tenpō-Era (1830–1844) Map of Matsumae-no-shima and the institutionalization of Tokugawa Cartography. Imago Mundi 70(2):183–198

Boyle E (forthcoming) Mapping Japan's north: material traces of an enduring encounter. In: Leca R, Storms M (eds) Mapping Japan: enduring encounters. Brill, Leiden

Braun B (2000) Producing vertical territory: geology and governmentality in late Victorian Canada. Ecumene 7(1):7–46

Capron H (ed) (1875) Reports and official letters to the Kaitakushi. Kaitakushi, Tokei (Tokyo)

Cattaneo A (2014) Geographical curiosities and transformative exchange in the Nanban Century (c. 1549–c. 1647). Études Épistémè 26. https://doi.org/10.4000/episteme.329

Cooper M (1965) They came to Japan: an anthology of European reports on Japan, 1543–1640. University of California Press, Berkeley

Frank AG (1998) ReOrient: global economy in the Asian age. University of California Press, Berkeley

Gong GW (1984) The standard of 'civilisation' in International Society. Clarendon Press, Oxford

Hasegawa S (1986) Hakodate Eigakushi Kenkyu [Research on the history of studying english in Hakodate]. New Country International, Tokyo

Hendrickson W (1961) Nineteenth-century state geological surveys: early government support of science. Isis 52(3):357–371

Hesselink RH (2002) Prisoners from Nambu: reality and make-believe in seventeenth-century Japanese diplomacy. University of Hawaii Press, Honolulu

Howell D (1994) Ainu ethnicity and the boundaries of the early modern Japanese state. Past Present 142:69–93

Imai I (1965) Reimeiki no Nihon Chishitsugaku: Senkusha no Shougai to Gyoseki [The beginning of Japanese geology: the lives and legacies of its pioneers]. Rateisu, Tokyo

Imanishi I (Ed.) (2008) Sekai Shisutemu to Higashi Ajia: Shokeiei/ Kokunai Shokuminchi/ 'Shokuminchi Kindai' [East Asia in the world system: small management/internal colonies/ 'Colonial Modernity']. Nihon Keizai Hyoronsha, Tokyo

Keene D (1969) The Japanese Discovery of Europe, 1720–1830. Stanford University Press, Stanford

Kikuchi I (2003) Ezoshima no Kaihatsu to Kankyo [Environment and development in Ezo]. In: Kikuchi I (ed) Ezo-ga-chishima to Hoppō Sekai [Ezo and the world of the north]. Yoshikawa Kōbunkan, Tokyo

Kim KN (2009) Oyatoi Gaikokujin Chishitsugakusha no Rainichikeii (4) Beijin Hakubutsugakusha Anchiseru –Kohen [Backgrounds to foreign geologists employed by the Meiji government coming to Japan (4) The American natural historian, Antisell—part 2-]. Sci. Move. Geol. Learn 62:48–62

Kues B (2008) Early geological studies in southwestern and south-central New Mexico. In: Mack G, Witcher J, Leuth VW (eds) Geology of the Gila Wilderness-Silver City area. New Mexico Geological Society, Socorro

Lyman BS (1873) On the importance of surveying in geology. Trans Am Inst Mining and Metall Eng 1:183–192

Lyman BS (1874) preliminary report on the first season's work of the Geological Survey of Yesso. Kaitakushi, Tokei (Tokyo)

Lyman BS (1877) A general report on the geology of Yesso. Kaitakushi, Tokei (Tokyo)

Morris-Suzuki T (1994) The technological transformation of Japan: from the seventeenth to the twenty-first century. Cambridge University Press, Cambridge

Oguma E (1998) 'Nihonjin' no kyōkai – Okinawa, Ainu, Taiwan, Chosen shokuminchishihai kara fukki undo made [The boundaries of the Japanese]. Shinyosha, Tokyo

Oka M (2016) Elusive Islands of Silver: Japan in the early European Geographic Imagination. In: Wigen K, Sugimoto F, Karacas C (eds) Cartographic Japan: a history in maps. University of Chicago Press, Chicago, pp 20–23

Pratt ML (1992) Imperial eyes: travel writing and transculturation. Routledge, London

Pumpelly R (1866) Geological researches in China, Mongolia, and Japan: during the Years 1862–1865. Smithsonian Institution 15(4)

Pumpelly R (1869) Across America and Asia: notes of a five years' journey around the world, and of residence in Arizona, Japan, and China. Leypoldt & Holt, New York

Pumpelly R (1918) My reminiscences. H. Holt, New York

Purchas S (1615) Hakluytus posthumus, or Purchas his pilgrimes: contayning a history of the world in sea voyages and lande travells by Englishmen and others, vol 3. Hakluyt Society (1905 Reprint), New York

Rudwick M (1982) Cognitive styles in geology. In: Douglas M (ed) Essays in the sociology of perception. Routledge, London, pp 219–241

Schütte JF (1962) Ignacio Moreira of Lisbon, cartographer in Japan 1590–1592. Imago Mundi 16:116–128

Secord J (1982) King of Siluria: Roderick Murchison and the imperial theme in nineteenth-century british geology. Victorian Stud 25(4):413–442

Sippel P (2006) Technology and change in Japan's modern copper mining industry. In: Hunter J, Storz C (eds) Institutional and technological change in Japan's economy, past and present. Routledge, London, pp 10–26

Stafford RA (1984) Geological surveys, mineral discoveries, and British expansion, 1835–71. J Imperial History 13(3):5–32

Stephan JJ (1969) Ezo under the Tokugawa bakufu, 1799–1821: an aspect of Japan's frontier history. Unpublished Ph.D. Thesis (SOAS)

Suzuki S (2009) Civilization and Empire: China and Japan's encounter with European International Society. Routledge, New York

Takakura S (1960) The Ainu of Northern Japan: a study in conquest and acculturation (trans. Harrison JA). Trans Am Philosophical Society 50(4): 1–88

Testa S, Whitney JD, Blake WP (2002) Conflicts in relation to California geology and the fate of the first California Geological Survey. Earth Sci History 21(1):46–76

University of Arizona (1910) The published writings of William Phipps Blake, 1850–1910. University of Arizona, Tucson

Unno K (1994) Cartography in Japan. In: Harley JB, Woodward D (eds) The history of cartography 2(2) cartography in the traditional east and southeast asian societies. University of Chicago Press, Chicago

W [unknown] (1877) Reviews-Lymans Geological Survey of Japan. Geol Mag 2(4):522–526

Walker BL (2001) The conquest of Ainu Lands; ecology and culture in Japanese expansion, 1590–1800. University of California Press, Berkeley

Willis B (1925) Memorial of Raphael Pumpelly. Geol Soc Am Bull 36(1):45–84

Winchester S (2001) The map that changed the world: William Smith and the Birth of Modern Geology. Penguin, London

Winichakul T (1994) Siam Mapped; A History of the Geo-Body of a Nation. University of Hawaii Press, Honolulu

Wittner D (2008) Technology and the culture of progress in Meiji Japan. Routledge, New York

Wood D, Fels J (1992) The power of maps. Guilford Press, New York

Wrigley EA (2016) The path to sustained growth: England's transition from an organic economy to an industrial revolution. Cambridge University Press, Cambridge

**Edward Boyle** is Assistant Professor at the Faculty of Law, Kyushu University. He holds a BA from the School of Oriental and African Studies in London and an MA and Ph.D. from the Faculty of Law at Hokkaido University. His doctoral thesis examined the incorporation of Japan's north into the space of the state during the seventeenth to nineteenth centuries, looking at the history of the cartography of the region as well as the concepts of territory that underpinned them. Edward's interests include the comparative history of early modern imperial mapping, contemporary practices of bordering and the multiscale nature of borders under globalization. He has established Kyushu University Border Studies (KUBS) as an interdisciplinary hub for all things border-related, and is investigating the multiscalar nature of contemporary border effects and their implications in projects on Japan, Georgia and Northeast India. His work intersects with political science, geography, history, and scholarship on international relations. In English, he has published articles and reviews in *Imago Mundi*, *Europa Regional*, *Eurasia Border Review*, *Journal of Borderlands Studies*, *Geographical Review of Japan (Series B)*, and others. He has also translated the key border studies text on Japan, by Akihiro Iwashita (*Japan's Border Issues: Pitfalls and Prospects*). He is editor-in-chief of *Border Bites*, and tweets at @BorderstudiesRM.

# Putting America's First Empire on the Map: American Early Efforts to Map the Philippine Islands

Eric H. Losang

**Abstract** In December 1884, John W Powell, second director of the U.S. Geological Survey (USGS), addressed the U.S. Congress seeking authorization to begin systematic topographic mapping of the United States. Alaska had just become an official district of the country, extending its territory to the most westerly point of the continent. The Westward Expansion (which officially ended 1912 when Arizona was admitted to the Union) therefore came to an end, and with it the subsequent contiguous mapping of US territory became the final act of nation-building. Cartographically, this was undermined by the publication of the *Atlas of the Philippine Islands* by the US Coastal and Geodetic Survey in 1899. As defined by Edney (2009), there is a difference between imperial cartography (maps used to create an image of the empire as a legitimate entity and to articulate a claim for territory) and colonial cartography (maps of varying sorts used for the immediate administration within a dependency); the former preceding the latter by defining the territory claimed. This chapter focusses on the development, processes and framework of the US survey and topographic cartography of the Philippines to suggest that US colonial topographic mapping, although connecting cartography with the exercise of imperial power, was more pragmatic and considerably different to the colonial and imperial cartographies of Europe, and reflects America's struggle to define itself as a colonial power.

## 1 Introduction

In 1896, Rand McNally & Company published the first edition of their *New Imperial Atlas of the World*. The first map showed 'U.S. Acquisitions' and depicted the familiar geobody of the United States, indicating the east–west growth of the

E. H. Losang (✉)
Leibniz Institute for Regional Geography, Leipzig, Germany
e-mail: e_losang@ifl-leipzig.de

© Springer Nature Switzerland AG 2020
A. J. Kent et al. (eds.), *Mapping Empires: Colonial Cartographies of Land and Sea*,
Lecture Notes in Geoinformation and Cartography,
https://doi.org/10.1007/978-3-030-23447-8_2

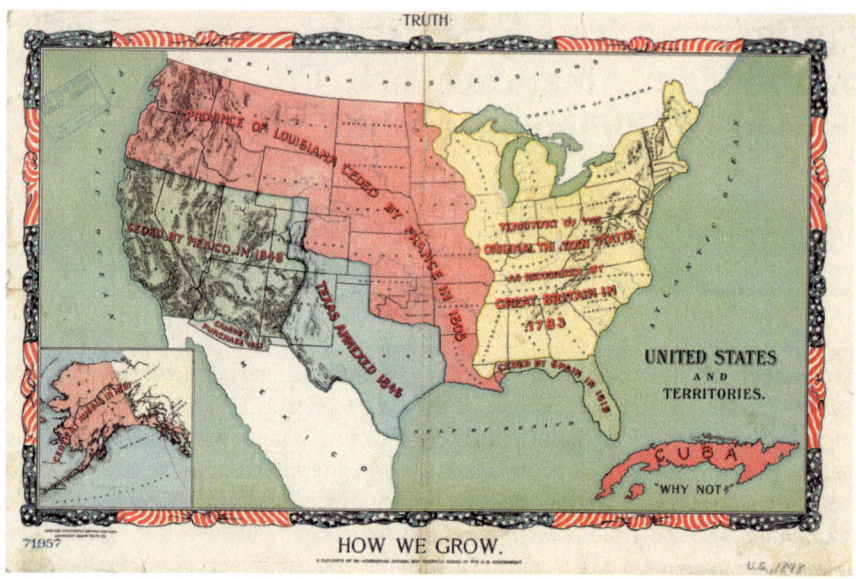

**Fig. 1** Map of US states and territories 'How We Grow' published in *Truth* (1898). Courtesy of NYPL Digital Collections, Image ID: 57081093

nation, complete with the 1853 Gadsden Purchase of today's southern Arizona. This map became a template for discussions on why and how the US could possibly extend her 'realm'. The wish to establish a European-style empire was fuelled by US neo-imperialism that had been economically induced after the Great Depression of 1873–1896 and the Panic of 1893. It was also driven by expansionist ideas as in Mahan's (1890) book *The Influence of Sea Power upon History* and by logical continuation of colonizing the North American continent (the 'Manifest Destiny').

In 1898, *Truth*, at that time a weekly full-colour humour magazine, published the map 'How we grow' (Fig. 1) as a centre-spread, calling it 'A duplicate of an interesting official map recently issued by the U.S. Government'. It shows Cuba, coloured as per other US states and territories, slightly rotated anti-clockwise and positioned very close to Florida. The date of inclusion into the Union (either by annexation or cession) is indicated for all territories, whereas Cuba has the words 'Why not?' written underneath. Although this did not reflect official government intentions,[1] it recapped public opinion, constantly heated by the yellow press.

In fact, the Cuban question led to the Spanish-American War of 1898. After several military uprisings to gain independence since 1868, the revolt of 1895 started with the death of the intellectual leader José Martí and marked the beginning

---

[1]This discussion goes as far back as 1823, when Thomas Jefferson wrote to James Monroe that to add Cuba 'to our confederacy' would 'round out our power as a nation' (Savelle 1967: 17). For details about the numerous associated attempts, see Hard (2003: 144).

of the Spanish retreat. A group of Cuban exiles (Cuban Junta) started to swing US popular opinion to the side of the rebels and sought financial support for the guerrilla war. The Spanish government 'introduced a policy of re-concentration, which involved the forced relocation of peasants from their homes and farms into fortified government-run cities and camps' (Tucker 2009: 163). In 1897, the desire to protect US commercial interests on the island led to the deployment of the battleship *USS Maine* to Havana, which exploded there in unexplained circumstances on 15 February.

Criticism of the ongoing brutality in enforcing the re-concentration rules led to a US blockade and finally to the declaration of war on 21 April by the US Congress. This was followed by the destruction of the Spanish fleet in the Philippines (1 May) and the establishment of a US military government in Manila (August 1898), the US conquest of Cuba and Puerto Rico and the defeat of the Spanish troops and naval forces (August 1898), and the establishment of an occupation government to secure a stable regime for protecting US economic interests. However, the negotiations that led to the Treaty of Paris signed on 10 December 1898 that formally ended the Spanish-American War also ended US dreams to add Cuba to the Union.

## 2 The Spanish-American War and Its Results: The Case of the Philippines

Although having won this 'splendid little war'[2] for the US, the outcome defined in the 1898 Treaty of Paris was ambiguous:

- The United States entered the club of colonial powers, established imperial footholds and appeared on the stage of world politics;
- Cuba gained independence (under US tutelage);
- Puerto Rico and Guam were ceded to the US as indemnity; and
- The Philippine Islands were ceded to the US for $20 million.

The case of the Philippines became increasingly complex. Having invited the leader of the Philippine independence movement to join the US effort against the remaining Spanish troops, Emilio Aguinaldo left his short exile in China and gathered 20,000 troops to fight against the Spanish. Soon after, he released the Philippine Declaration of Independence from Spain, and on 23 June set up a Revolutionary Government with himself as president. In August 1898, the US forces captured Manila and established a US Military Government, ruling out Aguinaldo, who was not allowed to take part in the talks of the Treaty of Paris.

---

[2]This phrase was coined in 1898 by John Hay, United States Ambassador to the United Kingdom, in a letter he wrote to his friend Theodore Roosevelt: 'It has been a splendid little war; begun with the highest motives, carried on with magnificent intelligence and spirit, favoured by the fortune which loves the brave' (Millis 1988: 335).

The Philippine Revolutionary Government recognized neither the treaty nor US sovereignty. The resulting tensions between US military forces and the Philippine revolutionaries led to several incidents, one of which, on 2 February 1899, finally led to the Philippine-American War. This lasted for two years and caused approximately 500,000 casualties,[3] most of whom were civilians having died of famine or disease, including cholera.

## 2.1  US Colonial Mapping of the Philippines: Key Cartographic Stakeholders

News of the sinking of the Spanish Fleet on 1 May by Admiral Dewey took some time to reach Washington. Various stories are associated with the reactions of the US Government, especially of President McKinley (Losang and Demhardt 2018: 102). In an interview with *The Christian Advocate* on 22 January 1903, after telling the story of his personal struggle to take the Philippine Islands, he described his eureka moment: 'I went to bed, and went to sleep, and slept soundly, and the next morning I sent for the chief engineer of the War Department (our map-maker), and I told him to put the Philippines on the map of the United States' (Rusling 1903).

The reason behind McKinley's demand is ambiguous. One interpretation is that he was responding to criticism of the decision to support the annexation of the Philippines in the wake of the war, thus taking on 'The White Man's Burden'.[4] Regarding the lack of reliable US maps, he might have also meant in his interview that the mapping and surveying institutions, namely the War Department, were called into action. Whatever McKinley had in mind, several institutions sent staff to the Philippines soon after the decision to embark on the colonial adventure. Five of these made a reasonable effort to put the Islands on US maps, although they employed different methods. These were:

- USASC (the US Army Signal Corps);
- USACE (the US Army Corps of Engineers);
- USC&GS (the US Coastal & Geodetic Survey);
- USGS (the US Geological Survey); and
- The National Geographical Society.

---

[3]The number of causalities is an estimate, compiled from different sources.

[4]*The White Man's Burden* is a poem by Rudyard Kipling that was published (on the eve of the Philippine-American War) on 4 February in London's *The Times* and on the 5 February in the *New York Tribune* and *Sun*. Written to encourage the American annexation, Kipling admonishes the addressees to risk the imperial adventure, but reminds them of the costs. The poem became a euphemism for imperialism and sketches the moral burden of the white race, which is divinely destined to civilize the brutish and barbarous parts of the world by encouraging economic, cultural, and social progress.

With the exception of the National Geographic Society, these were governmental institutions paid by Congress, but they lacked an effective coordination of their efforts. The reasons included: having different (and changing) remits; being assigned different tasks and operating with fluctuating levels of financial support; the lack of a central inventory of the maps acquired (hence, research efforts in different archives were often duplicated); and the rarity of exchange of information and base mapping between organizations (except in the case of the USASC and USACE).

### 2.1.1  The United States Army Corps of Engineers (USACE)

The USACE is one of the oldest operational units of the US Army and dates back to the American Revolution, when, in 1775 the Continental Congress authorized the first Chief Engineer to build the fortifications for the battle of Bunker Hill. Besides undertaking a range of obligatory military tasks, which involved building and improving transportation, infrastructure, fortifications and supply technologies, conducting surveys and reconnaissance, and directing siege operations, the Corps gradually became specialized in civil works, such as the building of hydropower plants and setting up flood-defence schemes. The most significant of these was their involvement in the construction of the Panama Canal.

For large-scale operations, maps, plans and charts were in constant demand. The Corps took over the battlefield and reconnaissance mapping by setting up mobile printing units and so it is not surprising that the USACE produced the first US maps of the Philippines. Most of these were simple battlefield maps of the Philippine-American War (Fig. 2). The sheer number of maps produced emphasizes their increasing importance in conducting modern warfare. Thus, the 1900/01 Annual Report to Congress mentions that 8,800 maps were produced and distributed throughout the division. In addition, the Corps organized and produced maps of the unsurveyed hinterland of Manila by supplying 'Instruments of various kinds [...] to officers in the division, for use in obtaining data for the compilation of maps' (Chief of Engineers, US Army 1901: 44). The report lists various tasks, of which the following indicates the strategic importance of maps made by USACE: 'Assembled material for pontoon trains, bridge construction, and railroad, repairs, and prepared maps for use of Major-General Lawton, Major-General MacArthur, and Brigadier-General Wheaton prior to their movements in Northern Luzon in October, 1899'. When the second battalion of Engineers set sail for the Philippines it 'was fitted out for service with astronomical and surveying instruments, reconnaissance and drafting materials, field photographic outfit [...] and a library of technical books' (Chief of Engineers, US Army 1902: 40). This might have included the new *Engineering Field Manual* with its chapters on surveys, reconnaissance and map compilation. This reinforces the idea that detailed Spanish maps of the main island (Luzon) did not exist, or, as will be seen, were inaccurate or outdated, such as the plans of Manila and its surroundings.

**Fig. 2** Map showing operations of the Second Division, 8th Army Corps, from 1 March to 31 May 1899; Major-General Arthur MacArthur commanding, United States Army Corps of Engineers, 1899. Courtesy of Library of Congress Geography and Map Division Washington, D.C. 20540-4650 USA dcu, Control Number 2009582412

By 1910, the USACE was heavily involved in surveying the interior of the islands. 'Engineer officers and troops are now engaged in making extensive military surveys […] while urgent requests for additional officers for duty in connection with the survey of the Philippine Islands have been received' (Chief of Engineers, US Army 1910: 5). Unfortunately, the USACE reports, although very detailed, do not provide details of the number and type of maps produced.

## 2.1.2 The United States Army Signal Corps (USASC)

The United States Army Signal Corps provides and manages communications and information systems to support the combined US armed forces. Although the technology in use changed, the Signal Corps' tasks remained the same and focussed in the Philippines on visual signalling (including heliography) and the supply of telephone and telegraph wire lines and cable communications. To manage these tasks, the requirement for detailed topographic information was essential. The Signal Corps employed photography and renewed the utilization of balloons to acquire spatial information while in combat. Despite having produced thousands of battlefield sketches and maps, the USASC published an overview map for its report, showing the progress of the construction of telegraph lines throughout the Philippines in 1900. This map is based on a Spanish hydrographic chart (Carta General del Archipiélago Filipino (en dos hojas) Levantada Principalmente por la Comisión Hidrográfica al mando del Capitan de navío D. Claudio Montero y Gay hasta el año 1870 con adiciones hasta 1875).

The Signal Corps' map was printed in an unusual format (107.95 × 52.07 cm) and matches the size of the montaged two-sheet original exactly, even showing minor fitting errors where the original map frame has been cut. A two-sheet version of the map was also available. The colour overprint, showing US military and telegraph lines and British underwater cable lines are drawn very crudely against the delicate design of the original. It reflects the lack of US maps of the Philippines at the time of the annexation. However, this very detailed map, like other hybrid maps, was replaced in subsequent years by a rather functional outline map.

## 2.1.3 United States Coastal and Geodetic Survey (USC&GS)

The USC&GS's tasks today are part of the National Oceanic and Atmospheric Administration (NOAA). Originally the United States Survey of the Coast in 1807, it was the first scientific agency of the government and started surveying by measuring and verifying the first baseline for setting up a triangulation network in 1817.

In 1836 the United States Department of the Navy was given control over the renamed United States Coast Survey, which led to a strong military influence that was not always beneficial to the long-range goal of surveying the US coast. Since the ships of the Survey belonged to the Navy and only Army and Navy officers were allowed onboard during missions, the agency strongly relied on trained staff that had been supplied directly from the Army and Navy academies. During wartime, both withdrew their officers, to replace them afterwards by fresh staff from the academies. This caused the constant change of staff, often for ongoing surveying operations.

The USC&GS was established in 1878 during the economic crisis and therefore lacked secure financial support from Congress. Finally, in 1900, it was re-established as an entirely civilian organization, solely responsible for its surveying ships and crew. It was at that time of reorganization when the Philippine task arose. After the loss of cruiser *USS Charleston* in shallow waters using Spanish pilot charts, the necessity for more accurate charts became obvious. Negotiations with the War Department and Philippine Commission led to the USC&GS obtaining the necessary legislation to start coastal surveying and a field office was established in Manila by 1900.

The surveyors and the hired Philippine staff faced multiple challenges during their work. These included constant fighting and guerrilla warfare even after hostilities had formally ended; weather conditions (e.g. typhoons with heavy rainfall); unstable coral reefs and sandbanks; the so-called 'wild men' of the non-Christian tribes (Samales) who often mistook surveyors for tax collectors; and salt-water crocodiles in the estuary waters.[5]

By 1907, five vessels were permanently engaged in hydrographic and topographic surveying, stretching the definition of 'coastal' to the maximum by covering complete Islands (e.g. Map 4714/1912: Southwest Luzon and Mindoro— 1:400,000 or Map 4305 Mindoro and vicinity—1:200,000) or stretching the survey along major rivers (e.g. Map 4260/62/1903 Cagayan River). By 1914 the USC&GS had mapped the complete archipelago, focussing on the principal harbours and bays (large scale) but also producing complete regional overview maps (smaller-scale maps, whose production had started in 1910) (USC&GS 1914). During different surveying campaigns (Fig. 3) basic data were collected from the southern tip of Taiwan to northern Borneo, which yielded 294 charts of the Philippine coastline and larger rivers, tide tables and sailing directions (coastal pilot charts). Regarding this production effort, the USC&GS outpaced the US Geological Survey (below) in providing reliable maps for public use due to the use of surveying vessels to facilitate mapping and charting activities.

### 2.1.4   The United States Geological Survey (USGS)

The USGS was established by an Act of Congress in 1879, charged with responsibility for the 'classification of the public lands, and examination of the geological structure, mineral resources, and products of the national domain' (Rabbitt 1975: 3). As a non-military, fact-finding agency, it was originally established to study the geological structure and economic resources of the national landscape. Contesting for public funds with the USASC and the USC&GS, the USGS became the main producer of topographic maps from 1894. In 1895, the USGS took over the survey to investigate the coal and gold resources of Alaska, which was followed by responsibility for surveying associated with the planned Nicaraguan Canal. After 1900, the work of the USGS encompassed four major science disciplines, i.e. biology,

---

[5]NOAA provides webpages on its organization's history, with a section called 'Philippine Tales' that includes eyewitness accounts of the surveying adventures of USG&CS staff (https://www.history.noaa.gov/philippine_tales.html).

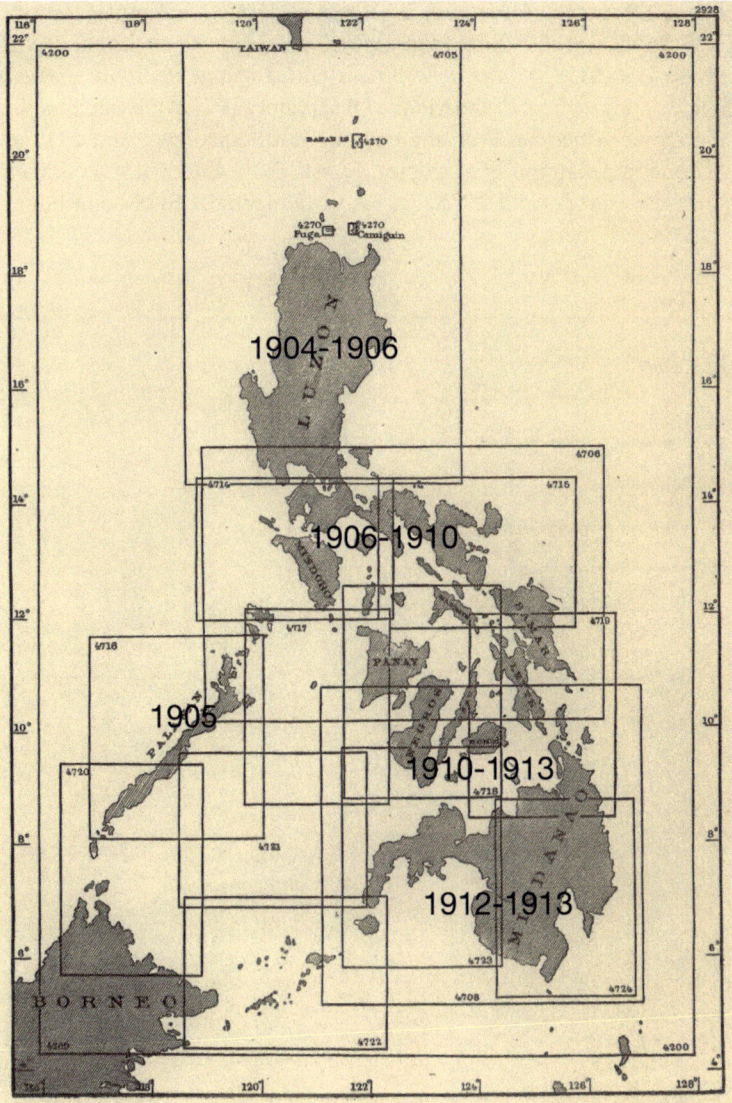

**Fig. 3** Progress of survey campaigns in the Philippine Islands by the US Coastal and Geodetic Survey. The base map is taken from US Coast and Geodetic Survey 1914, Catalogue of Charts (modified by the author)

geography, geology, and hydrology. It became responsible for surveying and producing the national topographic and geological map. However, only 26 per cent of the country, including Alaska, had been topographically surveyed by 1904.

As its first engagement in the Philippine Islands, the USGS commissioned the German geologist George Becker to provide thorough information on their geology

and mineral resources. Published as Volume 3 of the Annual Report of the
Geological Service (1899/1900), the report contained maps simultaneously published in the USC&GS's *Atlas de Filipinas (Atlas of the Philippine Islands)*, prepared by the cartographers of the Manila Observatory (Fig. 4). Even after the end of
the of Philippine-American War, the ongoing hostile actions against US advances
inland and the impenetrability of tropical forests did not allow for a comprehensive
mapping of the Islands until 1935. The most influential USGS contribution to the

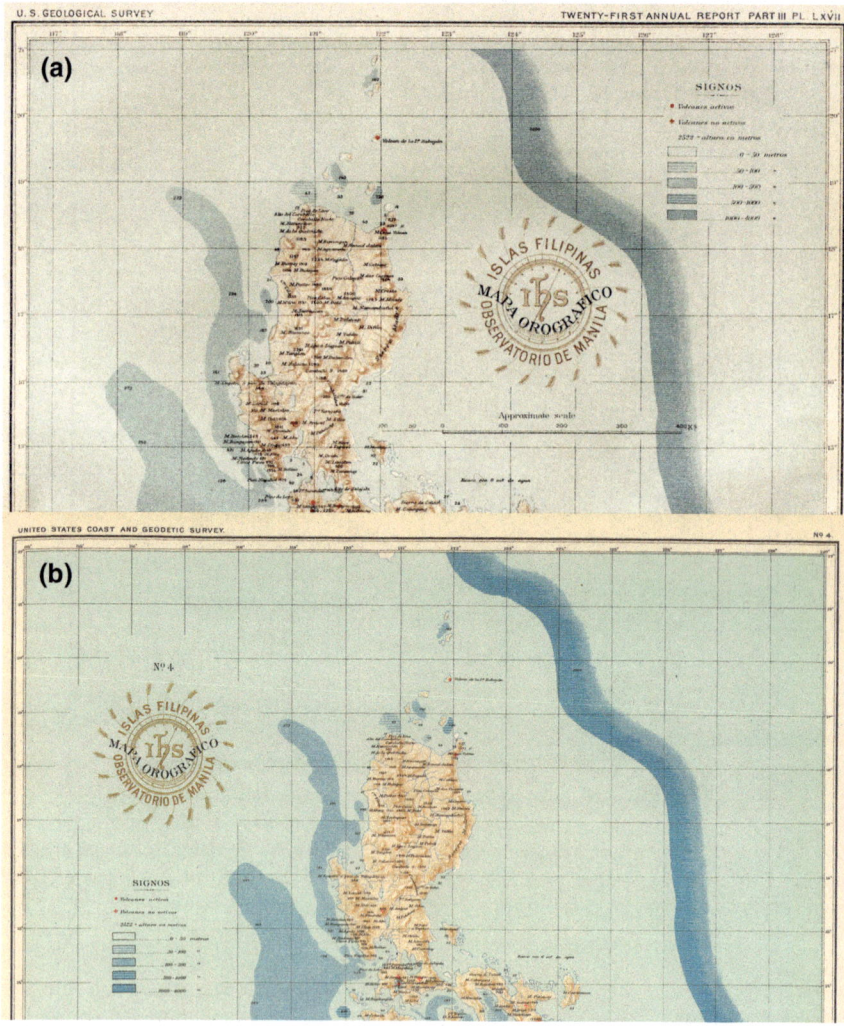

**Fig. 4** Details from the map of the physical geography of the Philippine Islands from **a** the Becker
Report (US Geological Survey) and **b** the *Atlas de Filipinas* (US Coastal and Geodetic Survey).
Courtesy of David Rumsey Map Collection: http://www.davisrumsey.com

early US maps of the Philippines was probably the appointment of the Principal Chief Geographer for the Geological Survey and Chief Geographer for the United States Census, 1890 and 1900 as Chief Geographer for the Philippine Census.

## 2.2  The Philippines Census

The first US Philippines Census took place soon after the approval of Public Act 467 by the US Philippine Commission. The Act provided for a census of the Philippines and the establishment of a Bureau of Census in the Department of Public Instruction (NLLRD 154). Although the Bureau was a temporary one, it closely cooperated with the US Bureau of Census and employed specialists such as Victor Olmsted and Henry Gannett, both having been Assistant Directors to the Cuban and Puerto Rican Censuses conducted in 1901 and 1902. Henry Gannett, later co-founder of the National Geographic Society (1888) and American Association of Geographers (1904), had been Geographer of the United States Census in 1880, 1890, and 1900. He introduced topographical mapping methods to precisely lay out the enumeration districts for the US Census and engaged in the production of several gazetteers, e.g. of Cuba and Puerto Rico. Anderson (1988: 30) sees colonial censuses as subjective texts, that can be interpreted (besides maps and museums) as an imagined cultural context for inhabited territories. This rather colonial interpretation may not fit the Philippine Census, whose main focus was to enable the US Government to establish a supervised democratic political system in the aftermath of the Philippine-American war (cf. Sloane 2002: 18).

Unlike the Cuban and Puerto Rican censuses, Philippine territory was hardly covered by reliable maps, so that establishing appropriate census districts became a crucial matter. To avoid the vague allocation of retrieved numbers and facts, Gannett had the idea to let the Philippine people locate themselves by providing a system in which 'presidents of the towns prepare diagrams of their respective municipalities, showing the relative location and approximate distance from the main barrio or seat of municipal government' (USBC 1905, vol. 1: 14). The method was tested but needed a sample map (Fig. 5) to explain the written instructions. The results enabled Census officials to divide municipalities having greater than 2000 inhabitants and to aggregate those having fewer than 1000. By using approximations of area sizes and distances to district centres, the method may be interpreted as the first 'thematic triangulation' of unknown territory and uncertain distance measurement to build comparable census units.

The 1903 Census comprised numerous maps compiled by scientists of the Manila Observatory, established and supervised by the Jesuit Order. Worth emphasizing are the maps on Philippine geology, earthquakes and volcanoes or the influence of tropic storms (cyclones) on the overall climate of the archipelago (USBC vol. 1: 172), with most topics already depicted in the *Atlas of the Philippine Islands*, whose circumstances of production have been outlined by Losang and Demhardt (2018: 112–116). The employment of approximately 7000 Filipinos as

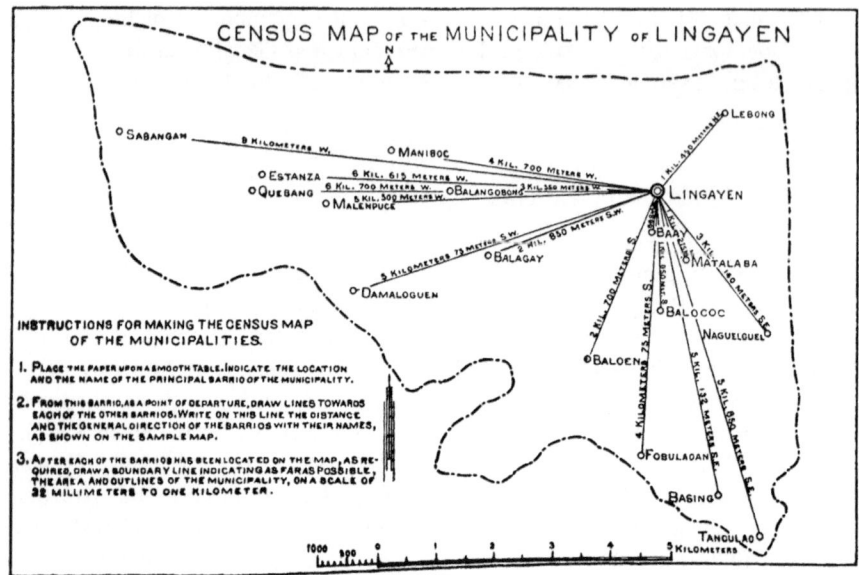

**Fig. 5** Mapping template for local communities, US Census on the Philippines 1902. From: USBC (1905, vol. 3, p. 15). Courtesy of archive.org (https://archive.org/details/ajb5835.0001.001. umich.edu)

supervisors and enumerators to conduct the Census and the use of Spanish as the language of the Census underline its purpose, as does the self-determined local-ization of barrios within communities.

# 3   Colonial or Imperial Mapping: Some Considerations

'To be or not to be' might be the question. 'To be or not to be mapped' is a question of power relations. Maps are, of necessity, selective representations of the envi-ronments their authors seek to portray, thus being intrinsically a reflection of the power of the interests that created them (Wood 1992: 71). The first years of US control after facing a serious insurrection, saw rather moderate mapping efforts. Wherever possible, these involved the local population, the educated elite and scientists, to replace the Spanish maps that continued in circulation. The publication of the *Atlas de Filipinas* as a co-production of the USC&GS and the Manila Observatory (later becoming the US Philippine Weather Bureau) reveals some aspects that suggest US mapping of the Philippines never reflected colonial claims but imperial strategy:

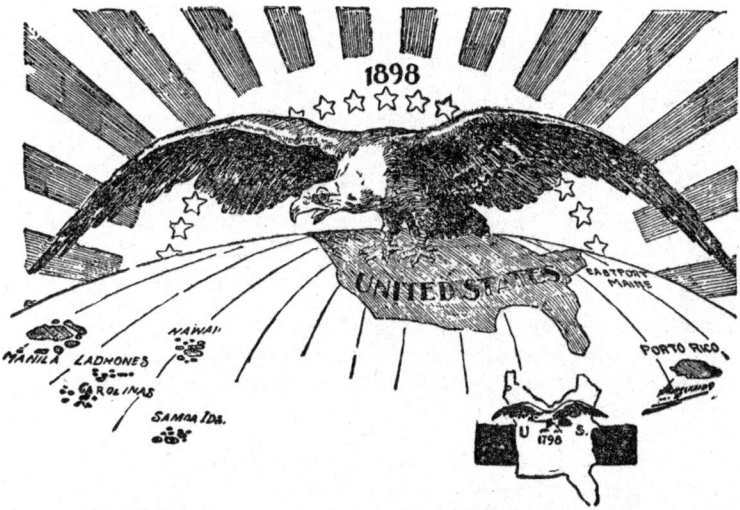

Ten thousand miles from tip to tip.—Philadelphia Press.

**Fig. 6** 'Ten thousand miles from tip to tip', published by Philadelphia Press. Image courtesy of Cornell University Library, Digital Collections, Persuasive Cartography—the PJ Mode Collection: Cornell University Library (https://digital.library.cornell.edu/catalog/ss:3293830)

– A decentralized administration seeking spatial information to improve the situation towards an elected (but controlled) government, thus incorporating federal principles;
– The use of existing federal institutions and procedures to gather this information (in particular the USC&GS and the Census); and
– The quality of integration of indigenous people in gathering information (Manila Observatory, participation in the Census).

In that sense, US Imperialism—and thus imperialist mapping—occurred as a direct continuation of the former Westward Expansion (Manifest Destiny) of the North American Frontier. In the following years, the domestic political clashes about being a colonial power vanished behind the imperial curtain on which the United States of America stretched ten thousand miles from tip to tip (Fig. 6).

# References

Anderson MJ (1988) The American census: a social history. Yale University Press, Newhaven
Chief of Engineers US Army (1901) Annual report of the Chief of Engineers, United States Army. Government Printing, Washington, D.C
Chief of Engineers US Army (1902) Annual report of the Chief of Engineers, United States Army. Government Printing, Washington, D.C
Chief of Engineers US Army (1910) Annual report of the Chief of Engineers, United States Army. Government Printing, Washington, D.C

Edney M (2009) The Irony of Imperial Mapping. In: Akerman J (ed) The imperial map: cartography and the mastery of empire. University of Chicago Press, Chicago, pp 11–48

Gannett H (1905) The Philippine census. Bulletin of the American Geographical Society 37 (5):257–271

Hard J (2003) Comparing empires: European colonialism from Portuguese expansion to the Spanish-American War. Palgrave Macmillan, New York

Losang E, Demhardt IJ (2018) Change of sovereignty and cartographic advance: cartographic implications of the Spanish-American war of 1898. In: Altić M, Demhardt IJ, Vervust S (eds) Dissemination of cartographic knowledge: 6th International symposium of the ICA Commission on the history of cartography. Springer, Cham, pp 99–128

Mahan AT (1890) The influence of Sea Power upon History, 1660–1783. Little, Brown & Co., New York

Millis W (1988) The martial spirit. Ivan R. Dee, Chicago

NLLRD (National Library Legislative Reference Division) (Philippines) (1918) Checklist of publications of the government of the Philippine islands September 1, 1900, to December 31, 1917. Bureau of Printing, Manila (Available at: https://quod.lib.umich.edu/p/philamer/AEZ4806.0001.001)

Rabbitt MC (1975) A brief history of the U.S. Geological Survey. U.S. Government Printing Office, Washington, D.C

Rusling J (1903) Interview with President William McKinley. The Christian Advocate, 22 January 1903, 17. In: Schirmer D, Shalom S (eds) The Philippines Reader. South End Press, Boston, pp. 22–23

Savelle M (1967) The origins of American diplomacy: the international history of Angloamerica, 1492–1763. Macmillan, New York

Sloane, FM (2002) The Philippines censuses of 1903 and 1939 and the representation of women's occupations. PhD thesis, James Cook University, Townsville Australia. Available at: https://researchonline.jcu.edu.au/86/

Tucker SC (ed) (2009) The encyclopedia of the Spanish-American and Philippine-American Wars: a political, social, and military history. ABC-CLIO, Santa Barbara

USBC (United States Bureau of the Census) (1905) Census of the Philippine Islands. Taken under the direction of the Philippine Commission in the Year 1903. In four volumes. United States Bureau of the Census, Washington, D.C.

USC&GS (United States Coast and Geodetic Survey) (1914) Catalogue of charts, sailing directions, and tide table of the Philippine Islands. Government Printing Office, Washington, D.C.

Wood D (1992) The power of maps. The Guilford Press, New York

**Eric Losang** studied Geography, History, Political Sciences and Economics at Trier University in Germany, followed by two years working in the international tourism industry. He joined the Leibniz Institute for Regional Geography in 1999 as a researcher and works on several atlas projects, such as the National Atlas of Germany and the Digital Atlas of Geopolitical Imaginaries of Eastern Central Europe. His research focuses on production modes and techniques in geovisualization, atlas concepts and cartographies. Recently, he has worked on developing pragmatic approaches to critical cartographies. He is Vice Chair of the ICA Commission on Atlases and a Visiting Lecturer at the Free University Berlin, the Research Academy in Leipzig and the University of Leipzig.

# The Exploration and Survey of the Outlying Islands of the Dutch East Indies

Ferjan J. Ormeling

**Abstract** The start of mineral exploitation and the gradual establishment of a network of government administrators in the outer islands of the Dutch East Indies in the 1850s made the establishment more conducive for systematic exploration and mapping. However, it took until the 1880s before this started—by that time the mapping brigades of the Topographic Survey in Batavia had finished their work on Java, and the triangulation, survey and mapping of Western Sumatra and Western Borneo had begun. All of this happened somewhat half-heartedly, as parliament in the Netherlands again and again tried to restrict these operations in order to economize. These restrictions were lifted when Van Heutz became Governor-General. His vision of the colony as a unitary state (not just as Java and a couple of vassal sultanates) meant imposing Dutch presence (Pax Neerlandica) and administration throughout the archipelago. In addition, as maps formed the basis for good administration, the funding of the Survey was more secure. Southern and Eastern Sumatra were tackled, as was southern Borneo, and the systematic mapping of Celebes commenced. A regular army exploration program for New Guinea was set up, and by the 1920s most of that island was no longer a white patch on the map. This chapter links changes in government policies towards mapping activities and the resulting map series of the outer islands that were produced by the Topographic Survey in Batavia from the 1880s until 1950, when the Survey was transferred to the independent Indonesian authorities.

## 1 Introduction

In 1883, heated discussions took place in the Dutch Parliament about the decision by the Governor-General of the Netherlands East Indies in the previous year to abort the proposed triangulation of Western Sumatra. The mapping of the central island of Java had been completed and the six mapping brigades of the

F. J. Ormeling (✉)
University of Amsterdam, Amsterdam, The Netherlands
e-mail: f.j.ormeling@uva.nl

© Springer Nature Switzerland AG 2020
A. J. Kent et al. (eds.), *Mapping Empires: Colonial Cartographies of Land and Sea*,
Lecture Notes in Geoinformation and Cartography,
https://doi.org/10.1007/978-3-030-23447-8_3

37

Topographical survey in Batavia (now Jakarta) were preparing to embark—triangulation personnel had been trained—when the Governor-General decided on 2 June 1882 (Governmental Decree No. 10) to remove the funding for the endeavour in next year's budget. This was proposed as an austerity measure, but there was more involved. Of the Dutch East Indies, it was only Java and some isolated strongholds like Padang, Makasar, Manado, Kupang and the Moluccas, that were directly under Dutch rule. The rest of the archipelago was under indirect rule, binding local rulers to the Netherlands by all sorts of treaties and contracts. There was a general consensus in the colony's administration to keep the Dutch involvement in the outer islands as little as possible. However, there were contrary forces, too: firstly, there was the army, involved in a war to subject the Aceh sultanate in northern Sumatra and aware of the need to have proper maps of the rest of the island; secondly, there were the trading companies eager to extend the tobacco and coffee plantations in Eastern Sumatra; and thirdly, there were mining companies keen to open-up coal mines in Western Sumatra and tin mines in Banka. Scientific institutions like the Royal Netherlands Geographical Society (Koninklijk Nederlands Aardrijkskundig Genootschap or KNAG) were clamouring for exploration and surveys, too (van den Brink 2010).

The debates in Parliament in The Hague—the place where almost everything was decided regarding the colony— ended in favour of the proposed triangulation and subsequent mapping of Western Sumatra, and a few months later triangulation started. Even so, in a sense it was a Pyrrhic victory; in order to bear the additional costs of the triangulation brigade, two surveying brigades had to be dissolved. Nevertheless, it meant the start of the systematic surveying and mapping of the outer islands, which, by the year 1942 (when the Japanese took over) or by 1950 (when the topographical survey was handed over to independent Indonesia), resulted in a country that had proper topographical maps for the homes of 95% of its population.

## 2 Unsystematic Mapping, 1816–1883

The above comments do not imply that the Dutch had not mapped any areas beyond Java before 1884. Since the restoration of Dutch authority over the colony in 1816, there had been an almost constant series of military actions to reassert Dutch sovereignty over outlying areas, as the Dutch colonial authorities had to remind local princes that the Netherlands government had taken over all responsibilities and rights of the Dutch East India company or defend what it saw as its sphere of influence. That frequently entailed military actions, which in turn were accompanied by topographic surveys. These military actions were quite bloody ones to sustain for a small country like the Netherlands. In the Java War (1825–1830) some 8000 Dutch soldiers died; the Padri wars in Western Sumatra (1821–1835) took 1600 Dutch casualties, and 12,000 Dutch and allied indigenous soldiers died in the

Aceh wars (1873–1905). The numbers of local victims must have been a multiple of these figures, for Aceh alone they are estimated at 100,000 (van 't Veer 1969).

The Padri wars started in 1821, when a number of Minangkabau princes in Sumatra's interior asked the Dutch at their coastal trading post in Padang for help against the Wahabi-inspired Muslims, who, on their return from the hajj to Mecca, had ousted them and set up a theocratic state under sharia law in west-central Sumatra. The Dutch military saw this as a good occasion to get a foothold in the interior, painted the event as something that threatened the whole colony, and, while sustaining heavy casualties, were able to build a chain of forts to subjugate an area the size of Scotland (Sumatra's West Coast of 50,000 km$^2$ and Tapanuli of 40,000 km$^2$). Reconnaissance maps were produced of the various principalities but not inter-connected ones, although a consensus was reached concerning the legend. In 1842, Captain LW Beyerinck arrived in Padang to correct and update the existing maps. As these did not match his standards of quality, he had to resurvey nearly all of them. In 1847, his officers had mapped the whole area at the scale of 1:50,000 in a series of partial surveys, supported by some astronomically determined positions. This series of 1:50,000 manuscript maps already bore the characteristics of later printed topographic maps, adopting the undulating brick-like symbol for paddy fields, the light green village areas with its black house symbols, and the dark green forests. In 1852, his overview map of the whole area at 1:500,000 was engraved and reproduced as *Kaart van het Gouvernement Sumatra's Westkust*. Beyerinck's maps were highly acclaimed by the Dutch military and were taken as muster for the subsequent systematic mapping of Java (1850–1884).

Other military actions linked to unsystematic surveys took place in Palembang, Jambi and Aceh on Sumatra, in Pontianak, Sambas/Montrado (against the Chinese gold miners that started an independent state), and Banjermassin (about the control over coalmines) on Borneo. Further surveys occurred in Boni on Celebes, in Lombok and in Bali, and maps of the contested areas were drawn by Dutch officers without using triangulation. The quality ranged from sketch maps and schematic maps of the distances between bivouacs to elaborate land-use maps, similar to the residency maps of Java (the *Kaart van Zuid-Celebes met uitzondering van het rijk Gowa, schaal van 1:200,000* of 1886). The only systematic mapping activities where the various surveys were linked and related to each other were the small-scale maps of the archipelago, like the well-documented map *Algemeene Kaart van Nederlandsch Oost-Indië* by G. von Derfelden van Hinderstein (Broeders 2007) and the atlases by Melvill van Carnbee and Versteeg (1853–1862) and Stemfoort and ten Siethoff (1883–1885), published at the instigation of the colonial government.

In 1873, this situation changed. In Aceh, a military conflict started that was to last at least 30 years. At first, the Dutch could only survey the area within a small perimeter around the capital Kutaraja (now Bandar Aceh), and this perimeter was gradually extended. It was only in the twentieth century that systematic mapping was possible beyond this perimeter, and even then, only in coastal areas. This is one of the reasons why the interior of the overview map of Aceh at 1:500,000 (the *Overzichtskaart van Atjeh en onderhoorigheden 1:500,000*), published in Batavia in 1901, is so devoid of names and symbols—the surveyors had no access there.

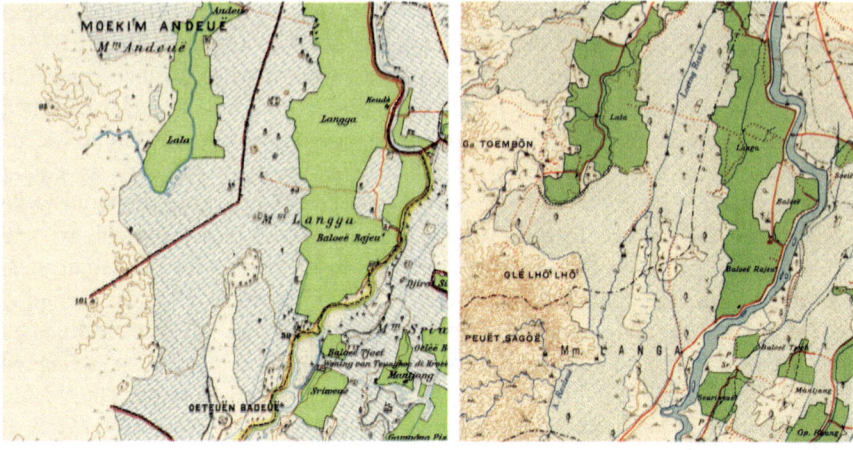

**Fig. 1** Details from the 1:40,000 map series Atjeh en Onderh: Moekim VII-Pidië: published in 1900 as Pidië sheet M (left) and published in 1916 as sheet IXa (right). Courtesy of Leiden University Library (shelfmarks KK 062-04-05 and KK 063-01-03)

The 1:40,000 coastal maps appeared in two editions, the first produced in the 1900s and the second a decade later. By then the situation had improved, in the sense that surveyors had better access (although they still had to be protected until 1912), which can be seen in the coverage of the maps (Fig. 1). The initially more dangerous mountainous areas were now mapped as well. In 1930, the government decided to extend the triangulation chain of Sumatra to Aceh.

Meanwhile, the political and economic situation had changed (Clemens and Lindblad 1989). In 1870, the Cultivation System (Cultuurstelsel), under which farmers had to plant one fifth of their fields with cash crops for the government (especially coffee), was abolished de facto[1] and this paved the way for the development of large-scale agricultural enterprises funded by Dutch or British private capital. While Java's high population density prevented the development of large-scale plantations, and instead saw the lease of part of the arable village lands for sugar or tobacco production, elsewhere, in the undeveloped and underpopulated outer isles like Sumatra, non-agricultural lands were opened up for plantations. Being under-populated, this caused a labour shortage and to safeguard the costs of bringing in agricultural labourers from China or Java, administrative measures and penal sanctions were proclaimed[2] that bound these indentured labourers. In order to guarantee the safety of people and property, the Dutch administration had to be

---

[1]The Sugar Law (Suikerwet) that was passed in 1870 ended the government's involvement in sugar cane cultivation and the Agricultural Law passed the same year made it possible for private companies to lease existing agricultural land while guaranteeing indigenous ownership. At the same time, uncultivated lands could be distributed as leasehold estates.

[2]The Koelie-ordonnantie (1880) imposed penal sanctions on imported labourers for breach of contract.

extended and estate boundaries were demarcated and mapped by topographic surveys. Accordingly, already before the turn of the twentieth century, the traditional non-intervention policy of the Dutch had been suspended within their sphere of influence. Around the year 1900, a spate of so-called 'pacification actions' imposed Pax Neerlandica all over the archipelago, with the exception of New Guinea.

# 3 Systematic Mapping Beyond Java

## 3.1 Western Sumatra

As initially stated, the triangulation of Sumatra started in Padang, half way along the west coast, in 1884. This was followed by a survey at 1:20,000 of the Padang lowlands and uplands—a region of 50,000 km$^2$. This consisted of a narrow coastal plain, a mountainous dorsal comprising two ranges with a plateau in-between, and part of the adjacent eastern lowlands. In 1867, the government had commissioned a geologist, de Greve, to prospect for minerals and to conduct a geological survey of Western Sumatra. In 1870, a major coalfield had been discovered in the latter at the upper reaches of the Indragiri river (Ombilin, Sawahlunto)—essential for a colony in the steamship age. A 1:100,000 geological map resulting from the survey was published in 1883.

In 1887, the government decided to build a railway over the mountains to the coast, with the aim of exporting the coal. A 1:20,000 series of the area was produced during 1889–1897, initially in black and white, but from 1890 onwards with brown contour lines added (these Sumatra sheets were the first in the colony to profit from this additional colour (Pannekoek 1946)). To begin to map an island the size of Sumatra (almost half a million square kilometres) at 1:20,000 required stamina if not hubris: the aforementioned territory required 160 sheets and for the whole island almost 6000 sheets would be needed. It was therefore decided to map the less economically developed areas at smaller scales such as 1:40,000 or 1:80,000. So-called 'military' maps at 1:40,000 were produced as well from 1893 onwards, which used an additional colour of green to indicate villages, and overview maps were produced at 1:80,000 or 1:200,000. The surveys continued until 1905, when the maps of Tapanuli residency, north of the Padang uplands, had also been completed. For the prosperous Padang uplands, a second edition of the 1:40,000 map was released in the 1930s in full colour, which was standard for the colony since 1898 (Fig. 2). In 1892, the Ombilin coalfields were connected by rail to the new port Emmahaven, just south of Padang.

## 3.2 South Sumatra

The Deli plantation owners in the north were clamouring to have their estates surveyed next. This could not be realized as the Batak areas located in-between, for

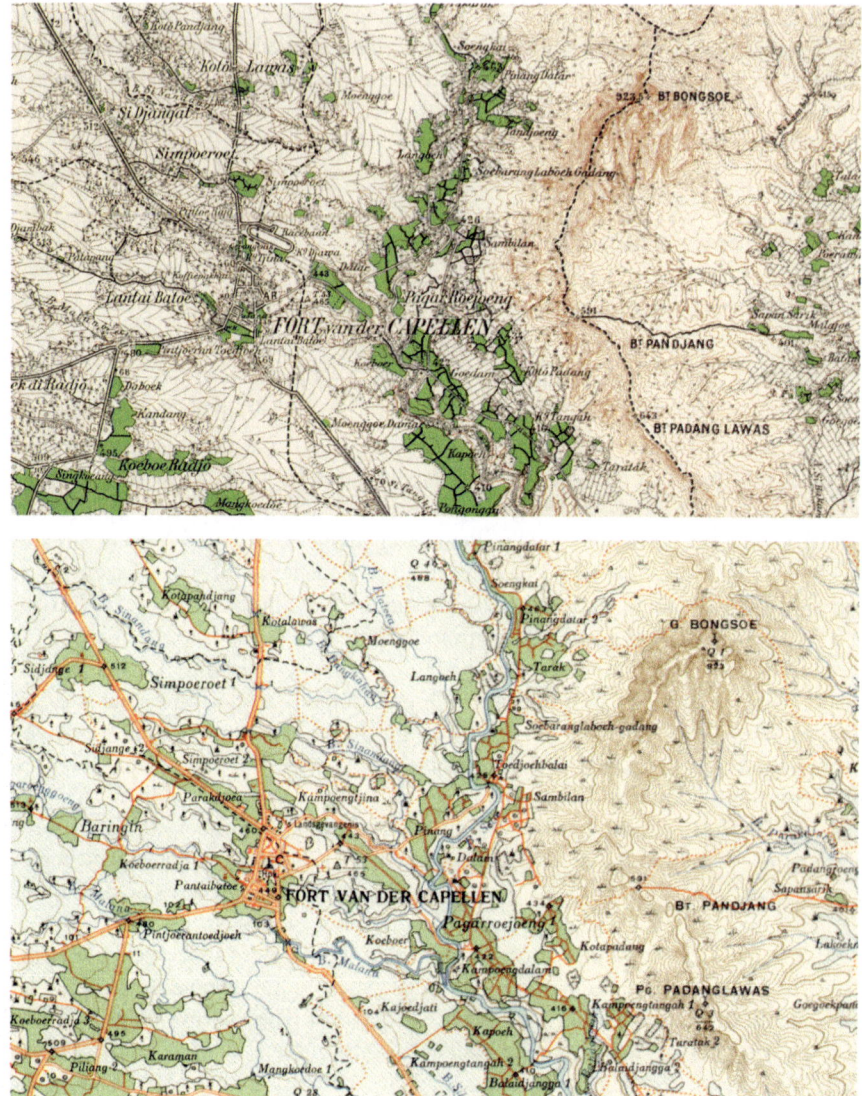

**Fig. 2** Details from the 1:40,000 series showing Sumatra's west coast, sheet 21: from 1896 (top) and from 1935 (bottom). Courtesy of Leiden University Library (shelfmarks KK 079-05-01 and KK ATL 222)

the time being, resisted Dutch pacification attempts and the proposed triangulation links with Deli were postponed. It was only after 1911 that these were established. The triangulation brigade now moved southwards instead, to link up with the Java primary triangulation net, so that the mapping of South Sumatra could start (1905).

This had been advocated since 1901 by the government, which was interested in building a railway line from Teluk Betung, the port closest to Java, to Palembang with its oil refineries and the Muara Enim coalfields. In 1905, the survey of Tapanuli was completed, and the two surveying brigades engaged there moved to South-Sumatra.

The parties interested in the construction of the South Sumatra railway had been asked to mark the route for the proposed line so that the route could be mapped at the scale of 1:25,000. However, after some areas had been mapped at that scale another route was chosen by the government. The Survey therefore decided in 1910 to complete the mapping of the area at the scale of 1:100,000. It was only in the 1930s that parts of the Lampung area, where Javanese and Balinese farmers were to be settled because of the transmigration (resettlement) policy, were mapped at the larger scale again. South Sumatra provided ample opportunities for the newly integrated indigenous surveyors to finish their training with some master's proof, like the mapping of volcanoes. A major problem in both the Palembang and Lampung residencies were the marshes: about 40% of the areas were seasonally inundated and 20% were permanent swamps, necessitating surveys by native crafts (perahus). The Survey actively participated in the planning of routes for roads and of settlements, and in its annual reports (Jaarverslag 1906–1940) criticized the local government's siting decisions (for example, the construction of Tanjungkarang as a communications hub in the 1915 annual report of the Survey).

## 3.3   Deli or Eastern Sumatra

In Deli sultanate (part of the residency of Eastern Sumatra), agricultural trading firms started to create tobacco plantations and tea estates in great numbers during the 1880s. As an area outside direct Dutch rule, it had not been mapped and it was up to the local princes to grant concessions for agricultural estates. Initially, these were not demarcated with sufficient precision. This led to a spate of boundary issues, and topographers complained persistently that they had to spend days negotiating and solving these boundary issues with the parties concerned instead of doing productive surveying work.

The impossibility to link the area of the Deli sultanate with the triangulation network in Western Sumatra, due to political unrest in the Batak area in between, necessitated an independent triangulation, for which a base measurement near lake Toba was taken in 1910. A fourth surveying brigade was created in 1911 consisting mainly of personnel engaged in ad hoc boundary surveys. The topographic maps produced here (1915–1925) after the triangulation were to a different scale (1:50,000) and had a different legend compared to the other series of Sumatra. They included an additional colour to indicate the height of the vegetation, and there were special symbols for the plantations, such as coffee, rubber, tea, copra and, later, oil palms. A distinctive feature of the maps—and not yet found elsewhere on Sumatra

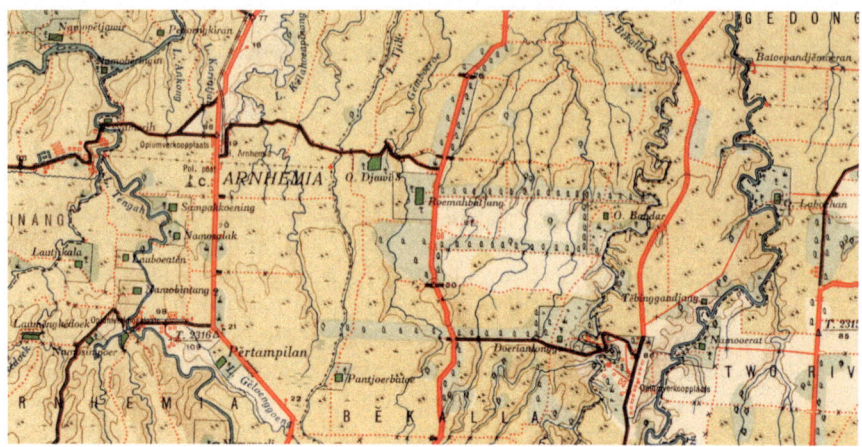

**Fig. 3** Detail from the 1:50,000 map series Oostkust van Sumatra, sheet 10D from 1915. The map detail shows three opium dens (opiumverkoopplaats). Courtesy of Leiden University Library (shelfmark KK 068-08-01)

—was the symbol used for opium dens (Fig. 3), which resembles the symbol for a public house on Ordnance Survey maps of Great Britain. The sale of opium was a government monopoly, as in British India, and the imported labour from China and from Java were an easy target for drug peddlers as the immigrants strove to cope with the harsh labour conditions.

In order to accommodate oil production refineries and shipping facilities of BPM, a Royal Dutch/Shell subsidiary, in northernmost Deli and adjacent Aceh, the 1:50,000 mapping was extended northwards to Pangkalanbrandan and Aru Bay. Gradually, improved safety conditions made it possible to map the eastern half of Aceh at this scale too. Agricultural interests, for oil palm plantations, pushed for an extension of the mapping programme in a southerly direction, turning the practically uninhabited marshy coastal part of East Sumatra (Asahan, Bengkalis) into productive areas as well.

## 3.4 Bangka

The reasons for the mapping of the isle of Bangka at a scale of 1:25,000 (1930–1936) were completely different again: Bangka was being mined for tin, like the adjacent island of Billiton. To codify the mining claims and plan the necessary infrastructure, this larger scale was chosen (Fig. 4). It was the first application of aero-triangulation. The aerial photographs were paid for by the mining companies, as the survey could not afford it during the depression. These were also interpreted for the exploration of tin and for assessing the forestry reserves suitable for charcoal production for the smelters used to process the tin ore. Due to the importance

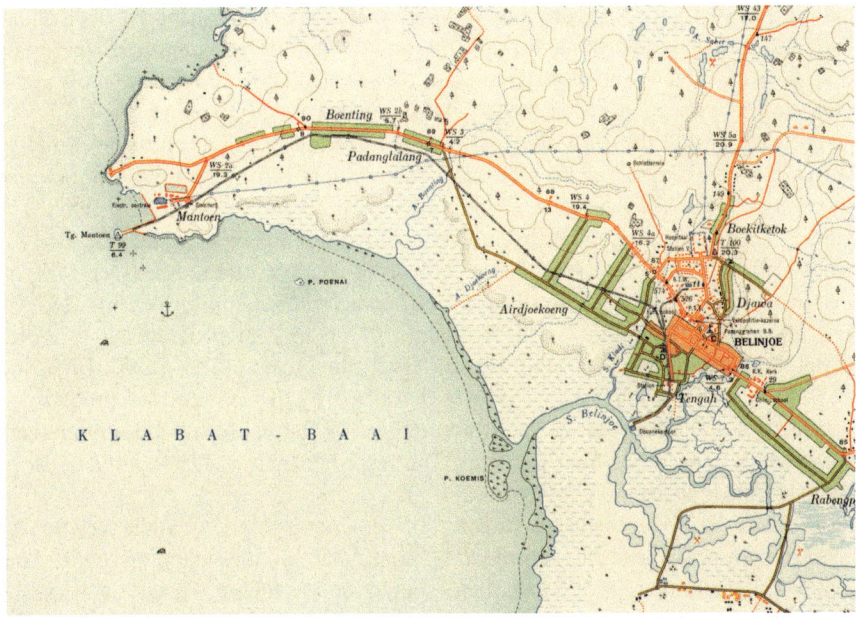

**Fig. 4** Detail from the 1:25,000 map series Res. Bangka en Onderh. (sheet 33/XXIIIq, from 1934), with the mining company town Belinyu, port facilities, power plant, tin smelter, railway, schools, company headquarters, churches, jail, government guesthouse, police barracks and shooting range, customs office and railway stations. Courtesy of Leiden University Library (shelfmark KK 083-07-02)

attributed to an accurate rendering of the microrelief, additional contour lines were rendered (at 3-metre intervals). A secondary levelling was also executed for accurate assessment of soil composition, to allow for a systematic study of the tin ore deposits.

Thus, the mapping of the outer islands certainly also served the interests of the trading, mining and logging companies. On topographical maps apart from the boundaries of the estates/plantations ('onderneming') the mining and the forestry claims were drawn in, too; the maps of Southern Borneo also testify to this.

## 3.5   West Borneo

The main reasons for the mapping of West Borneo were political. The encroaching of British adventurers like the Brookes or the British North Borneo Company on what was regarded as the Dutch sphere of influence, necessitated a demarcation of boundary lines. The subjugation of the Chinese goldmining kongsis or unions that had established independent states in the Sambas River area required military actions that should be supported by accurate maps. When the mapping of Java had been completed on 16 January 1886, the government decreed that two surveying brigades were to move to West-Borneo (a vast lowland area drained by the Kapuas River and its tributaries) in 1887, for its triangulation and mapping.

The mapping of this vast area also was the making of JJK Enthoven (Ormeling 1996), who organized a river-borne triangulation during which astronomical observations and the water-borne transport of time pieces were elemental. The Kapuas River and many of its tributaries could still be navigated and provided the means of transportation (the latter with ever-smaller crafts for over 900 km upstream). Ships with a depth of up to 3 m could navigate up to Sintang (465 km from the mouth) and those with a draught of up to 2 m could reach the town of Putussibau (902 km from the mouth). Over nine years (1886–1895), Enthoven worked in an environment that is best described by Joseph Conrad in *Heart of Darkness*, with a group of 12 surveyors and locally hired helpers and porters, and surveyed an area larger than England. The resulting map at 1:200,000 was to be printed from 5 plates.

In 1897, Enthoven became director of the Survey and as such saw to the establishment of a central coordinating committee in the mapping field. That committee, in turn, decreed that there should be no more overlap between the surveys of the various mapping establishments in the country, like those of the forestry department, the irrigation department, the railways, the cadastre, the hydrographic, geological and topographic surveys. Enthoven also homogenized the sheet system, integrated the land-use survey into the topographic service and, together with Colijn, sped-up the mapping of the outer isles.

## 3.6   Southeast Borneo

In these coastal areas, a situation existed that was similar to that of Bangka. Mining interests called for large-scale maps and were willing to pay for part of the expenses. The area in between Bandjermasin, Balikpapan and Samarinda, where coal was found and exploited since the 1850s, got an additional boost by oil strikes at the beginning of the twentieth century. The fields were acquired gradually by BPM, a subsidiary of Royal Dutch/Shell and in order to safeguard their infrastructure, as well as for strategic purposes, the area was mapped in the 1930s (Fig. 5). This included the major oil field of Tarakan Island, of which maps were not published. Parts of upland Borneo, on the banks of subsidiaries of the Barito River were mapped to allow for *transmigrasi* projects, the colonization by landless Javanese and Balinese farmers. This project in the Kandangan area was not mapped by one of the surveying brigades, but by one of the land rent brigades from Java.

**Survey of Indonesia's outer islands 1900–1942**

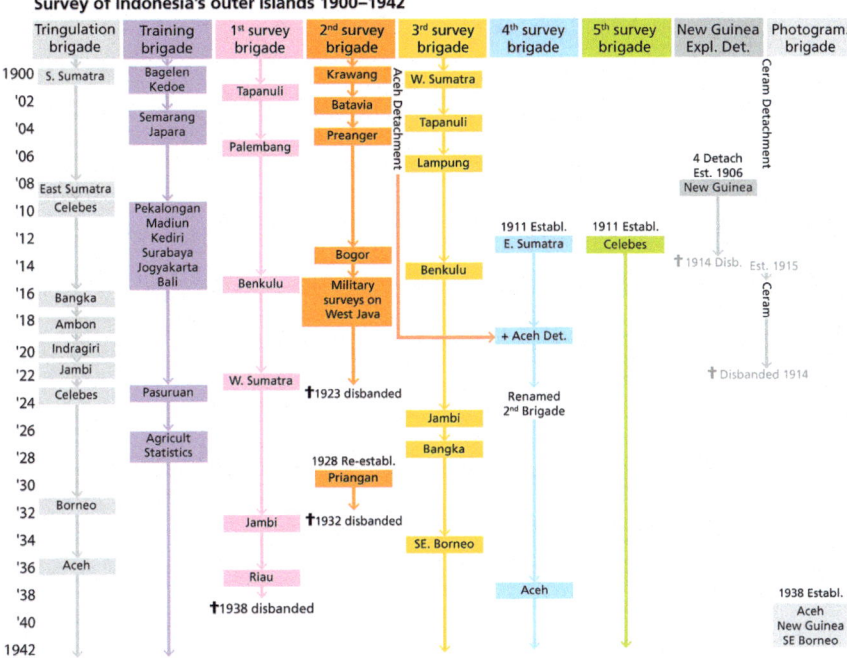

**Fig. 5** Activities of the mapping brigades of the Topographic Survey in the Outer Islands (1900–1942)

## 3.7 Lesser Sunda Isles

The latter policy also applied to the Lesser Sunda Isles. Bali and Lombok had only been brought under Dutch sovereignty recently by bloody campaigns (Lombok in 1896 and Bali in 1907), and to make amends and show the advantages of Pax Neerlandica, many infrastructural projects were undertaken for the benefit of the inhabitants. For a proportional distribution of the land rent, village holdings would be mapped by land rent brigades from Java as well. In 1918, an independent land rent surveying detachment was set up in Den Pasar, in charge of land rent mapping at 1:5000, and it was from these land rent maps that the topographic maps 1:50,000 of Bali (1924–1934) and Lombok (1926–1933) and parts of Sumbawa were produced. They differ in the legend from the maps of other Indonesian islands by the incorporation of symbols for Hindu temples (poera) and Hindu graveyards/cemeteries (Fig. 6), but otherwise cannot be discerned from the maps produced by the surveying brigades. This is a sign of the fact that the land rent brigades and the surveying brigades were being assimilated. Actually, on Java, the land rent brigades and the map revision brigade were merged into a new system under which three revision brigades were set up, for the provinces of West, Central and East Java instead of for the residencies. These revision brigades were meant to operate apart

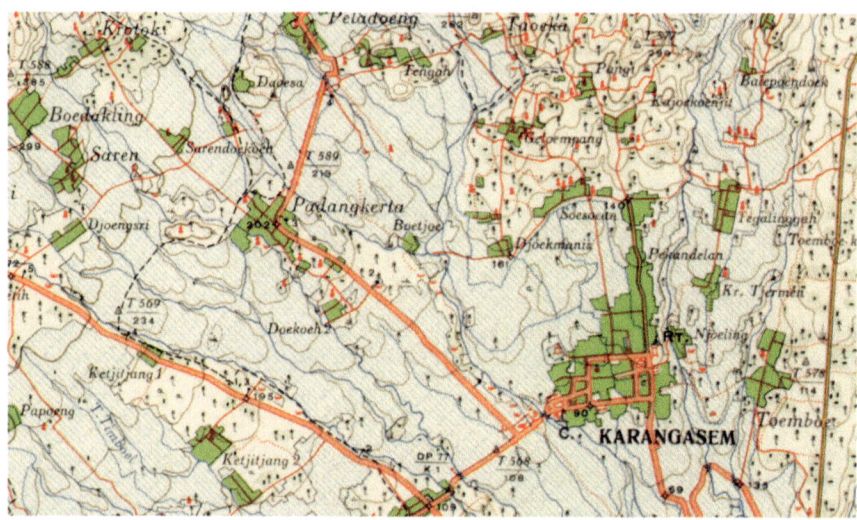

**Fig. 6** Detail from the 1:50,000 sheet from series Kl. Soenda-eilanden (Bali), sheet XLIV-63A, surveyed 1924–26 and published in 1928. Courtesy of Leiden University Library (shelfmark KK 101-07-09)

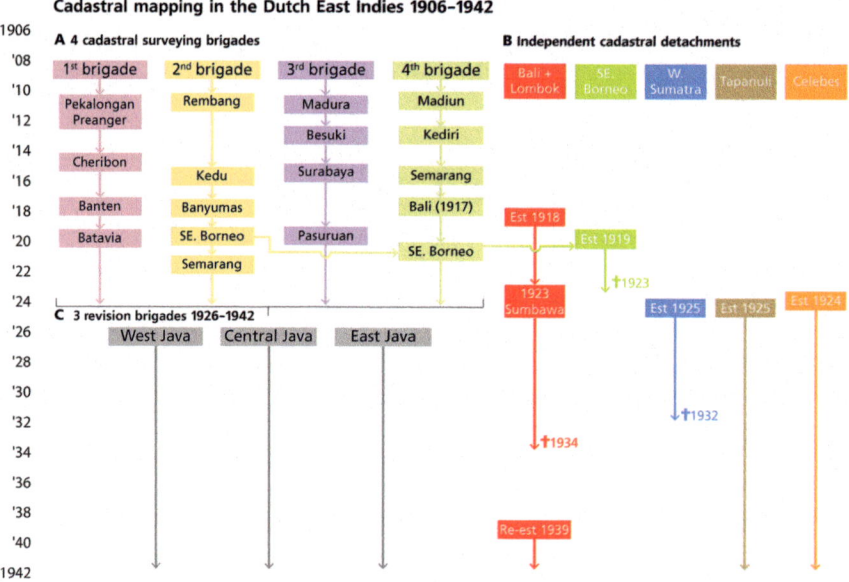

**Fig. 7** Activities of the land rent brigades on Java (left) and in the Outer Isles (1906–1942)

from the residencies, which were responsible both for revision of the topographic maps at 1:25,000 and 1:50,000 as well as for updating the village land rent maps at 1:5000 (Fig. 7).

## 3.8    Southwest Celebes

From 1908 onwards, Celebes was safe for surveyors and a systematic survey was started after the triangulation brigade had established its network, with a base measurement at Jeneponto. In 1911, a new (fifth surveying) brigade was set up for this purpose, and its priorities were set by the irrigation service. This can be seen in the various areas deemed fit for irrigation, where levelling pillars and bolts with their values are included on the map. The scales were 1:25,000 and 1:50,000; the first for the economically more promising agricultural areas near Makassar (1922) and the Tempe Lake area (1928–1931). Both the triangulation brigade and the surveying brigade moved northwards, the first to link up with the northern chain extended from the Minahasa area. By 1942, the topographers had reached the Toraja-area in the interior as well as the Malili transmigration fields at the head of the Gulf of Bone.

## 3.9    Smaller Outer Isles

Progress in mapping these territories can be summarized as follows:

Halmahera: surveyed 1912–1929, 1:100,000 maps published 1924–1934
Ceram: surveyed 1915–1922, 1:100,000 maps published 1919–1923
Ambon: surveyed (to be determined), 1:50,000 maps published 1924–1925
Timor: surveyed 1921–1940, 1:100,000 maps published 1924–1931 (west), 1938–1940 (central)
Sumba: surveyed 1923–1935, 1:100,000 maps published 1925–1938 (symbols for indigenous cult sites).

## 3.10   New Guinea: Military Exploration

In 1906, Governor-General Van Heutz commissioned Colonel Colijn (who wrote a famous guide for civil administrators on how to deal with newly pacified areas, and was to be Prime Minister of the Netherlands later on) to set out a policy for the mapping of New Guinea. In his report, dated 4 February 1907, Colijn advised the government to systematically explore the Dutch part of New Guinea. There was an immediate follow-up with a Military Exploration programme starting that same year. It comprised four exploration detachments: South New Guinea (Merauke), West New Guinea (Fakfak), North New Guinea (Mamberamo River) and Humboldt Bay, each with a specific exploration task. The programme stopped in 1915, when, due to the perceived threat by World War I, the military was withdrawn and allocated to areas that were more vulnerable. From the reports of the exploration

**Fig. 8** Two samples of the sketch map series of New Guinea at 1:250,000: sheet 21, Kokenau (1939) (left) and sheet 45, Boven-Digoel (second edition 1940, 1947 reprint) (right). Courtesy Leiden University Library (shelfmarks KK 150-02-05 and KK 150-01-06)

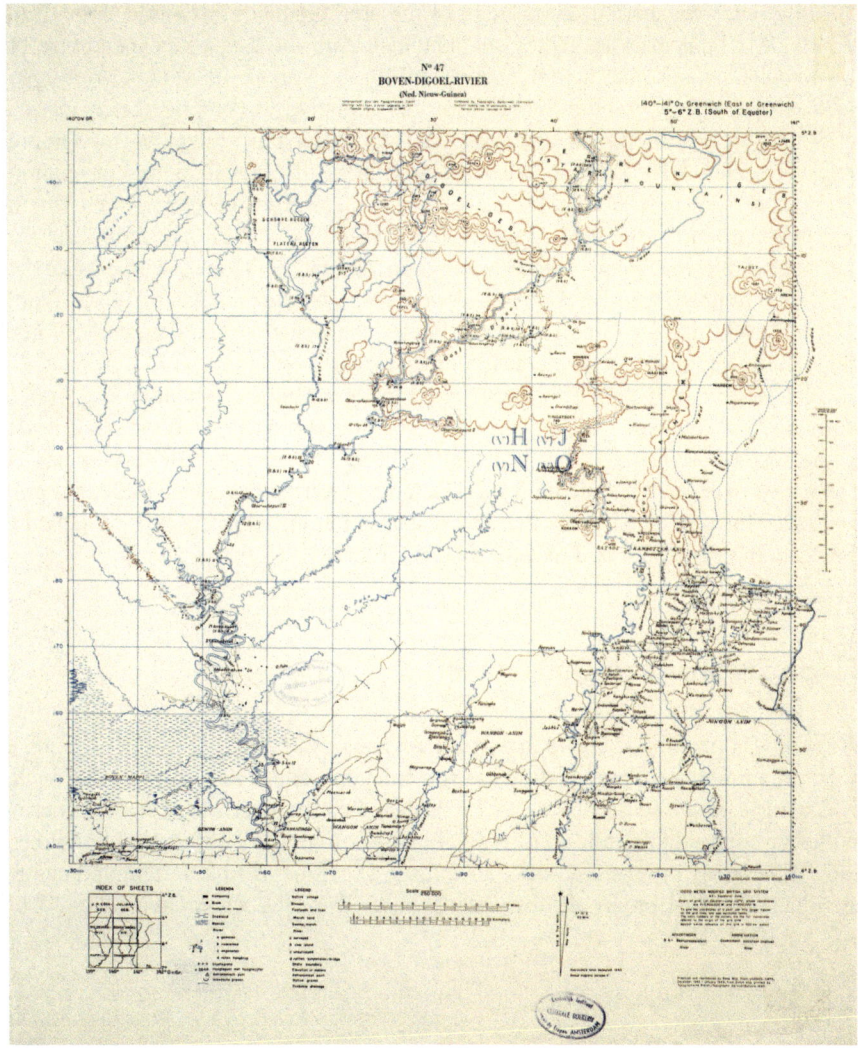

**Fig. 8** (continued)

programme[3] in combination with hydrographic charts, a sketch map of New Guinea was produced at 1:1,000,000. The image of New Guinea provided by this *Schetskaart van Nederlandsch Nieuw Guinea op schaal 1:1,000,000* remained practically unchanged until 1935.

_____

[3]Verslag van de Militaire Exploratie van Nederlandsch Nieuw Guinea van 1907–1915, published in 1920.

In the inter-war period, knowledge of the area gradually increased through the reports of military patrols, geologists, administrators and scientific expeditions, but this knowledge was both heterogeneous and local. In about 1934, the Cartographic Section of the Topographic Survey took the initiative to compile sketch maps at 1:200,000 or 1:250,000 for all areas that lacked regular systematic topographic maps at scales 1:100,000 or larger. New Guinea also fitted in this programme (Pannekoek 1940). The data for these sketch maps were needed as well for the International Map of the World (at 1:1,000,000) and later for air navigation charts at the same scale that were needed even more urgently. Of the 54 map sheets at 1:250,000 required for New Guinea, 33 were ready by the outbreak of World War II and published in easy-to-add-to sepia sheets. These sheets would contain all geographical information available at the time (Fig. 8).

Such sheets were not only produced for New Guinea, but also for parts of Borneo (23 sheets) and Celebes (43 sheets), Wetar (11 sheets) and for smaller isles in the eastern archipelago that had not yet been surveyed at 1:100,000 or larger. Thus, well over one hundred of these sketch maps were produced at 1:200,000 prior to 1942. This continued under the new dispensation in Indonesia in 1950, albeit at the scale of 1:250,000, and excluded New Guinea.

# 4  Similarities and Differences

When it was mentioned above that systematic surveys began in the outer islands in 1884, this point should be qualified to some degree. These surveys were regionally systematic—in the sense that up to 150 sheets were produced of regions, adhering to the same standards, rules and regulations—but between different regions there were differences in legends, scale, sheetlines, sheet numbering system and central meridian. The first proposal for a sheetlines valid for the whole archipelago dates only to 1924 (Schepers 1925).

## 4.1  Triangulation

From 1884 onwards, when the systematic mapping of West Sumatra began, to well into the twentieth century, there was no central system, legend and no overall sheetlines for the colony. The legend, scale and sheetlines of the series for West Sumatra differed from those used for East Sumatra or South Sumatra. As already mentioned, it was only in 1924 that a standardized sheet system for the whole archipelago was chosen (Schepers 1925). On 9 December 1924, the Permanent Council for Mapping and Surveying in the archipelago (Permanente Commissie voor kaarteerings- en opnemingswerkzaamheden) decided to adopt a one-sheet index system for the whole archipelago, based on 20-min quadrangles and the Central Meridian of Batavia. So finally, the various partial systems each based on their own central meridian were to be unified.[4] In Sumatra three different geodetic bases existed: that in Padang based on a baseline measurement (1884), that in East Sumatra (1910), and that in South Sumatra linked to the triangulation network of Java. There was also the astronomical observations-based network in Southeast Sumatra, as no there were no mountains convenient to use. These networks were merged and recomputed into one system in 1931.

## 4.2 Scale

Gradually, the insight was reached that it was neither necessary nor desirable that the same scales should be used for the whole area under consideration, but rather to adapt the scale to the economic potential, population density, or resources of the areas concerned. The 1:20,000 scale used for West Sumatra was dropped in favour of 1:40,000 as soon as the surveyors moved into the more northerly Tapanuli residency. Likewise, it was used for Aceh, which was covered by three different scales (1:40,000, 1:80,000 and 1:200,000). For the less densely populated parts of Southern and Eastern Sumatra, 1:100,000 was deemed sufficient too, except for the proposed route of the South Sumatra railway, which merited 1:25,000 maps. The

---

[4]According to Schepers (1925), there are multiple reasons that led to the introduction of a general sheet line system for the whole archipelago: The existing partial index sheets did not answer to any set system; some were numbered from the north in an east-westerly direction, some from the south, some in a west-easterly direction, some according to the date of their being ready, thus lacking a logical sequence altogether. Confusion reigned especially on Sumatra. Here the central meridian of the maps for Western Sumatra was supposed to be at Padang, 6° 26' 42" West of Batavia, while for the maps of Southern Sumatra the central meridian of 3° 15' West of Batavia was adhered to. Both central meridians were not multiples of 20', so map sheets of the two systems did not match but overlapped or showed gaps.

[De redenen, die hebben geleid tot het invoeren van een dergelijken, algemeenen bladwijzer voor den geheelen archipel, zijn velerlei. De bestaande partieele bladwijzers hadden geen vast systeem; enkele waren van het noorden genummerd in de richting O - W, andere weer vanuit het zuiden, bij weer andere was de volgorde W - O, terwijl bij de opneming van Sumatra's Westkust de bladen eenvoudig genummerd zijn geworden in de volgorde, waarin zij gereedkwamen, zoodat daar van een regelmatige volgorde heelemaal geen sprake is geweest. Vooral op Sumatra was de verwarring, zelfs voor de ingewijden, hopeloos, waarbij nog kwam, dat daar twee systemen van bladindeeling bestonden, waarbij de nulmeridiaan van het systeem van West-Sumatra geacht werd te loopen over het westelijk uiteinde A van de basis nabij Padang en gelegen was op ongeveer 6° 26' 42" West van Batavia, terwijl voor Zuid-Sumatra als nulmeridiaan was aangenomen de meridiaan, gelegen 3° 15' West van Batavia. Beide nulmeridianen waren dus noch onderling, noch van den nulmeridiaan van Batavia een geheel veelvoud van 20' verwijderd, met het gevolg dat de bladen van beide systemen elkaar overlapten en niet aansloten.]

economically interesting parts of Eastern Sumatra were rendered at 1:50,000. On Celebes, the agriculturally promising Tempe Lake plains were rendered at 1:25,000, the remaining lowlands at 1:50,000, and the mountainous interior at 1:100,000. For town maps, a scale of 1:5000 was established (van Diessen and Ormeling 2003).

## 4.3  Legend

Local differences referred to the portrayal of different cash crops and parts of the infrastructure that stood out locally. In Bali, special symbols were used for showing Hindu temples, and in Sumba or Nias, for indigenous places of worship. In areas with important irrigation projects, the levelling markers were added. The plantation area in East Sumatra showed the height of the original vegetation with areal colours: blueish for high rising and yellow for shrubby vegetation. On the maps of East Sumatra, a special dotted line would mark the boundary between the permanent wetlands and the areas that dried up in the dry monsoon. Plantation roads that were fit for oxen but not for trucks in the dry season ('plantwegen') had a special yellow colour not found elsewhere as well.

## 4.4  Cultural Differences

The cadastral mapping that went together with topographical surveying in many areas not only gave the surveyors problems with languages, as they were unable to speak the Bugi or Arafura languages in the eastern part of the archipelago, but also problems with different social customs pertaining to land rights. In Java and most of Sumatra, land rent was based on the agricultural area belonging to the inhabitants of a village, and the total of the land rent taxes imposed on a village would be apportioned locally. In Celebes, up to six different types of landed property could be discerned, with extreme fragmentation of the holdings between several villages, so that other ways to impose the land rent had to be devised. In the Batak lands on Northern Sumatra, this led to the assessment and mapping of individual land holdings. It was so expensive however, that it almost equated the government's tax income from this area.

## 4.5  Updating

Only Java and a few areas in Western Sumatra saw a revision and thus a second or —for some strategic areas on Java—even a fourth edition of the 1:40,000 or 1:50,000 maps.

# 5   Geopolitical Background

The reasons for mapping the various areas in the Outer Isles were different: mining interests were behind the mapping of East and South Sumatra, Bangka, and, after its pacification, of Aceh, and of South-eastern Borneo. After the demise of the system of forced farming (Cultivation System or Cultuurstelsel) and the subsequent opening up of the territory for private enterprise and the pacification policy that enabled the government to guarantee the safety of persons and goods all over the archipelago, a tempestuous increase in the production of cash crops and agricultural commodities occurred. This plantation economy was behind the mapping of Eastern Sumatra (Deli) and Southwest Celebes. Strategic reasons first led the Dutch to organize the mapping of Western Borneo and to send military exploration detachments to New Guinea.

Opening-up the country also meant that it was more exposed to outside influences. The years between 1905 and 1914 saw more change in the Outer Isles than in the three preceding centuries. Cartographically, this involved the military exploration of New Guinea and the creation of new surveying brigades for East Sumatra and Celebes. The influence of World War I and the subsequent fear in the colony of potential enemies led to the surveying endeavours outside Java being halved. It also brought a stop to the exploration of New Guinea. In the annual reports of the Topographical Survey from this period, there are even descriptions of meetings of the Sarikat Islam on Sumatra (1915) and of the Samin sect in East Java (1914: Blora). In 1914, half the surveyors in the Outer Isles were recalled to Java in order to help with the military surveys that were deemed necessary there. In 1918, the Survey's annual report described the surveying parties and the local population alike suffering from the Spanish Influenza. After the war, in the 1920s, when surveyors returned to the Outer Isles again, the surveying operations in Eastern Sumatra were influenced by strikes of the indigenous personnel of the DSM (the Deli railway company). In addition, the rubber boom led to such an increase in earnings that the survey could not find imported labour or assistants to help them as porters. Consequently, the survey of Jambi was interrupted from 1927–1931. In 1925 and 1926, the Survey sacked (i.e. honourably discharged) indigenous personnel that participated in communist rallies as a reaction to the communist uprising in the colony. The communist party was outlawed after the 1926 insurrection.

The 1930 report described in positive terms the fact that the rubber boom was over, and that imported labour could again be hired as porters for reasonable terms. This concealed joy, however, was soon to be reversed as the depression led to seven years of retrenchments. The training brigade was dismembered as training new recruits stopped, and it was only at the end of 1937 that matters were reversed and the Survey could expand again.

In 1939, when the Survey celebrated its 75th anniversary (Vijfenzeventig jaren 1939) most regular surveying activities came to a halt in view of the deteriorating global political situation. Henceforth, strategic considerations prevailed in setting out priorities for surveys or resurveys. Maps of areas with strategic resources, like

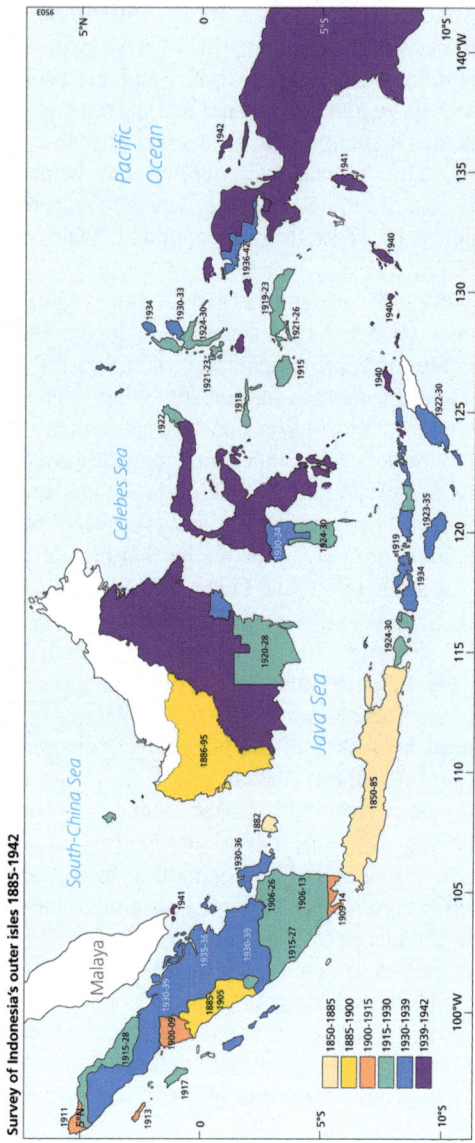

**Fig. 9** The sequence of topographic mapping in the Outer Isles. From 1950–1962, most of Netherlands New Guinea was mapped at 1:100,000 (see Fig. 10)

Tarakan and Balikpapan with their oil refineries, were updated, as were those areas of Java that were likely to be used in fending off landing troops. The Riau Archipelago close to Singapore was mapped as well (1941), but the maps were not published (Fig. 9). The surveys made were as follows:

| | |
|---|---|
| Minahasa (N Celebes) | 1:100,000 (1941) |
| Towuti lakes (SE Celebes) | 1:100,000 (1940) |
| Sketch maps Celebes | 1:200,000 1939–41 |
| Northern Dutch Timor | 1:100,000 (1939–1941) |
| Aru Islands | 1:200,000 (1940) |

All this was rather in vain, however, because as soon as the colonial government was drawn into the hostilities, almost all troops were recalled from the outer islands to Java. It was there that the army made its last stand and surrendered in March 1942 to the invading Japanese troops. When Dutch troops returned in 1946, Dutch authority was re-established over most of the archipelago, except for Central and Eastern Java, which remained under the control of independent Indonesia. The Topographical Survey resumed its activities beyond Java until after the transfer of sovereignty at the end of 1949, when the organization and premises of the Topografische Dienst were transferred to the independent Indonesian authorities in June 1950.

This did not end Dutch involvement in the topographic mapping of the Outer Isles, however. Dutch New Guinea had not been part of the transfer of sovereignty to independent Indonesia to allow for the autonomy of the Papua ethnic group. The Netherlands were engaged in the production of a 1:100,000 topographical map

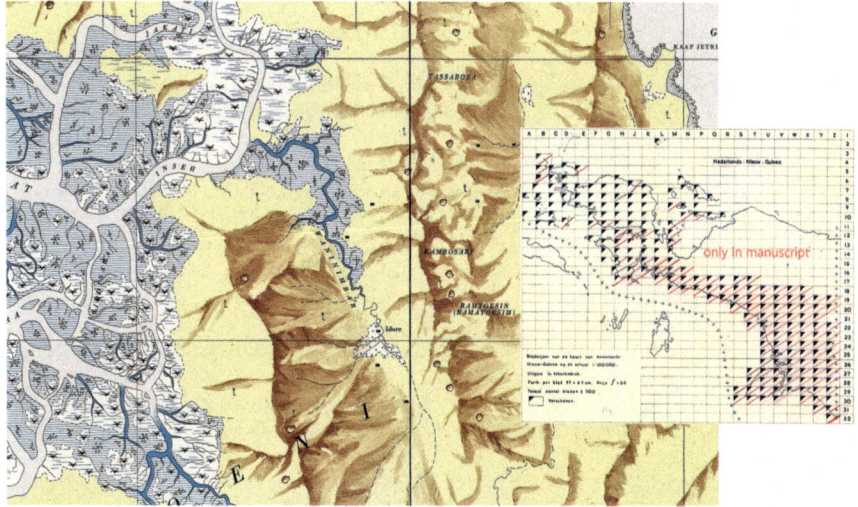

**Fig. 10** Index sheet of the 1:100,000 map series of Netherlands New Guinea 1950–1962, with a detail from sheet K12, Wendehsi (Topografische Dienst, Delft 1958). Courtesy of Utrecht University Library

series of that area until 1962 (Fig. 10). Based mainly on photogrammetric surveys and produced in the Netherlands (Ormeling 1952; Kint 1954; Bramlage 1959), about two-thirds were finished when sovereignty over New Guinea was transferred to the United Nations, which in turn handed the area over to Indonesia. Copies of the printed maps and the manuscript versions of those not printed yet are hopefully still lingering in corners of some Dutch archives, being a souvenir of 360 years of Dutch involvement in overseas cartography.

**Acknowledgements** For the chapter's structure, as well as for his comments, the author is indebted to Paul van den Brink, with whom he originally intended to write it.

# References

Bramlage JH (1959) De kaartering van Nederlands Nieuw-Guinea (The mapping of Netherlands New Guinea). In: Kartografie, Mededelingen van de Kartografische Sectie van het KNAG 4, TAG LXXVI -3-1959, pp 307–316

Broeders PWA (2007) Gijsbert Franco, Baron von Derfelden van Hinderstein 1783–1857. Leven en werk van 'eene ware specialiteit' in kaart gebracht. 't Goy-Houten: Hes & De Graaf

Clemens AHP, Lindblad JT (eds) (1989) Het belang van de buitengewesten: economische expansie en koloniale staatsvorming in de buitengewesten van Nederlandsch-Indië, 1870–1942 (The importance of the outer isles: economic expansion and colonial constitution on the Dutch East Indies, 1870–1942). NEHA, Amsterdam

Jaarverslag van den Topografischen Dienst in Nederlands-Indië (1905–1939). Topografische Dienst (Annual reports of the Topographic Survey in Batavia), Batavia, 1906–1940

Kint A (1954) Kaartering [van Nederlands Nieuw-Guinea] In: Klein WC (ed) Nieuw-Guinea, de ontwikkeling op economisch, sociaal en cultureel gebied in Nederlands en Australisch Nieuw-Guinea. Staatsdrukkerij, Den Haag, pp 1–30

Melvill van Carnbee P, Versteeg WF (1853–1862) Algemeene atlas van Nederlandsch Indië. Van Haren Noman & Kolff, Batavia

Ormeling FJ Sr. (1952) De groei van de kaart van Westelijk Nieuw-Guinea. Tijdschrift van het Koninklijk Nederlandsch Aardrijkskundig Genootschap van April 1952, pp 199–244

Ormeling FJ Sr (1996) JJK Enthoven (1851–1925) Inspirerend promotor van de kartering in voormalig Nederlands-Indië. Kartografisch Tijdschrift 22(2):7–14

Pannekoek AJ (1940) Nieuwe schetskaarten van Nederlands-Indië. Jaarverslag van den Topografischen Dienst in Nederlands-Indië 1939, Batavia, pp 101–119

Pannekoek AJ (1946) Enige aantekeningen over Indische kaarten. TAG LXIII 1946:627–639

Schepers JHG (1925) Het algemeene systeem van bladindeeling voor de kaarten van den Ned. Ind. archipel. In: Jaarverslag Topografische Dienst in Nederlandsch-Indië, Batavia [the general sheetlines for maps of the Indian archipelago], pp 111–119

Stemfoort JW, ten Siethoff JJ (1883–1885) Atlas van de Nederlandsche bezittingen in Oost-Indië. Topographische Inrichting, The Hague

van den Brink P (2010) Dienstbare kaarten. Een cartografische geschiedenis van het Koninklijk Nederlands Aardrijkskundig Genootschap en het Tijdschrift 1873–1966. Houten: Hes & De Graaf

van Diessen JR, Ormeling FJ (eds) (2003) Grote Atlas van Nederlands Oost-Indië. Asia Maior (contains facsimiles of most of the map series discussed in this paper), Zierikzee

van 't Veer P (1969) De Atjeh-oorlog. Amsterdam: Arbeiderspers [The Aceh war]

Verslag van de Militaire Exploratie van Nederlandsch Nieuw Guinea van 1907–1915. Weltevreden, Landsdrukkerij, 1920 [Report of the military exploration of Netherlands New Guinea 1907–1915]

Vijfenzeventig jaren Topografie in Nederlands-Indië (1939) Jubilee issue, introduced by director MT van Staveren [75th anniversary of the Topographical Survey in Batavia]

**Ferjan Ormeling** held the Chair of Cartography at Utrecht University from 1985 to 2010 and since then has been a member of the Explokart research group at the University of Amsterdam. His research focuses on atlas cartography, toponymy and the cartographic history of the Indonesian archipelago. He was one of the editors of the national atlases of the Netherlands and contributed to the *Comprehensive Atlas of the Dutch East India Company*. He was vice-chair of the United Nations Group of Experts on Geographical Names from 2007 to 2017.

# A View from Inside: Chinese Mapping of the World Against the Backdrop of Colonial Experience

Laura Pflug

**Abstract** From the mid-nineteenth century, a fragmented landscape of formal colonies, such as Hong Kong, and other structures of foreign influence, such as the International Settlement in Shanghai, spread over parts of Chinese soil. This colonial patchwork encompassed centres of transcultural encounters, translation, publishing, and commerce, and it also brought forth a vibrant strata of cultural intermediaries, such as translators, teachers, and journalists. Among these were social actors who employed cartography as a means to communicate knowledge about the world around them. Thus, while the outside world was gaining spheres of influence on Chinese soil, certain members of the Chinese elite turned their attention towards distant lands and created world maps and atlases. Based on cartographic visualizations from the mid to the late nineteenth century, this study explores a view from inside a country that was partially exposed to colonial influences. It looks at the impact of the colonial condition on individual Chinese mapmakers and the topics that they deemed important when presenting the world outside of China to their fellow countrymen.

## 1 Introduction

Qing China's (1644–1911) realm was never fully dominated by powers that came from abroad. The Chinese empire's defeat in the Opium War (1840–1842), however, laid the ground for the successive colonial expansion on its territory. These colo-

This paper presents research from the project 'Maps of Globalization: The Production and the Visualization of Spatial Knowledge' at the Leibniz Institute for Regional Geography. The project is part of the Collaborative Research Centre (SFB) 1199: 'Processes of Spatialization under the Global Condition', which is funded by the German Research Foundation (DFG).

L. Pflug (✉)
Leibniz Institute for Regional Geography, Leipzig, Germany
e-mail: L_Pflug@ifl-leipzig.de

© Springer Nature Switzerland AG 2020
A. J. Kent et al. (eds.), *Mapping Empires: Colonial Cartographies of Land and Sea*,
Lecture Notes in Geoinformation and Cartography,
https://doi.org/10.1007/978-3-030-23447-8_4

nized spheres, albeit entailing situations of conflict and tension,[1] also brought about a new interest in and outlook on the outside world. They created new cosmopolitan spaces that were integrated within global flows of goods, people, and ideas. The colonies, as the German Sinologist Klaus Mühlhahn has described them, 'became fluid unstable zones of contact where identity lines and cultural borders were blurred through exchanges and transfer' (Mühlhahn 2012: 38).

The period between the end of the war in 1842 and the 'global scramble for colonies' (Goodman and Goodman 2012: 1) by the end of the century marked a phase of transition in Chinese cartography. On the one hand, old-established patterns of visualizing the world beyond China as a marginalized frame for the grandeur of the empire were still present as material as well as mental maps.[2] On the other hand, however, some Chinese sought to produce cartographic images that allowed their fellow countrymen to take a closer and more 'realistic' look at the world, and give them an opportunity to use that knowledge to their advantage, e.g. for political support, trade, travel, or migration. This is not to say that prior to that, fairly accurate, European-style world maps had not been produced at all in late-eighteenth and early-nineteenth-century China,[3] but they were rather scarce, while in the latter half of the nineteenth century, the need for information about the outside world took on a new quality and urgency. The maps from that era that will be presented in this paper did not serve for mere aesthetical pleasure. The themes, which their creators chose to highlight on them, were mostly of practical use and mirror a pragmatic approach in highly politicized times.

This study will examine maps that were produced or co-produced by three educated Chinese whose lives were strongly affected by the presence of foreign powers on Chinese soil. While neither of them was making a living as a cartographer—one was a government official, one was a translator and one was, amongst others, a teacher, author and publisher—they left behind maps of the world that are connected to their names. Besides examining their works, the study will also take a look at the people behind these cartographic images and ask for the motivations that can be deduced from the focal points of their maps.

---

[1]For diverse aspects of colonialism in China see Goodman and Goodman (2012). For a discussion of colonial violence and humiliation in nineteenth-century China see Hevia (2003).

[2]For nineteenth-century Chinese maps representing 'all under heaven' (*tianxia* 天下) see Pflug (2019: 249–252).

[3]An example is the world map in two hemispheres in Li Mingche's 李明徹 (1751–1832) work *Huantian tushuo* 圜天圖説 (Illustrations of Encompassing Heaven) from the year 1819 (Smith 1996: 61–62), which was based on an earlier Chinese map from 1794 (Mosca 2013: 211).

# 2   Xu Jiyu and His Short Account of the Maritime Circuit, 1848

> Geography without maps is not clear. Maps [that are drawn] without making observations are not detailed. The world has a [fixed] shape and it cannot be extended and constricted [as one] wishes. Westerners[4] are good at traveling afar [and their] sails and masts encircle the Four Seas. [When they reach their] destinations, they always take out their brushes and draw maps. Therefore, only their maps can be relied upon. (Xu 1848: preface 1.8a)[5]

In 1848 Xu Jiyu 徐繼畬 (1795–1873), the then governor of the Chinese coastal province of Fujian, completed a cartographic and geographic project that he had already started in 1843, shortly after China's defeat in the Opium War. The outcome of this project was a treatise on the countries of the world called *Yinghuan zhilüe* 瀛環志畧, Short Account of the Maritime Circuit, which comprised forty-two maps printed in black and white, as well as detailed accompanying texts. Except for the map of Japan and Liuqiu (Ryūkyū), all the cartographic images drew upon foreign maps (Drake 1975: 44, 60). This treatise was to become an important step in nineteenth-century Chinese world geography.[6] During the Opium War, Xu had already been in the coastal province, where he had witnessed British military domination. Before China's defeat, overseas trade had been restricted to the city of Canton, but now, more harbour cities such as Shanghai and two ports in Fujian were forcedly opened, Hong Kong was ceded to Britain as a Crown Colony, and foreign presence in China's coastal areas increased. Shocked by the events he had witnessed in Fujian, Xu stated that he could 'neither eat or sleep, trying to think of ways to help' (Drake 1975: 1). Being convinced that the outcome of the war was a consequence of insufficient knowledge about the outside world, he began to gather information from non-Chinese materials as well as Chinese sources. In the coastal province, he had met the American Missionary David Abeel (1804–1846), through whom he had access to a book with world maps. Xu copied about ten of the maps and Abeel translated the titles for him, as Xu did not understand their language. Later, Xu gathered more foreign materials, including works that had been written in Chinese by non-Chinese authors (Han 2014: 12), and he used the possibilities to learn about other countries whenever he met foreigners. On this process, he said in his foreword:

---

[4]The term 'Westerner' that is used here is a translation of the Chinese term *taixi ren* 泰西人 (literally 'person from the utmost west').

[5]This is the opening passage of Xu Jiyu's own foreword (dated 1848) of the *Yinghuan zhilüe*. The 1848 edition of the *Yinghuan zhilüe* that I have looked into has been digitized by the Bavarian State Library in Munich. For the original passage see Bayerische Staatsbibliothek München, 4 L. sin. D 136-1/6, image no 840, urn:nbn:de:bvb:12-bsb11128998-8. For another English translation see Drake 1975: 60.

[6]The other well-known Chinese work on world geography from the 1840s is Wei Yuan's 魏源 (1794–1856) *Haiguo tuzhi* 海國圖志 (Illustrated Gazetteer of the Maritime Countries), which was first printed in 1844 (Mosca 2013: 274).

I repeatedly sought and obtained several types [of Western books]. [If] these books were unpolished and unrefined [so that] someone deeply cultured and refined could not look at them, then I collected them and made a selection. I got a piece of paper and also kept a record of [these materials], I did not discard them. Whenever I met a Westerner, I always opened [my] volumes [so that he could] examine and verify the current conditions of all the foreign countries and the topography beyond our borders. Gradually I gained [knowledge about] their outlines and then, relying on maps, I set forth my views and gathered what was credible in all the books. I expanded this and created sections. For a long time I accumulated [these sections] and they turned into a book. Every time I obtained a [new] book or received new information, I always revised and enlarged my draft. Overall, I changed it several tens of times. From the year *guimao* [i.e. 1843] until today, passing through five winters and five summers and in addition to my official duties, I did only this to while away the time and I did not stop for a single day. (Xu 1848: preface 1.8a–8b)[7]

After finishing his *Yinghuan zhilüe*, Xu did not fare well, as he met with resistance from conservative forces in Chinese government who criticized his lenient attitude towards the foreigners (Smith 1996: 70). In 1851, he was dismissed from his post in the coastal province. Nevertheless, the importance of his treatise in Chinese world geography was acknowledged by some of the scholar-officials (Drake 1975: 49, 56) and in the 1860s, it even served as educational material for Chinese officials in foreign affairs that were trained in Beijing (Matten 2016: 134). It finally gained broad recognition and, as Wagner notes, 'became the most widely read world geography for the next decades in China as well as Japan' (Wagner 2017: 23).

The Short Account of the Maritime Circuit begins with a map of each hemisphere, followed by a little over nine pages of textual information. The cartographic images merely give a general, rather sketchy introduction to world geography, which, apart from some mountains, hardly includes topographic information. As can be seen on the map of the eastern hemisphere (Fig. 1), differing conceptions of bordering were employed here. The Chinese empire is subdivided into its provinces whereas other countries are defined by their national boundaries and Africa is largely demarcated only by the borders of the continent.

The texts give basic information on the globe, its two hemispheres, and its continents. The next map is an image of Qing China, which is drawn in much more detail than the representations of other countries in the treatise, while the accompanying text emphasizes the empire's grandeur. The rest of the work is classified into the continents of Asia, Europe, Africa, and America, always starting with one or more maps followed by textual information, with Europe covering the largest part among these four. As Xu had witnessed Britain's forceful intrusion during the Opium War, it does not surprise that he devoted much of his attention to this continent. The maps, as can be seen by the depiction of the British Isles (Fig. 2), appear rather sketchy. The bulk of information, however, is not to be found in the images, but in the texts accompanying the cartographic depictions of the continents

---

[7]For the original passage in Xu's foreword see Bayerische Staatsbibliothek München, 4 L.sin. D 136-1/6, images no 840 and 839, urn:nbn:de:bvb:12-bsb11128998-8. For another English translation see Drake (1975:54).

**Fig. 1** Map of the Eastern Hemisphere from the *Yinghuan zhilüe* 瀛環志畧, 1848 (Bayerische Staatsbibliothek München, 4 L.sin. D 136-1/6, images no 823 and 822, urn:nbn:de: bvb:12-bsb11128998-8)

and their countries. They contain geographic, historical, cultural, and economic knowledge. In the section about the German state of Saxony Xu for example wrote about the religious reformer Martin Luther as well as about the annual book fair in the city of Leipzig (Xu 1848: 5.20a).[8] But not least, the work mirrored the strong political motivation underlying it and included information such as the following, which was referring to Britain's domination over China's neighbour India:

> In 1755 Bengal was annexed, and taking advantage of their victories the English stealthily encroached on the various states like silkworms eating mulberry leaves. The various parts, scattered and weak, could not resist, and consequently more than half became British colonies. (Suzuki 2009: 59)

Nevertheless, information about Britain in the Short Account of the Maritime Circuit is not only grave or grim, and some of Xu's descriptions also bear unintentionally amusing elements, for example when stating that 'the [British] men always obey the orders of the [British] women, [and] the whole country likewise

---

[8]Bayerische Staatsbibliothek München, 4 L.sin. D 136-1/6, image no 424, urn:nbn:de: bvb:12-bsb11128998-8.

**Fig. 2** Map of the British Isles from the *Yinghuan zhilüe* 瀛環志畧, 1848 (Bayerische Staatsbibliothek München, 4 L.sin. D 136-1/6, image no 225, urn:nbn:de: bvb:12-bsb11128998-8)

[proceeds like that]' (Xu 1848: 7.50a),[9] an idea that perhaps arose from the fact that the British throne was held by a woman.

The *Yinghuan zhilüe* is conceptualized in the tradition of the Chinese local gazetteers, the *difang zhi* 地方志 or *fang zhi* 方志.[10] An older designation for this genre is *tujing* 圖經, which can be translated as 'maps and treatises' or 'illustrated guides'. The local gazetteers comprised texts as well as maps about Chinese administrative divisions of various scales, and were mostly, but not solely, used for administrative matters. Their texts covered various topics such as famous buildings, local products, local customs, famous people, travel accounts, poems, and so on. Thus, with his Short Account of the Maritime Circuit, Xu had transferred this traditional Chinese approach to a global level, in the quest of bringing knowledge

---

[9]Bayerische Staatsbibliothek München, 4 L.sin. D 136-1/6, image no 188, urn:nbn:de: bvb:12-bsb11128998-8.

[10]For the genre of Chinese local gazetteers see Moll-Murata (2001).

about the world to his fellow—well-educated and literate—countrymen. In a highly condensed form, this approach can also be seen on the two following maps.


## 3   Chen Xiutang and the Complete Map of the Globe, 1855

While by 1850 Europeans still were a rather rare sight in the harbour cities of Fujian (Waley-Cohen 1999: 154), they were much more common in another part of China. Hong Kong Island, which had been occupied by Britain in 1841 and then formally declared a colony in 1843, grew steadily after surmounting initial difficulties in the 1840s. In the 1850s, the overseas trade increased and many Chinese, among them many merchants (Carroll 2007: 29–30), settled in Hong Kong after fleeing a rebellion that spread over parts of China from 1851. The colony also attracted foreigners from diverse professions, such as missionaries, physicians, printers, and journalists. Furthermore, Hong Kong's geographic location not only promoted the flow of goods, but also the flow and circulation of information from around the world (Sinn 2013: 21).

Chen Xiutang 陳修堂 (?–?), whose name and place of origin is to be found on the *Diqiu quantu* 地球全圖, the Complete Map of the Globe from the year 1855 (Fig. 3), was a native of the province of Guangdong in the hinterland of Hong Kong. Unfortunately, only few details are known about him. In Hong Kong, he cooperated with the British medical missionary Benjamin Hobson (1816–1873) on medical translations that were published between 1851 and 1858 (Andrews 2015: 118). Apparently, he also had some medical knowledge himself, as he coauthored the works together with Hobson and another Chinese (Elman 2005: 287). The Complete Map of the Globe is what seems to be another Sino-British cooperation involving Chen Xiutang. The second name, which is given on the map, is that of a certain De Chen 德臣. De Chen is stated as the producer of the map while Chen Xiutang was responsible for its text. The identity of De Chen is nebulous, but it is possible, that it was Andrew Scott Dixon (?–1873), as De Chen was Dixon's Chinese name. Andrew Scott Dixon was a man of Scottish origin who had been trained as a printer. Thus, his professional abilities would have enabled him to print a map. During the 1850s, he worked in Hong Kong. He came to the colony together with another Scotsman called Andrew Shortrede (?–1858), who founded the English-language newspaper *China Mail* in Hong Kong in 1845. Dixon was first the assistant editor and manager, and then became the editor in chief of the paper, which in Chinese was called *De Chen xibao* 德臣西報 (De Chen's Western Newspaper) or simply *De Chen bao* 德臣報 (De Chen's Newspaper) after him. In his articles, Dixon fought corruption in the government of Hong Kong and defended the rights of the Chinese in the colony. He stayed in China until 1863, when he had to return to his country because of an illness (Chen 2005: 38, 46). Before that, he was actively taking part in and promoting the cultural life in Hong Kong, and he associated with Chinese intellectuals (Vittinghoff 2002: 75, 87), which makes him a likely candidate for a Sino-British cooperation in the mid-1850s.

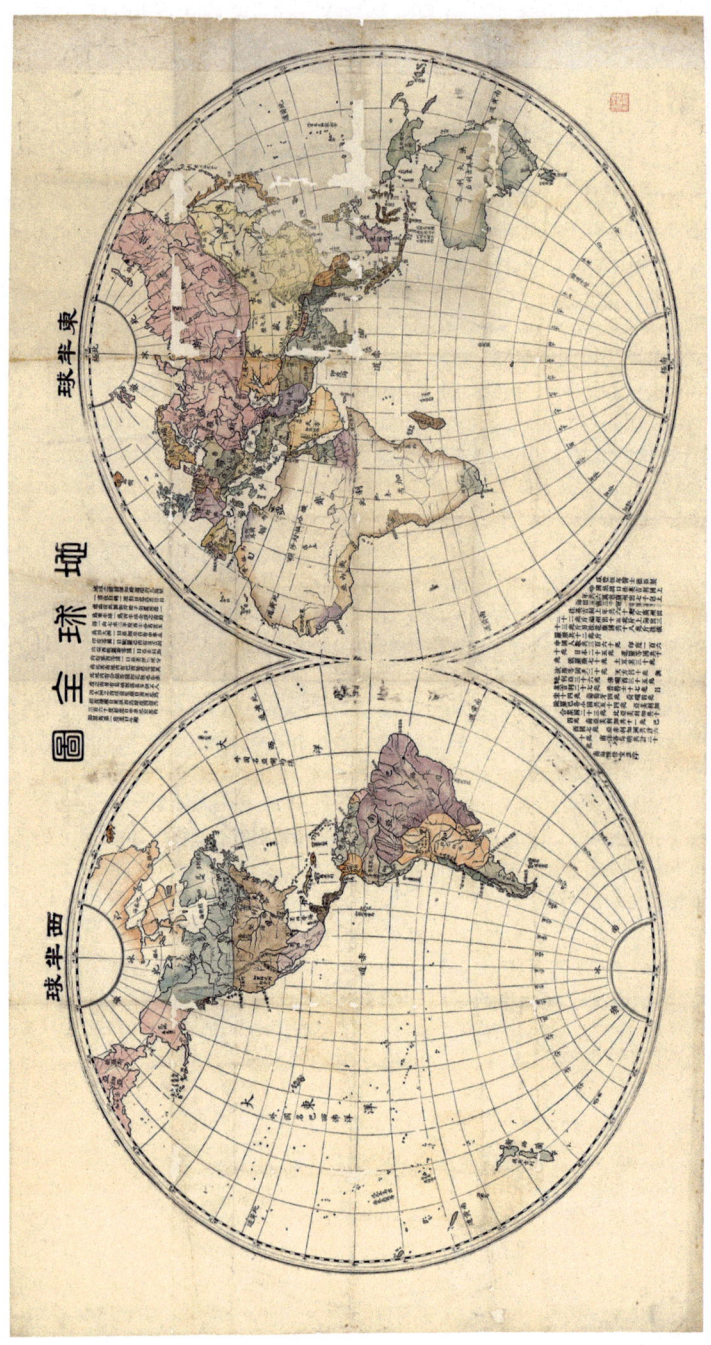

**Fig. 3** The complete map of the Globe, *Diqiu quantu* 地球全圖, 1855 (Harvard-Yenching Library, Harvard Library)

The coloured map, which he might have cooperated on with Chen Xiutang, is a world map in two hemispheres. According to Harvard Library, its size is 93 × 70 cm.[11] The map was published only seven years after Xu Jiyu's work, but compared to the latter, it looks far more 'modern'. Other than on Xu's map of the Chinese empire from the *Yinghuan zhilüe*, the designation of China on this world map is not associated with the Qing dynasty, but bears the title *Zhonghua guo* 中華國 (The State of the Central Florescence), which is another name for the country of China. And while De Chen's and Chen Xiutang's Complete Map of the Globe associates India with British rule, the Crown Colony Hong Kong is not associated with Britain on this map. The states are each depicted in a different colour, and topographical information, such as mountains and rivers, are added to the image. Although it does not look as sketchy as the cartographic depictions in the *Yinghuan zhilüe*, the map still has a rather plain design. A thematic purpose is embedded in the texts that are printed on the world map. The text on the upper half reflects the cartographic image, informing about the general shape of the earth and about the continents of Asia, Europe, Africa and America. The text on the lower half, however, is moving beyond the graphic information. On the left side, it lists the numbers of inhabitants of certain countries such as China, India, and Britain, as well as of the four continents stated above. On the right side, the text states the amount of Chinese silk and Chinese tea exported to Britain in the previous year, and it also indicates export numbers from individual provinces, as well as the amount of tea exported from China to America and Russia. Thus, the textual information is adding an economic theme to the map. Economy and trade were topics of interest on Chinese world maps from the second half of the nineteenth century, as the next cartographic image confirms.

## 4   Kuang Qizhao and the Complete Map of the Five Continents of the World, 1875

While the image of the Complete Map of the Globe from the year 1855 appears rather plain, the Complete Map of the Five Continents of the World, *Diqiu wu dazhou quantu* 地球五大洲全圖 from 1875 (Fig. 4) is packed with information. As its predecessors, it combines depictions and texts. Its producer was an interesting, but rather little-known Chinese cultural intermediary called Kuang Qizhao 鄺其照 (1836?–1891?). Like Chen Xiutang, Kuang was a native of Guangdong province in the hinterland of Hong Kong (Chan 2013: 228, 252–253). Although he was not a trained cartographer, he used maps as one medium among others to convey knowledge. In the early 1860s, Kuang moved to Hong Kong to study English in the Government Central School (Chan 2013: 228). From the mid-nineteenth century, the Crown Colony had developed into what Sinn has called a 'space of flow', a port

---

[11]See http://id.lib.harvard.edu/alma/990094062810203941/catalog, last accessed 30 November 2018.

city importing and exporting commodities as well as people who brought and took with them material as well as immaterial entities such as knowledge and culture (Sinn 2008: 13). The Gold Rush of that era drew Chinese from the harbour of Hong Kong to California and Australia. Kuang was among the Chinese who sought their luck abroad. He went to Australia, where he set up a business trading Chinese medicine and where he also improved his proficiency in the English language. He lived in Melbourne for five years before returning to China (Chan 2013:228). Around this time, the Chinese government made efforts for technological modernization and the enhancement of knowledge in various areas in the quest to strengthen the empire's position vis-à-vis the foreign powers. This so-called Self Strengthening strategy also included sending students abroad. In the mid-1870s Kuang Qizhao accompanied a group of students that went to America. Before that, he had worked as an English teacher in a school in Shanghai that prepared the teenage boys for this endeavor. The approximately 120 boys that were sent to America mainly came from Canton and Shanghai (Rhoads 2011: 40, 35, 14–20). In America, Kuang lived in Hartford, Connecticut, where he engaged in cultural and social life and even met Mark Twain (Chan 2013: 242).

Apart from teaching English and producing the Complete Map of the Five Continents of the World, Kuang wrote A Dictionary of English Phrases, which he compiled while staying in Hartford (Wong and Wong 2017: 1478). The dictionary displays Kuang's social and political commitment, as it contained phrases like 'He cheated me out of my wages', or 'They were lying in ambush' (McCunn 1988: 21), to be used by his fellow countrymen living and working abroad. He was shocked when he encountered anti-Chinese sentiment in America and publicly condemned the unfair treatment of his fellow countrymen. After returning to China in 1882, he developed further material for Chinese who wanted to learn English (Chan 2013: 244–245, 247–248). Moreover, he established a newspaper in Canton in the 1880s. In this newspaper, he took a clear political stance and promoted the anti-imperial cause, especially vis-à-vis Britain and Russia (Wong and Wong 2017: 1491–1493).

Kuang Qizhao's Complete Map of the Five Continents of the World displays an interconnected world. According to Harvard Library, the original size of this coloured map is 91 × 97 cm.[12] It combines an image of the world for which the Mercator projection had been used with a depiction of the two hemispheres on the lower half of the map. Kuang, who had travelled wide distances himself, shows the numerous shipping routes connecting America and Europe, and the not so numerous routes connecting East Asia and America, e.g. from Shanghai and Hong Kong to San Francisco via Yokohama. On the lines that designate the shipping routes, the distances between the respective destinations are given. The explanatory text in the middle of the image indicates that he took western maps as a model for his work, but unfortunately, he does not name any titles. Kuang Qizhao's map shows some resemblance to Hermann Berghaus' Chart of the World, especially

---

[12]See http://id.lib.harvard.edu/alma/990147735230203941/catalog, last accessed 30 November 2018.

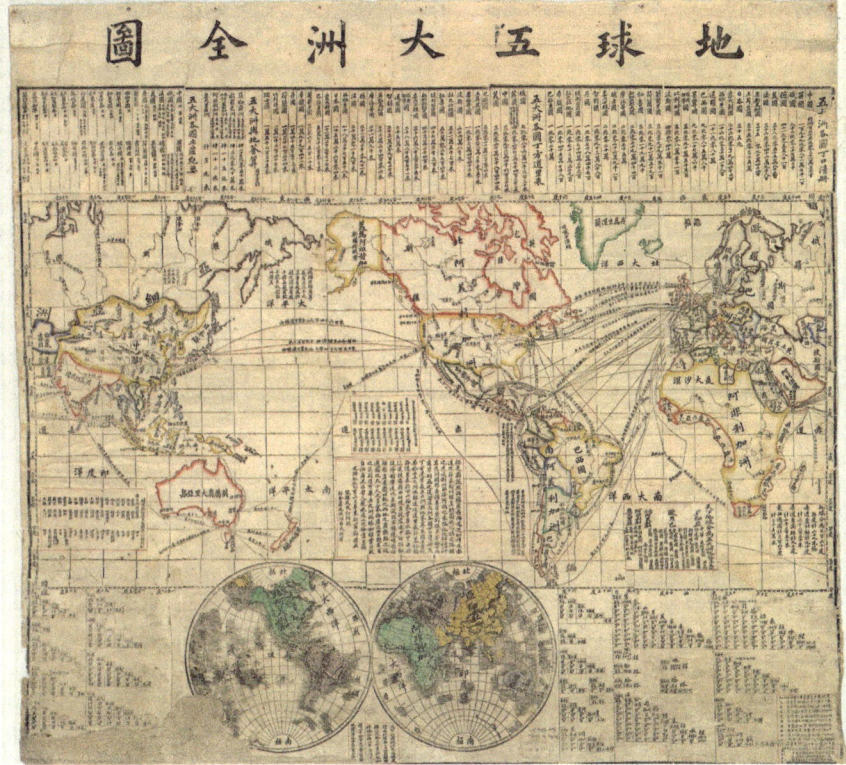

**Fig. 4** The complete map of the five continents of the world, *Diqiu wu dazhou quantu* 地球五大
洲全圖, 1875 (Harvard Map Collection, Harvard Library)

when compared to the 1871 edition of that map. Berghaus' work had been copied
and turned into a Japanese version titled Yochi shinzu 輿地新圖 in 1874.[13]

The text on Kuang's Complete Map of the Five Continents of the World is
spread all over the map and creates a visual pattern in addition to the cartographic
representation. It gives various information of practical use.[14] On the upper part, the
text lists the population figures of various countries, as well as the sizes of the
continents. It further names local products of various countries. For China, for
example, it lists silk, tea and porcelain as local products, for America, it lists cotton

---

[13]The Japanese Version of Berghaus' Chart of the World can be found in the National Diet Library
in    Tokyo,    http://iss.ndl.go.jp/books/R100000002-I000009243046-00?locale=zh&ar=4e1f.
Accessed 30 November 2018.

[14]Kuang also listed information that he gives on this map in a text that was included in the first
volume of the *Xiaofanghuzhai yudi congchao* 小方壺齋輿地叢鈔 (Collected Geographical
Writings from the Xiaofanghu Studio). This multi-volume collection was compiled by Wang Xiqi
王錫祺 (1855–1913) between 1877 and 1897.

and flour, and for Italy, it lists corals and olive oil. It also names the local products of certain cities. It says, for example, that the Old Gold Mountain (*Jiu jinshan* 舊金山) produces gold, silver and flour. Old Gold Mountain was the name the Chinese used for San Francisco and, more generally, for California. The text further states, that the New Gold Mountain (*Xin Jinshan* 新金山), which means Melbourne, produces gold, and the Xueli Gold Mountain (*Xueli jinshan* 雪梨金山), which means Sydney, produces gold as well as coal. All these Gold Mountains reflect the gold rush since the mid-nineteenth century, which triggered a flow of Chinese leaving their country. The texts on the other parts of the map mainly address shipping routes. One text for example informs about travel durations. According to the map, travel duration between Shanghai and England lasted fifty days, between Shanghai and France forty-four days. Thus, Kuang Qizhao's map presents an interconnected world full of economic possibilities.

## 5    Conclusion

The presented cartographic images that were produced against the backdrop of the nineteenth-century colonial condition in China served as educational media, allowing for an educational experience. They informed about the basics of world geography, as well as about the world's economy, connectivity, travel and trade.

The biographies of their creators are all strongly connected with spaces that were shaped by this condition, and they reflect multiple cross-cultural influences: The official Xu Jiyu drew on materials and knowledge that he acquired from foreigners in China, the translator and co-author Chen Xiutang cooperated with the medical missionary Benjamin Hobson and possibly also with the printer and editor Andrew Scott Dixon, and the reform-minded and multi-talented Kuang Qizhao lived and worked in Australia and America, mingling with intellectual circles in New England. Both Xu and Kuang were politically active and they wanted to better the situation of their fellow countrymen by providing them with knowledge. In their cartographic works, these cultural intermediaries addressed world geography by drawing on the Chinese tradition of local gazetteers, combining visual with textual information, while at the same time utilizing information they had acquired in the cosmopolitan spheres in outside of China.

## References

Andrews B (2015) Blood in the history of modern Chinese medicine. In: Chiang H (ed) Historical epistemology and the making of modern Chinese Medicine. Manchester University Press, Manchester
Carroll JM (2007) A concise history of Hong Kong. Rowman & Littlefield, Lanham
Chan BA (2013) A forgotten Qing Era progressive: Kwong Ki Chui—lexicographer, interpreter, textbook author, newspaper publisher. J. R. Asiatic Soc. Hong Kong Branch 53:227–261

Chen M 陳鳴 (2005) Xianggang baoye shigao 香港報業史稿, 1841–1911 [A history of the Press in Hong Kong, 1841–1911]. Huaguang baoye you xian gongsi 華光報業有限公司, Hong Kong

Drake FW (1975) China charts the World. Hsu Chi-Yü and his Geography of 1848. Harvard University Press, Cambridge, MA

Elman BA (2005) On their own terms. Science in China, 1550–1900. Harvard University Press, Cambridge

Goodman B, Goodman DSG (2012) Introduction. Colonialism and China. In: Goodman B, Goodman DSG (eds) Colonialism and China. Localities, the everyday, and the world. Routledge, London

Han Y (2014) Kenntnisse der Chinesen von Deutschland in den 1840er Jahren. In: Leutner M, Steen A, Kai X, Jian X, Kloosterhuis J, Wanglin H and Zhongliang H (eds) Preußen, Deutschland und China. Entwicklungslinien und Akteure (1842–1911). Lit Verlag, Berlin

Hevia JL (2003) English lessons. The pedagogy of imperialism in nineteenth-century China. Duke University Press, Durham

Matten MA (2016) Imagining a postnational world. Hegemony and space in modern China. Brill, Leiden

McCunn RL (1988) Chinese American portraits. Personal Stories 1828–1988. Chronicle Books, San Francisco

Moll-Murata C (2001) Die chinesische Regionalbeschreibung. Entwicklung und Funktion einer Quellengattung, dargestellt am Beispiel der Präfekturbeschreibungen von Hangzhou. Harrassowitz, Wiesbaden

Mosca MW (2013) From frontier policy to foreign policy. The question of India and the transformation of geopolitics in Qing China. Stanford University Press, Stanford

Mühlhahn K (2012) Negotiating the Nation. German Colonialism and Chinese Nationalism in Qingdao, 1897–1914. In: Goodman B, Goodman DSG (eds) Colonialism and China. Localities, the everyday, and the world. Routledge, London

Pflug L (2019) From 'All Under Heaven' to 'China in the World': Chinese visual imaginations from the nineteenth and early twentieth centuries. In: Storms M, Cams M, Demhardt IJ, Ormeling F (eds) Mapping Asia: cartographic encounters between east and west. Springer, Cham

Rhoads EJM (2011) Stepping forth into the world. The Chinese Educational Mission to the United States, 1872–81. Hong Kong University Press, Hong Kong

Sinn E (2008) Lessons in openness. Creating a space of flow in Hong Kong. In: Siu HF, Ku AS (Eds) Hong Kong mobile. Making a global population. Hong Kong University Press, Hong Kong

Sinn E (2013) Pacific crossing. California Gold, Chinese migration, and the making of Hong Kong. Hong Kong University Press, Hong Kong

Smith RJ (1996) Chinese maps. Images of 'All under Heaven'. Oxford University Press, New York

Suzuki S (2009) Civilization and empire. China and Japan's encounter with European International Society, Routledge, London

Vittinghoff (2002) Die Anfänge des Journalismus in China (1860–1911). Harrassowitz, Wiesbaden

Wagner RG (2017) 'Dividing up the [Chinese] Melon, *guafen* 瓜分': the fate of a transcultural metaphor in the formation of National Myth. Transcultural Stud. 1:9–122

Waley-Cohen J (1999) The sextants of Beijing. Global currents in Chinese history. Norton & Co, New York

Wong S, Wong V (2017) The role of the Guangbao in promoting nationalism and transmitting reform ideas in Late Qing China. Modern Asian Studies 51(5):1469–1518

Xu J 徐繼畬 (1848) Yinghuan zhilüe 瀛環志畧 [A short account of the maritime circuit]. Bayerische Staatsbibliothek München, 4 L.sin. D 136-1/6, urn:nbn:de:bvb:12-bsb11128998-8. Accessed 24 November 2018

**Laura Pflug** is a research fellow at Leibniz Institute of Regional Geography in Leipzig in the research group 'Maps of Globalization: The Production and the Visualization of Spatial Knowledge' from the Collaborative Research Centre (SFB) 1199: 'Processes of Spatialization under the Global Condition'. Within this project she analyzes Chinese cartographical visualizations from the mid-nineteenth century until today. She holds an MA in Sinology as well as History and Society of South Asia from Humboldt University Berlin, where she has taught courses on modern and classical Chinese as well as Chinese culture and history after graduation. She has submitted her Ph.D. thesis focusing on Chinese historical geography to Ruhr University Bochum.

# The Middle East and India

# French Cartographic Services in the Levant: Putting Syria and Lebanon on the Map of the Empire

Louis Le Douarin

**Abstract** This chapter looks at the history of the establishment of French carto-graphic services in Syria and Lebanon from 1919, contextualizing it within a larger institutional, imperial and scientific framework. More specifically, it focuses on the circulation of imperial cartographic knowledge and practices between imperial realms and inside the French empire. The establishment of the League of Nations Mandate coincided with the immediate post-war years that led to a reconfiguration of the power balance at the world scale and saw the growth of new international platforms for political but also scientific cooperation. Against this tensed context, the cartographic project for the Levant and the first surveys were thus entangled in complex and connected levels, which in turn helps to temper the idea of an omnipotent and performative imperial cartography.

## 1 Introduction

In October 1919, General Hamelin, commander of the *Troupes françaises du Levant* (TFL or French Levant troops) expressed his concerns about the state of French cartography in the Levant to general Bourgeois, head of the *Service géographie de l'Armée* (SGA or Geographical Service of the Army). He claimed that 'whereas in Madagascar, in the South of Oran and in Morocco, only after a few months, and with fewer resources' than he had himself in the Levant, 'the troops and the administration had sketches, topographic and traverse surveys essential for civil and military operations'. Now, his army only possessed 'after one year of presence, the English 1:250,000 map, which [was] but a poor reconnaissance map.' 'Drawing a 1:100,000 map' he concluded 'should precede all other military and public works, and I particularly regret this absence.'[1]

---

[1]Hamelin to Bourgeois, 6 October 1919, Service Historique de la Défense, Vincennes (SHD) 4 H 12/8.

---

L. Le Douarin (✉)
European University Institute, Florence, Italy
e-mail: louis.ledouarin@eui.eu

© Springer Nature Switzerland AG 2020
A. J. Kent et al. (eds.), *Mapping Empires: Colonial Cartographies of Land and Sea*,
Lecture Notes in Geoinformation and Cartography,
https://doi.org/10.1007/978-3-030-23447-8_5

77

If Hamelin's frustration illustrates the importance he gave to cartography in the process of conquest, control and development of an overseas territory, this quote, too, reveals the problems encountered in quickly mapping Syria and Lebanon. After World War I, the end of the Ottoman Empire and the partition of the Middle East, the French mandate was established in the Levant. Organizing their power in the region, the French shaped new polities and established infrastructures and administrations in order to effectively administer and control the territory. At a political level, the mandate system, organized for Ottoman Syria at San Remo in April 1920, was thought to be a transitory system, used to prepare peoples not yet 'able to stand by themselves under the strenuous conditions of the modern world' to become independent nation-states. However, if this system did allow international oversight, setting fragile bases for a new era of international cooperation (Pedersen 2015), scholars have also seen it as a camouflage for colonial domination. The High Commissioner, chief of administrative services but also at the center of all political, legislative and military power, 'became the one and only channel that accumulated all the instruments state-sovereignty: in Syria and in Lebanon at the beginning of the 1920s, State power was the power of the mandate' (Mizrahi and Méouchy 2013: 40). This particular form of imperial domination has attracted significant academic attention, but many of its dimensions, including cartography, have yet to be examined. The renewal of interest in the historiography of imperial cartography that explored the complexity and the ambiguity of every imperial cartographic project, if it has extended to French-dominated territories over the last ten years (Blais et al. 2011), has indeed left the French Levant out.

This lack is in part explained by the scarcity of the material available to historians: unlike the documents of other services of the mandate, the archives of the *Bureau topographique du Levant* (BTL) were not transferred to Europe after Independence and were divided between Beirut and Damascus. The history of both these cities since 1975 explains the obstacles historians have to face when trying to locate and access this material. By drawing on the reports published by the *Service géographique de l'Armée* and on other archival material, this chapter presents the context that saw the installation of French imperial cartographic services in the Levant. If a *Bureau topographique du Levant* was indeed created as soon as 1918, Hamelin's frustration, a year later, illustrates his difficulties on the ground, but also the tensions between different layers of imperial authority. Hamelin's dependence on British maps, illustrates not only the vivid competition at play between two imperial powers, but also the importance kept by material produced before the war for the period at the beginning of the mandate. Maps and geographical knowledge on the region actively circulated between different imperial powers, but also different types of actors, questioning the continuity of practices and knowledge beyond the official 'imperial' chronology. His complaint being addressed to Bourgeois, head of a central service that managed, from the rue de Grenelle in Paris, the mapping of France, its empire and other parts of the world, this quote also illustrates Hamelin's dependence on a central administration, which, during the difficult period of the reconstruction, did not always give priority to Syria. The Levantine case thus provides an example of how a given imperial system in perpetual evolution, namely the French overseas cartographic services, came to face and adapt to the inclusion of a new territory inside the empire, in a tense and changing international, national and local context.

## 2 The Extension of a Cartographic Empire

### 2.1 Imperial Institutions and Military Occupation

As the latest additions to an already extensive colonial empire, Syria and Lebanon inherited models and practices designed before the war in other geographical contexts. With the Mediterranean expeditions of the early nineteenth century, the constitution of topographic brigades to accompany troops sent to a foreign land was generalized. This expeditionary cartography differed from the regular work as it was conceived on the mainland. Unable to conduct regular and definitive geodetic and topographic work in a systematic manner, cartographic services conducted itineraries and reconnaissance surveys alongside the progression of troops, and according to their needs, dealing with a certain number of practical and scientific issues: language, security, climate and an ability to read the landscape. Blais has therefore shown that if the institutional frame, and the general ideal of systematic covering, were directly inherited from the metropolitan model, specific practices soon developed inside the military cartographic services for these expeditions (Blais 2014). The French *départements* of Algeria offered the first durable experience of imperial topography and, with time, Algeria became a true laboratory where exploration cartography coexisted with a regular systematic topographic project, and where officers were formed before being sent all over the expanding empire. As the director of the geographic services of the French West Africa put it one century after the Algiers Expedition, Algeria had become 'the great school of French military geography in the nineteenth century' (De Martonne and Martin 1931: 11).

With the growth of the colonial empire in the nineteenth century, several new services of this type were thus created in the colonial capitals.[2] As in Algeria, a first bureau was generally set up with small personnel and resources in order to answer the direct needs of the army, then followed by a larger service, answering both military and administrative needs, taking part in the making of the territory, and in its conquest, control and economic development. When, at the end of the summer of 1918, the French Detachment for Palestine and Syria was gathered in Beirut following the Allied conquest of that region, it was only logical that a first small cartographic service was created. This *Bureau topographique* was then officially

---

[2]In Algeria: *Brigade topographique* (1830), *Section topographique* (1831 and 1838), *Brigade topographique de l'Algérie* (1861); in Tunisia, *Service géographique de Tunisie* (1881); in Indochina, *Bureau topographique de l'état-Major* (1886), *Service géographique de l'Indochine* (1899), and so on.

installed in Beirut, and its status was confirmed in March 1919.[3] Before the summer of 1920 and the start of the regular geodetic work, this first topographic bureau operated alone in the Levant, where the last battles of the global conflict in the region were taking place.

In November 1919, the departure of the British from the northern occupied territories left the Nationalist Arab government of Damascus to face the French troops alone (Frémeaux 2016). After several months of tensions between Arab nationalists and French authorities, the situation accelerated in the spring of 1920. In March, the Syrian National Congress denounced the compromise signed by Faysal and Clémenceau in January 1920, declaring the independence of the Kingdom of Syria. Meanwhile, in Sèvres and San Remo, the European powers were shaping the mandate regime and assigning the Arabian territories of the late Ottoman Empire to France and the United Kingdom. The conflict between these two incompatible principles, national and imperial, proclaimed in Damascus and San Remo, thus fostered tension. In July 1920, General Gouraud, a fervent advocate of French historical rights in the Levant, who had been appointed both High Commissioner and Commander of the Levant Army in November 1919, issued an ultimatum to the Syrians. After the refusal of the nationalists, the Battle of Khan Maysalun saw their defeat on 24 July and in September, Gouraud proclaimed the French mandate on the Levant States from Beirut. In the meantime, French troops had also occupied parts of Cilicia, where they faced a violent and prompt resistance, backed by the nationalist movement of Mustafa Kemal. Besides the conflict with the Turks and the Syrian nationalists, the French also faced violent oppositions to their occupation, in the coastal range and in the wide plateaus of the hinterland. The 1918 Armistice of Mudros and the Allied victory in Syria were, therefore, far from putting an end to clashes in the region (Neep 2012).

In this turbulent context, the role of the BTL[4] was therefore to accompany the troops' operations on all these different grounds. Answering the direct need of the military, it had to 'supply maps, optical and topographic instruments to the occupation forces; execute sketches and special prints for the staff; keep the existing maps up to date […] execute transverse surveys […] and establish temporary maps for regions presenting a special interest.'[5] The first mission of the bureau in the first half of 1919 was therefore to gather all existing cartographic material in order to print useful documentation. The most famous small-scale maps, which were published for a large audience at the end of the nineteenth century, constituted a first resource, but only because they were easily accessible.[6] However, other material,

---

[3]Ministère de la guerre to Commandant des Troupes françaises du Levant 2/03/1919, SHD 4 H 12/8. Hamelin also asked for the creation of a similar office in Adana, the *Bureau topographique de Cilicie*, but it would never be fully functional, because of the military situation in Cilicia.

[4]The topographic bureau had different official names between 1918 and 1939. For this chapter this name *Bureau topographique du Levant* or BTL is kept, which is the one commonly used in the sources.

[5]Bourgeois to Hamelin 22 August 1919, SHD 4 H 12/8.

[6]Kiepert's maps of Asia Minor (Anatolia, modern Turkey) were an important resource.

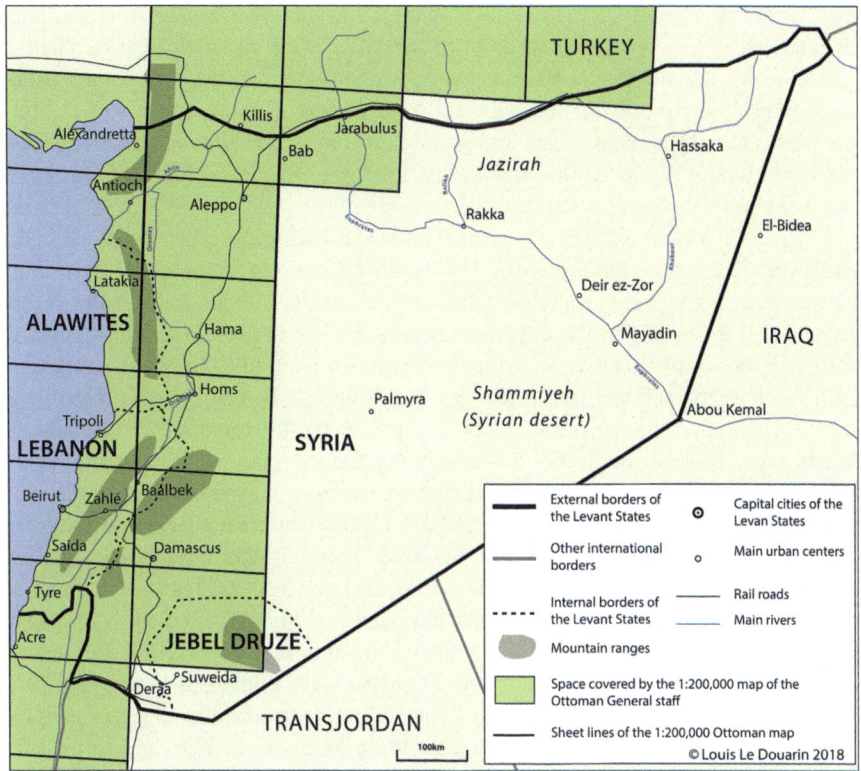

**Fig. 1** The Levant States created under French mandate in 1920. Only the coastal regions were covered by the Ottoman General Staff 1:200,000 map, used to draw the BTL's reconnaissance map (map drawn by the author)

which came from military institutions, was deemed more precise and reliable.[7] Sheets 27–49 of the 1:250,000 'Eastern Turkey in Asia' series, published between 1915 and 1917 by the War Office, along with other British maps like the Western Palestine map of the Palestine Exploration Fund (1881), were therefore acquired as early as May 1919.[8] The main reference for the BTL however, was the precious map of the Ottoman General staff. These were drawn under the supervision of Djemal Pasha, between 1908 and 1916, with help from German officers. In July 1919, the French topographic Bureau in Istanbul acquired a copy that it sent to the SGA in Paris, which translated it for the BTL (Fig. 1).[9] The French military was

---

[7]Duraffourd 'Rapport concernant l'énumération des cartes, plans et documents topographiques existants sur la Cilicie' SHD 4 H 12/8.

[8]Lemoine to SGA 18 May 1919, Messire to SGA 9 July 1919, SHD 4 H 12/8.

[9]The Istanbul bureau was created inside the expeditionary force created to occupy the city after the Ottoman defeat.

also aware of several surveys, drawn during the war by German officers. They included especially regular surveys of to the 1:25,000 of strategic regions such as the Bekaa Valley, the Houla Bassin or the Aleppo region. Along with more general maps published by the German General Staff during the war, these surveys represented some of the most recent and precise information available. However, if the officers in charge knew about, and looked for these precious documents, in 1919, they were still nowhere to be found.[10]

If the BTL was thus a new service set up in the midst of war, by the French, the actual cartography of the region in 1919 built on existing knowledge, illustrating the transnational or trans-imperial nature of these maps. The gathering process thus became a central stake in the negotiations taking place between French and British staff after the September 1919 joint convention organizing the British departure from the French mandate territories, and at the end of November, the British had indeed given part of their cartographic material to the French. These included British and Ottoman maps but documents on the German surveys in the Bekaa Valley.[11] Collection centre, the BTL played its most important role in the conception and printing of maps, plans and sketches to different scales and their distribution to the mandate services. The first maps were printed in Paris, by the SGA, but after 1920, the BTL acquired new presses: between 1920 and 1927, the 33 sheets of the Ottoman map concerning the whole of French and British mandates were translated and reprinted. Likewise, in the same period, a 1:100,000 version was designed from the same map, and 23 sheets were printed in Beirut, this time only for the French mandate territory. The BTL also produced rapid traverse and reconnaissance surveys in order to keep certain maps up to date and to produce temporary plans for the most important urban areas.

## 2.2  Diverging Priorities

The limitation of the BTL's work on preliminary surveys and compilation work was, however, soon contested by senior officers of the occupation forces, who were aware of the shortcomings of approximate and small-scale maps. This is why, in the letter in introduction, General Hamelin complained, in October 1919, that no regular cartographic work had begun, and that the army had to make do with old and imprecise documents. He asked for the regular topographic work to start. Early in 1919, with Captain Lemoine, first director of the BTL, the General thus proposed to Paris a complete program for the regular map of the Levant. The idea was to cover, as quickly as possible, the whole territory to the scale of 1:100,000, starting from

---

[10]'Rapport concernant l'énumération des cartes, plans et documents topographiques existants […]', Duraffourd, 10 July 1919, SHD 4H 12/8.

[11]Guerre to Armée du Levant 1/11/1919 and Armée du Levant to OETA North command 1/11/1919, SHD 4 H 12/8.

the coast. This very ambitious program, aimed at finishing the 'forty-eight most needed sheets' in four years, with two brigades operating each year around Beirut and Alexandretta, in the mountains and on the coast, alternatively. To lead the operations, they wanted 'senior officers who had been operators on the 1:100,000 in North Africa' to be transferred to the BTL, in order to lead the missions and train other topographers.[12] Hamelin and Lemoine were thus eager to establish, as fast as possible, a complete geodetic grid that could answer the needs of the different administrations of the mandate: Public Works, Cadastre, Army, and so on. This project also established the necessity of learning from the experience of the Army in other parts of the empire, and most of all to 'prevent the same hesitations that marked the progressive extension of surveys in Algeria, in Tunisia and in Morocco, in order to commence as of today, to prevent useless works and expenses, the map of the future'.[13]

This program, expression of a direct need for a good map but also of a larger scientific ideal, the 'map of the future', was, however, refused by the Parisian headquarter, which favoured financial and scientific pragmatism. The quantity and quality of information already available on Syria and Lebanon was indeed substantial, and the SGA, also recovering from the war, did not have sufficient personnel. Geodetic work and surveys in Syria and Lebanon would thus have to be postponed. Hamelin was to wait for the official cartographic program to be elaborated in Paris by the officers in charge, and for the regular brigades, which would be sent from Paris in due time: 'the geodetic operation on which the topographic survey will be based needs to be prepared with care and done by experienced topographers. A hasty and poorly grounded work would not have any value and would have to be repeated. The BTL does not have neither the experience personnel nor the material resources needed to do this kind of work.'[14]

These tensions actually related to the larger institutional question of the organization of cartographic services inside the French empire. If personnel working on the Empire's maps did circulate between all the different French possessions, fostering an intense circulation of practices and institutional models, not all imperial cartographic services answered to the same authorities. On the eve of the integration of the Levant into the empire, two different models coexisted. In Indochina, Madagascar and French West Africa, cartographic services were independent and answered directly to the local governors. However, in the rest of the Empire, and in the whole of North Africa, it was the central *Service géographique de l'Armée* that was in charge, from Paris, of organizing the cartography of all the different territories. If the missions were planned according to the requests of the colonial authorities, the undertaking depended on Paris both administratively and technically. In these territories, small local bureaus were also maintained on the ground as extensions of the SGA, to answer the everyday needs of the authorities.

---

[12]Hamelin to Guerre 19 February 1919 SHD 4 H 12/8.

[13]See Footnote 12.

[14]Bourgeois à Hamelin 22 August 1919 SHD 4 H 12/8.

The conflict between Bourgeois in Paris and Hamelin in Beirut can thus be understood as a debate on the institutional frame to be given to the Beirut bureau, with Hamelin pushing for a more independent service, capable of quickly answering the needs of both the army and the civil services of the mandate. In contrast, for Bourgeois, it was the North-African model that was to be applied to the Levant and the BTL was indeed legally put in place only as a local bureau, and explicitly copied from the Moroccan bureau, itself inspired by the Algerian one. If this local service was under the command of the *Armée du Levant*, and therefore of the High Commissioner, it was technically an extension of the Parisian *Service géographique*, and most of the cartographic work would indeed be conducted by officers sent from Paris. The regular maps would also be printed in rue de Grenelle, in the headquarters of the SGA, from which the mandate authorities therefore depended on for precise geographical knowledge.

## 3   Defining a Cartographic Project

With military control relatively secured after 1920, the SGA could start mapping the territory of the mandate. The idea of producing a regular topographic map of the whole territory corresponded to a certain extent to the application to different parts of the empire of the cartographic practices that were elaborated in France in the eighteenth and nineteenth centuries (Pelletier 2002). This conception, which strongly associated scientific representation and state sovereignty, was based on a claim for a technical grasp of the land, expressed through a systematic and exhaustive cartographic endeavour. In the Levant, this paradigm was partly transferred with the idea that France, as a mandatory power, had to draw the first modern map of Syria and Lebanon, newly 'independent' entities, states, and future nation-states. However, the elaboration of the cartographic project for Syria and Lebanon corresponded to a specific moment, after World War I, when the SGA took part in the efforts to promote common cartographic standards in the world, while continuing to work in different places, in the mainland and across the world. The elaboration of this project itself illustrates different tensions, with the need to adapt to different standards and situations.

## 3.1   Projection: Standards and Practical Uncertainties

While defining the project, the first obstacle the officers of the SGA encountered was the question of defining the actual extent of territory under the French mandate. Indeed, if during the summer of 1920, Gouraud defeated the Arab Kingdom of Syria and captured Damascus, it was still not certain if French officials would to be able to hold their claims against the British in the South and the East on the one hand, and against the Turks in Cilicia and the Nationalists in Damascus on the other. In December 1918, Clémenceau had already accepted a modification of the Sykes-Picot Agreement, renouncing the Mosul province and accepting a British Palestine. In 1920, when the cartographic project was discussed, only the coastal strip from Alexandretta to Beirut was secured and this political and territorial uncertainty directly influenced the technical characteristics of the map, delaying technical choices. To map the whole of Syria, one needed to know where it started and where it ended. More directly, these territorial doubts delayed the choice of projection: a coastal strip, which extended from north to south, would not be suitable for the same projection as a wide zone going from the sea to the Euphrates.

Ludovic Driencourt, officer of the SGA responsible for the report on the project, thus summarized the issues at stake: 'As far as Syria is concerned, [...] it is the Gauss projection that seems to be the most suited if we only consider the region located between the coast and the parts next to the Anti-Lebanon in the east. But one should also keep in mind that the hinterland spreads far to the east, and that [...] it is the Lambert projection that suits the best the shape of the country.'[15] If 'useful' Syria (*Syrie utile*), stretching along the coast, thus perfectly fitted the Gauss projection that limited distortion from North to South, the inclusion of the large eastern deserts would have supposed major deformations, hence the need for the Lambert projection, more suited to spaces extending both in latitude and longitude. However, it should be noted that this choice of projection, which was also adopted for Morocco in the same year also corresponded with a more general adoption of the Lambert by the SGA, across the empire, but also in the mainland. The Lambert projection had indeed replaced the Bonne projection in 1915 to answer the needs of the war and was applied to the regular '*carte d'état major*' after 1918 (Levallois et al. 1988).

The question of the actual extent of the mandate territory also made it difficult for the officers of the SGA to define the centre in order to establish the projection tables. As the Lambert projection entails distortions at the latitudinal extremities, to the North and South, the idea was to determine a zero point in such a way that the most 'important', strategic, regions of the territory would not be too concerned by the distortion. If it was impossible to determine this point without the eastern borders being defined, a temporary system would be put in place with a zero point at the centre of the coastal range, between the towns of Hama and Homs. This choice, which entailed major distortions in the representation of the eastern part of Syria was thus primarily a result of the economic priorities for the French in Syria, which focused on the fertile and populated occidental plains, but also on the political and military context as, at the time, only these regions were secured and administrated.

---

[15]Driencourt, 'Instruction au sujet du système de projection et des coupures des cartes définitives du Maroc et de la Syrie'. 30 August 1920, 8–9. Archives of the geodetic section of the Institut Géographique National, (IGN) 206 02 01.

## 3.2    Gridlines: The Levant and the Map of the World

As far as the division of sheets was concerned, the officers at the *Service géographique* dealt with another issue that also reveals the overlapping of different levels of scientific stakes. The post-war period was indeed the occasion for different cartographic services to adopt, or not, some of the standards promoted by the conferences of the International Map of the World (IMW) (Pearson and Heffernan 2015). In the SGA, the cartographic section for instance pushed for the new index tables of Morocco and the Levant to be based on the sheetlines of the 1:1,000,000 sheets of the IMW, which followed parallels and meridians, starting from the Equator and Greenwich. However, this was not the idea of the geodetic section, led by the charismatic Georges Perrier who defended a less international and more French division based on meridians and parallels, but was counted in grads and started from the Paris meridian.

In the end, Lucien Driencourt asked the geodetic section to forget its national pride. If 'France managed to make its views prevail in a number of questions, metric system' and so on, it should accept being 'less fortunate' in other matters, and accept the international sheetlines, 'with the enormous advantages that should ensue'.[16] These advantages referred of course in part to the compatibility between different cartographic systems but also to the material issue of the use of the map on the ground. The maps drawn from by the IMW gridlines, with their division in degrees (4° or 6° for the 1:1,000,000-scale sheets), were deemed easier to manipulate and fold than the large sheets proposed by the geodetic section.

As early as 1920, before any geodetic work had started, the *Service géographique* had derived an index table for the map of Syria and Lebanon, illustrating the nature of the project: to create an exhaustive map of a coherent territory, symbolizing the extension of French political and technical control over space, as well as the unity and coherence of a new part joining an already-existing Empire. Syria and Lebanon were thus symbolically covered and framed by a rigid grid of fractions of meridians and parallels, organized in a complex yet complete nested structure, each part of the country being framed in a specific sheet with a precise reference, according to the nomenclature initiated for the International Map of the World.[17]

---

[16]Ibid. 11.

[17]The Zahle sheet was thus NI36XII4d: from left to right and from bottom down: 9th zone from the Equator; 36th sector; 12th 1:200,000 sheet, 'Beirut'; 4th 1:100,000 sheet, 'Zahlé' and 4th 1:50,000 sheet. See Fig. 3.

## 3.3 Scales: The Triparition of the Mandatory Space

Whereas in Madagascar, Indochina and West Africa, the independent colonial geographic services answered directly to the local administration and conducted different types of work by themselves, in the rest of the empire, all the global cartographic work, if planned in Paris, was divided between the local and permanent bureaus on the one hand and by the survey parties sent punctually from Paris on the other. The local bureaus had to answer the basic needs of the administration and their activity mainly consisted of compiling and reprinting reconnaissance maps for different colonial services and realizing temporary surveys in areas where information was lacking. In contrast, the regular brigades focused exclusively on the production of the regular map: geodesy, detailed surveys, levelling, astronomical observations, and drawing and printing of the maps.

In the Levant, once the project had been agreed at the end of 1919, a similar division of labour was thus put in place: regular brigades were to be sent each year by the SGA, to do the regular geodetic and topographic work, while the BTL continued to compile reconnaissance maps and to conduct rapid surveys in strategic regions. As was the case in the Maghreb and other parts of the empire, this functional partition also entailed a strict spatial dichotomy. On one side, the regular Parisian brigades focused their work on the coastal plains and the first slopes of the coastal ranges, where the major urban centres and sedentary populations were located. The work of the *Service géographique* was therefore mainly focused on the 'useful' regions of Syria and Lebanon, namely the coastal plains of Beirut and Latakia and the regions of Damascus, Aleppo and the Orontes valley. There, the Parisian officers benefited from the security and the infrastructures needed for their precision work. On the other hand, the *Bureau topographique*, apart from its regular printing mission that was realized in Beirut, mainly operated in the hinterlands, in the different mountainous regions and more generally east of the Anti-Lebanon, in the arid plateaus and the desert, where the populations, as well as the landscapes, were deemed more hostile and still needed to be 'discovered' and properly charted.

As far as the scales of these maps were concerned, the original idea was to cover the whole territory at a scale of 1:100,000—a practical solution that was suited to the lack of finance and manpower that followed the war. Cheaper to make than the 1:50,000 that was used for the regular map of Algeria, the 1:100,000 was still more precise than the 1:200,000-scale, already accessible from the Ottoman documents. In the French practice of overseas cartography, the scale of 1:100,000 was more generally used for areas where control was not fully assured, or simply for large territories, where a more precise map was of use only in certain strategic centres.

This 'conquest scale' was for instance used for the regular maps of Indochina and Madagascar.[18] However, the scale of 1:50,000 finally proved to be more suited for the Levant and was chosen for the regular topographic map, to be drawn by the SGA brigades. This choice was mainly a consequence of a major divergence the SGA and the military on the ground came to observe, between, on the one hand the relatively important amount of cartographic information already available at a smaller scale thanks to the Ottoman General staff map, and, on the other, the absence of any precise surveys on the mot strategic regions. Thus, there was an urgent need for precise topographic surveys on the coastal region, which would be performed at 1:40,000 for a 1:50,000 map. The Great Syrian Revolt, that introduced certain modification in the way the French administered the region, also influenced the general cartographic project. The uprising, also called the Druze Revolt, had started in the southern Druze (or Suweida) State in 1925, and had expanded to several regions of southern and central Syria and Lebanon. The violence of the repression showed to the world the nature of French mandatory power in Syria. But it also exposed a lack of geographic information on the eastern regions, where the hostility against the mandatory power was substantial. From 1927, the BTL was therefore put in charge of a new 1:100,000 series. The same officers were responsible for mapping some strategic points of the eastern desert and preparing the work of the eastern border boundary commission, with the objective of better controlling tribal movement between Syria, Iraq and Transjordan.[19]

The choice of scales can thus be understood as a compromise between an ideal initial scheme that favoured affordability and rapidity and the reality on the ground, building a fragmented geography according to different imperial priorities: development, military control of hostile regions and general coverage, overseeing and organization of the territory. In the end, the French mandate was thus partitioned lengthwise into three parts: the useful and densely populated coast and coastal ranges in the West were mapped at 1:50,000; the intermediary mountains and desert fringes (populated but hard to control) were surveyed at 1:100,000; and the eastern peripheries were only represented at 1:200,000, which was used as the general reference map for the whole of Syria and Lebanon (Fig. 2). This tri-partition, although unseen before in this specific form, was, in fact, similar to that which had been designed for other colonies of the empire, especially in North Africa, where the same gradient can be seen between the coastal Tell, the 'intermediary' high

---

[18]In Indochina, the territory was divided between the more populated and strategic regions that were mapped at the 1:25,000 scale and the mountainous hinterland, mapped at the 1:100,000. This scale was then generalized as a reference scale and applied to the whole Union, while in parts of Cambodia, the 1:50,000 was also used. In Madagascar, after a different initial project, the 1:100,000 was adopted for the regular map in 1907. See De Martonne and Martin (1931).

[19]Captain Delienne thus led several missions in the east in 1928–1929 before participating on the joint French-British reconnaissance mission in the border region in 1931–1932.

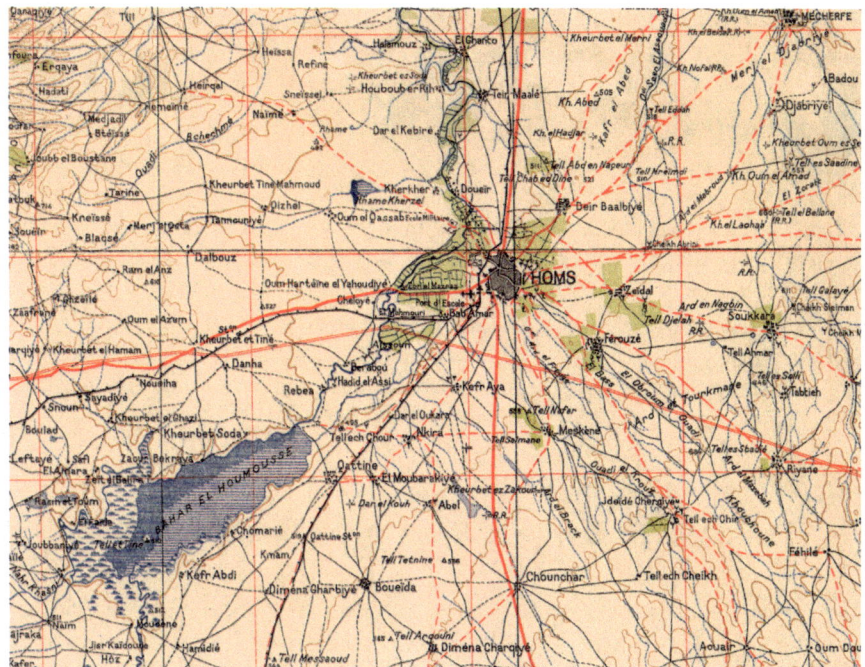

**Fig. 2** Extract from the Homs sheet (NI-37-XIII, 1:200,000) of the provisional reconnaissance map. The map was a compilation of different elements: a rapid survey made by the BTL in 1932 in the north of the Anti-Lebanon; the regular topographic surveys made by the SGA for the Homs sheet of the 1:50,000 series; and the revision of the Ottoman General Staff map. BTL 1937 (Courtesy of the Maison de l'Orient de la Méditerranée)

plateaus and the Saharan Desert.[20] If some circulation of staff between the regular services of the SGA and the BTL happened, the division of labour also meant that different people were in charge of different spaces and different scales: the 1:50,000 to the more experienced, mainly superior officers of the SGA, and the reconnaissance 1:100,000 and 1:200,000 to the less-trained officers of the BTL.[21]

---

[20]If the Algerian and Tunisian costal range was also mapped at 1:50,000, different types of scales and surveys were applied to the intermediary zone (1:100,000 and 1:200,000) while the desert was mapped at 1:200,000 in Tunisia and parts of Algeria, and at 1:500,000 for most of the southern part of Algeria.

[21]If some members of the BTL helped the regular SGA brigades on the ground, the officers generally did not belong to the two groups. Only the BTL's directors usually possessed extended experience in the field and might have taken part in the annual missions before joining the local bureau, as did Captain Delienne. SHD GR 8 YE 126927.

## 4   Covering Syria and Lebanon: Local Resources and Conflicting Strategies

During the first months of French occupation, the work of the geographical services became fragmented: the elaboration of an ideal and efficient project in Paris and the compilation and preliminary surveys on the ground to answer the direct needs of the conquest. Only in August 1920 was a final project agreed upon and the first geodetic brigade was sent to Beirut. This was led by Colonel Georges Perrier himself, at the time director of the Geodetic Section of the SGA, and he brigade measured the first baseline of the future geodetic network in the Bekaa Valley, along with two first signals.[22] If, on the eve of World War II, most of the territory would indeed be covered by one of the new cartographic series of the mandate, this process developed with a complex and chaotic chronology.

### 4.1   The Ksara Observatory: Heart of the Geodetic Network

The French were not the first Europeans to choose the Bekaa Valley to conduct geodetic work. Between 1917 and 1918, a German brigade had indeed measured a first geodetic baseline of more than a kilometre between Rayack and Deir Zinoum, viewed seven points, and surveyed the region at 1:25,000 with the aim of improving irrigation. The choice of the region as the centre for the geodetic network of the Levant by the French and the Germans can be explained by geographical and technical reasons. Firstly, the Bekaa 'Valley', or plateau, is a large and mostly flat area surrounded by two mountain ranges: the Lebanon in the West and the Anti-Lebanon in the East. This situation thus offered multiple observation possibilities for triangulation, with several summits clearly visible from the bottom of the valley. The centre of the plateau was also relatively close to Beirut, allowing a rapid extension of the triangulation chain to the strategic coastal areas. The valley was also crossed by the Beirut-Damascus road and railroad, which secured a connection to some of the main urban centres of the region and granted good accessibility to the baseline, allowing men and instruments to travel easily from Beirut to the baseline, and to other parts of Syria.

The main asset the Bekaa had to offer, however, was its infrastructures, namely the Jesuit observatory of Ksara (Udias 2003) (Fig. 3). Created in 1906 by the Reverend Fathers of the *Université Saint-Joseph* of Beirut, it had been directed by RF Berloty since its opening. Important works were conducted in the building in order to make astronomic calculations and Berloty managed to import several

---

[22]Five signals were built in total by this first mission: the southern end of the baseline (altitude 858 m, south of Bar Elias), northern end of the baseline (altitude: 912 m, south of Houch Hala), intermediary point of the baseline (altitude: 874 m), Djebel Kneysseh (altitude: 2091 m) and Djebel Sannine (altitude: 2548 m).

**Fig. 3** Detail from the Zahle sheet (NI-36-XII-4d, 1:50,000) of the regular topographical map of the Levant States. The survey for this sheet was conducted in 1926 by Captain Baudry's topographic brigade. It was the second sheet of the regular map series to be completed following the one of Beirut and it was printed in 1927. In this extract, the Ksara Observatory and the three terms of the original Bekaa baseline, south of Bar Elias, Tell ouasi and Haouch Hala, can be seen (Courtesy of IGN)

instruments from Europe. Until World War I, the Jesuit Fathers thus realized the first astronomical and topographical works around the Observatory, while Berloty tied links with several scientific institutions in France and over the world (Berloty 1921).[23] The calculation of longitude, a key focus of the director's work, took him several times to Europe to convey the Observatory's experiences and introduced him to Gustave Ferrié, who helped him to acquire a radio station in 1912.

In the spring of 1919, when Lemoine and Hamelin designed their first project to map the Levant, it was therefore natural that they turned to the Ksara Jesuits, whom they asked for a preliminary report on the possibilities the Bekaa could offer in terms of geodetic work.[24] However, the nature of the relationship between the French geographic services and the Observatory staff changed scale when Berloty travelled to France in May 1919. Between the end of 1914 and December 1918, he

---

[23]Private archive of the Université Saint-Joseph, Beirut (USJ) 1C3.
[24]Armée du Levant to SGA, 5 May 1919 SHD 4 H 12/8 and Berloty (1921).

had indeed left Ksara for Egypt, while during the war the Observatory had been partly demolished and looted. In France, Berloty was looking for help from the new mandatory power in order to rebuild, and quickly resume its work. The opportunity was seized on by Georges Perrier, who saw the benefits he could gain from the knowledge and experience of the Jesuits. This collaboration was acted on October in a letter from Bourgeois, head of the SGA, to Hamelin.[25] It was then confirmed by the expedition of several scientific instruments to Ksara. In anticipation of the first geodetic mission, a radio station was sent to the Observatory. Perrier's first geodetic mission thus arrived on site at the end of August 1920 and was hosted at the Observatory. Berloty and Combier, his assistant at the Observatory and a former lieutenant in the French army, took an active part in the work of the first brigade, led by Perrier, that chose and measured the baseline.[26] Meanwhile, the Observatory itself was at the centre of all elementary calculations: Captain De Volontat determined the longitude from a pillar built at the Observatory while the second brigade proceeded to the levelling of the baseline, connecting Ksara to the medimarimater installed in Beirut.[27] The Fathers' Observatory, thanks to its localization, its history and its infrastructure, had thus become an essential and strategic resource for the mandatory cartographic services, and as a consequence, the centre of the Levantine geodetic chain, which would then spread to the rest of the country. The history of the Observatory also helps to grasp the transforming nature of French (and European) imperialism in the Levant. The role played by Ksara and Berloty is indeed an example of how French influence developed in the region during the end of the Ottoman era through commercial, educational, religious—and also scientific —institutions and networks, that were reactivated and transformed after 1920 (Arsan 2016). In Ksara, Bourgeois and the mandate's officers thus found and used allies and infrastructures that preceded them in introducing and producing formalized, European, geographical knowledge to and on the Levant.

## 4.2   Competing Agendas: The Fragmented Geography and Chronology of the Covering of the Mandate

After 1920, the first step of the general cartographic enterprise was to extend the geodetic network north and south in order to establish the Beirut-Aleppo Meridian chain. The 43 signals were built and viewed between 1920 and 1927 along the coastal range, from the Palestinian to the Turkish border. The original idea was that the proper surveying would closely follow the extension of the geodetic network, in

---

[25]Bourgeois to Hamelin, 22 August 1919. SHD 4 H 12/8.

[26]Berloty and Combier also helped to measure the second baseline in Bab, near Aleppo, in the summer of 1923.

[27]The third and last brigade, led by Captain Schmerber, built and observed the first signals in the Lebanese mountains.

order to process in the regular coverage, by a rate of two or three sheets per year. This progressive and orderly enterprise would have covered coastal Syria in only a few years. However, from the early years of the mandate, this ideal plan devised in Paris contrasted against the reality on the ground in Syria.

Priority was not given to cartography, but to economy. For the experts of the High Commissioner's office, the emergency was to measure, delimitate and register land plots in order to start a large land reform that would restore Syria's antique prosperity (Achard 1922). Consequently, the first topographic brigade that was supposed to be sent to survey the first sheet was cancelled. Instead, and for several years, the High Commissioner insisted that the *Service géographique* should focus mainly on the development of the geodetic network, in order to accelerate the work of the Cadastre. This explains the rapid development of the network towards the north and into the Orontes Valley, at a time when the Beirut sheet of the regular map had not even been published. After a first extension between Damascus and Homs, not far from the baseline, work began in the Aleppo region, where the *Régie du Cadastre* needed to complete its survey. The distance from the original network forced the 1923 mission to start new primordial operations, independently of the Bekaa reference, and measure a new baseline in Bab, near Aleppo. This region was also not on the same longitude as the southern geodetic chain, and the new base was approximately a hundred kilometres further to the east than Ksara. The next year, it was thus decided to quickly fill the gap between the two networks in order to include the Bab baseline into the Beirut Meridian. Between July and December 1924, three brigades duly built and viewed fourteen new signals between Hama and Aleppo (Service géographique de l'Armée 1932: 25). This priority given to land reform and to the *Travaux publics* services of the mandate in general also explains the rapid extension of the geodetic network in the Jazirah region, between the Euphrates and the Turkish and Iraqi borders. Besides the western strip where urban centres and fertile plains were concentrated, two thirds of the mandate territory were indeed constituted of desert regions where nomadic and semi-nomadic populations raised certain specific issues for the French authorities in terms of tribal control and development of the territory. In the middle of the 1920s, several political and military solutions were designed to secure the control of the Jazirah, but the region was mostly subject to a large colonization and development plan, based on irrigation, the settlement of local tribes, and the promotion of migration of Kurdish and Christian refugees (Velud 2016).

In 1934, the parts of Syria where economic development was deemed possible were thus included in an L-shaped geodetic chain, between the Hauran at the Palestinian border and the Jazirah at the Turkish-Iraqi border (Fig. 4). The two parts of the network, however, did not answer the same objective as the Euphrates-Jazirah region was not to be concerned by the regular topographic map. To the contrary, the Aleppo Meridian, although its development was in part determined by the land reform agenda, was destined to be the base for the 1:50,000.

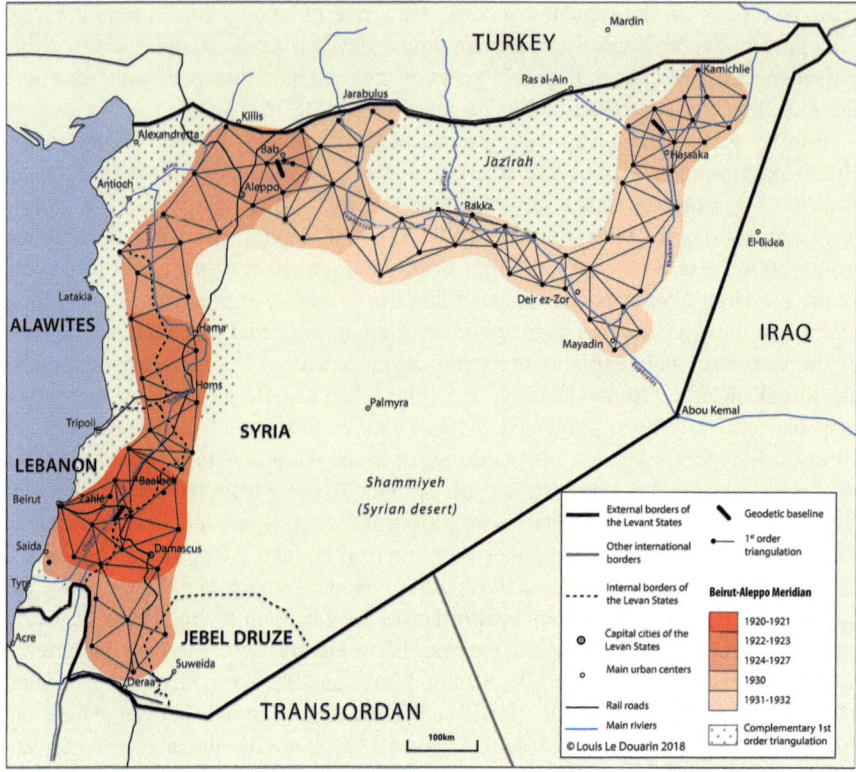

**Fig. 4** Progressive extension of the Levantine geodetic network (map drawn by the author)

The progress of the regular map was therefore strongly linked to the priorities set by the civil administration, and between 1920 and 1939, only some specific sheets had been surveyed, in the western part of the country (Fig. 5).[28]

While the Parisian brigades' work indeed mainly served colonial development purposes, the role of the topographic bureau was much more linked to operations of control of the territory, and to the military. Between 1925 and 1927, this direct military role was confirmed when, after the end of the clashes, the human and financial resources of the bureau were increased in order to fill in the blanks on the map. Between 1927 and 1930, several reconnaissance missions were sent in marginal spaces, in south-eastern and north-eastern Syria. As already mentioned, this also corresponded to the emergence of a need for a new map series at 1:100,000, which mainly concerned the Druze states and the Damascus region.

---

[28]In 1939, approximatively 35 sheets were surveyed, and 25 sheets had been published. The concerned regions were the major urban centres and their surroundings, the Aleppo plain, the Bekaa and the Homs gap and Orontes valley.

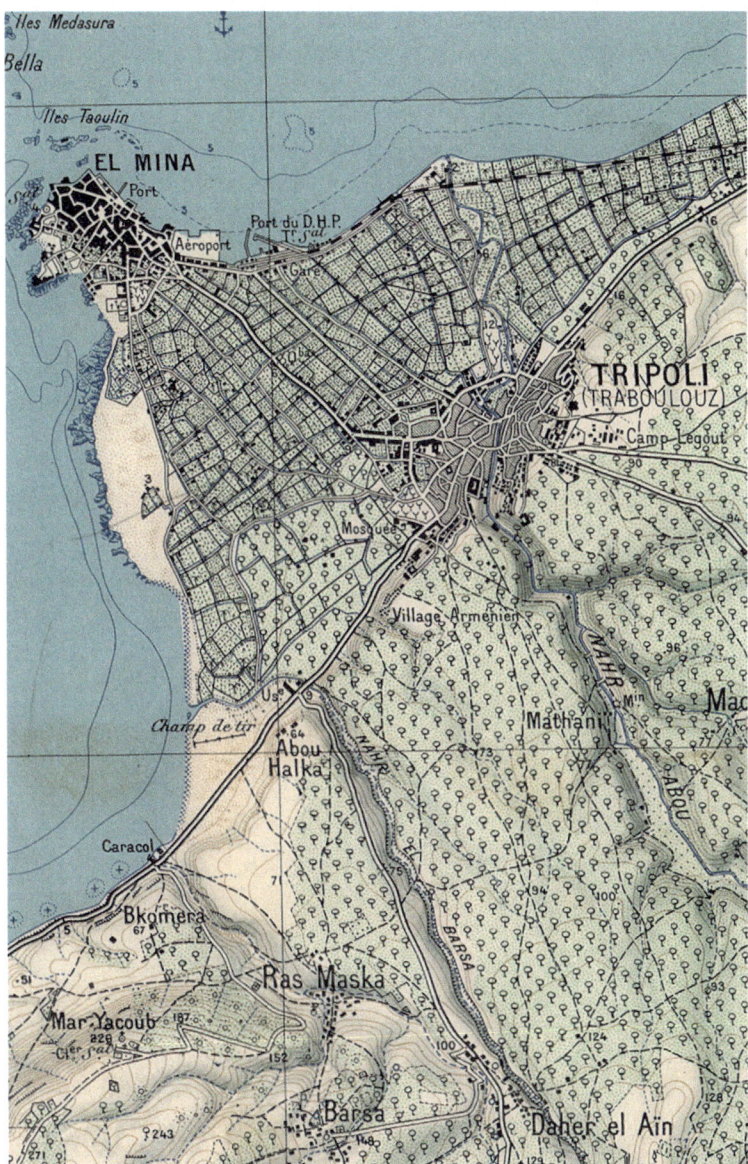

**Fig. 5** Extract from the Tripoli sheet (NI-36-XVIII-2d, 1:50,000, 1933) of the regular topographical map, representing the strategic assets of the city: the citadel, the old and the modern port, the train line, the airport and the gardens surrounding the city. The survey for this map were made in 1931 by Captains Houdmont and Juilland. Service Géographique de l'Armée, 1933 (Courtesy of the Maison de l'Orient de la Méditerranée)

The difficulty of controlling these regions was partly illustrated by distance, one
of the main obstacles that colonial powers have to face, despite their modernist
discourses and their claim to comprehensiveness. The substantial network of roads
and railroads that covered parts of Syria and Lebanon, and that extended with the
mandate, allowed topographers to reach certain areas more rapidly. However, the
bureau's operations in the east went outside of these networks. The development of
air transportation was therefore an important asset for these topographers in order to
reach their fields of operation. This specific use of aerial transportation was particularly brought to light in July 1924, when two officers of the SGA, Henri Renaud
and Louis Govin, were killed in an airplane accident while flying to Deir ez-Zor to
conduct astronomical observations. The literature has shown the symbolic and
effective importance of aviation for the enforcement of the European domination in
the Middle East during the inter-war period. First used as a coercive tool for
repression and control, aviation found a new function in intelligence and cartography (Thomas 2013). Aviation thus helped the topographic bureau as early as
1919, as a provisional solution. Nevertheless, the cooperation reached its full
development only after 1927, when the 39th *Régiment d'Aviation* became a central

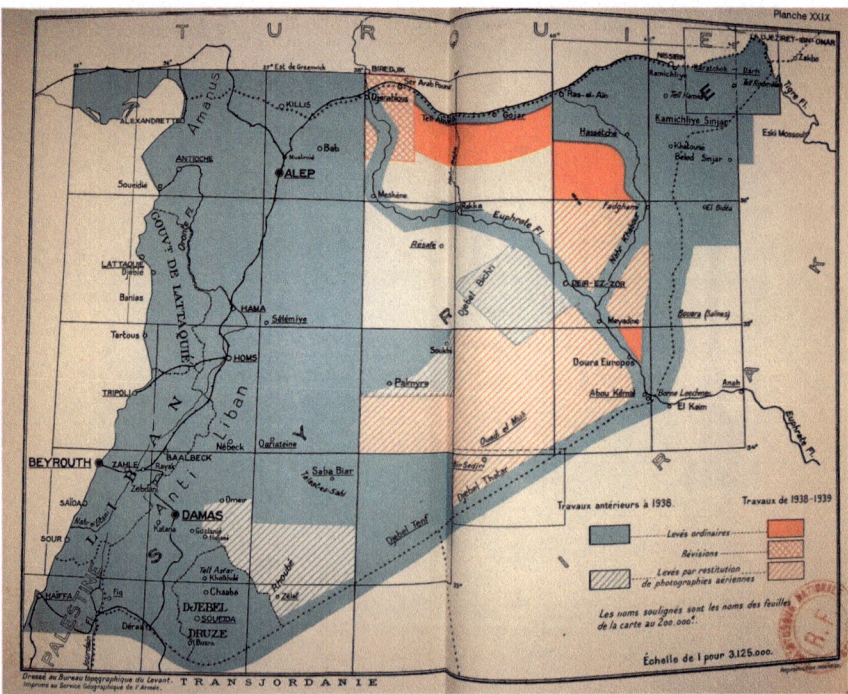

**Fig. 6** 'Reconnaissance surveys and 1:200,000 map'. Progress of the surveys made by the BTL to
complete the new edition of the exhaustive 1:200,000 series, respecting the IMW gridlines and
progressing from the coast towards the east. Service géographique de l'armée, 1939 (Courtesy of
BnF)

actor in the bureau's coverage of the east, and especially of the border regions. Eventually, these surveys were included in a specific project: the publication of a new reconnaissance map at 1:200,000, which was based on surveys and aerial photography, in order to replace the Ottoman map.[29] The main objective of this map was exclusiveness: from the coast, the coverage progressed in orderly fashion, towards the hinterland, along the meridians that marked the division between each sheet. They reached the 37th meridian in 1933, and the 38th in 1935. Progressively, the blanks on the map were thus filled, thanks to the bureau's eastward push. This dimension was particularly emphasized in the official reports of the SGA, with maps where a swathe of dense colour gradually covered the white peripheries, recounting the almost teleological tale of the inevitable coverage of the Syrian land (Fig. 6).

# 5 Conclusion

The addition of Syria and Lebanon to the map of the French Empire and of the world coincided with the post-war moment, which saw a reconfiguration in the balance of power and a growth in international cooperation. To a certain extent, this was intended as a new way to deal with conflict, which gave birth to institutions, platforms and norms. The mandate regime was one of these tools, aiming at reforming imperial practices. The norms of the International Map of the World and other international standards were others that reformed cartographic practices. When in both fields the French adhered to their own objectives, without a doubt this new international frame played a role in the way these territories were designed and mapped.

Other factors played a major role in the evolution of imperial cartographic practices during the inter-war period, determining the ways in which the general project was adapted to the reality of the ground. If the total coverage of the Levant states was therefore completed through the 1:200,000 map, it was only thanks to the help of aerial photography. Because of this technical innovation, the military answered the issues it encountered in mapping an uncontrolled land in a new way, different from the traverse surveys of the nineteenth century: flying over the land, they could apply an exhaustive scheme to regions where no soldier had set foot.

While standards for scales of the world and of the rest of the empire were discussed and set in Paris, the occupation troops relied on their experience, technical innovation and on existing resources. In that sense, this Levantine example

---

[29]Commencing in 1932, this new series was also the opportunity to apply the IMW gridlines to the reconnaissance map, in line with the 1:50,000 sheets.

also questions the continuity of practices between the end of the Ottoman era, where French and European interest led cartographers to take interest in the region, and the mandate period. This continuity, impersonated by Berloty and the Jesuits in Ksara, is also questioned by the use French officers made of the 1862 *Carte du Liban* in 1919. Drawn to 1:200,000 from the survey made during the Syria campaign of 1860–1861, it was the cartographic expression of Napoleon the Third's imperial glaze on the Levant. In 1919, this map was the first to be reprinted, with almost no modification, serving renewed, yet different, imperial plans.

# References

Achard E (1922) Études sur la Syrie et la Cilicie. Le coton en Cilicie et en Syrie. L'Asie Française 3:19–62

Arsan A (2016) There Is, in the Heart of Asia, … an entirely French population. France, Mount Lebanon, and the Workings of Affective Empire in the Mediterranean, 1830–1920. In: Lorcin P, Shepard T (eds) French Mediterranean transnational and imperial histories. University of Nebraska Press, Lincoln, pp 76–100

Berloty B (1921) Annales de l'Observatoire de Ksara 1921. Imprimerie catholique, Beirut

Blais H (2014) Mirages de la carte: L'invention de l'Algérie coloniale. Fayard, Paris

Blais H, Deprest F, Singaravelou P (2011) French geography, cartography and colonialism: introduction. J Hist Geogr 37:146–148

De Martonne É, Martin J (1931) La carte de l'Empire colonial français. Imprimerie de G Lang, Paris

Frémeaux J (2016) Les interventions militaires françaises au Levant pendant la Grande Guerre. Guerres mondiales et conflits contemporains 262(2):49

Levallois J-J (1988) Mesurer la terre: 300 ans de géodésie française. Presses de l'Ecole nationale des ponts et chaussées, Paris

Mizrahi J-D, Méouchy N (eds) (2013) France, Syrie et Liban 1918–1946: Les ambiguïtés et les dynamiques de la relation mandataire. Presses de l'Ifpo, Beirut

Neep D (2012) Occupying Syria under the French mandate: insurgency, space and state formation. Cambridge University Press, Cambridge

Pearson A, Heffernan M (2015) Globalizing Cartography? The International Map of the World, the International Geographical Union, and the United Nations. Imago Mundi 67(1):58–80

Pedersen S (2015) The guardians: The league of nations and the crisis of empire. Oxford University Press, Oxford

Pelletier M (2002) Les Cartes des Cassini : La science au service de l'Etat et des provinces, 2nd edn. Editions du Comité des travaux historiques et scientifiques, Paris

Service géographique de l'Armée (1932) Mémorial du Service géographique de l'armée, Tome VI: Description géométrique des États du Levant. Imprimerie du Service géographique de l'armée, Paris

Thomas M (2013) Markers of Modernity or Agents of Terror? Air Policing and Colonial Revolt after World War. In: Baxter C, Dockrill M, Hamilton K (eds) Britain in global politics volume 1, security, conflict and cooperation in the contemporary world. Palgrave Macmillan, London, pp 68–98

Udias A (2003) Searching the Heavens and the Earth: the history of Jesuit observatories. Springer, Dordrecht

Velud C (2016) La Politique Mandataire Française à l'égard Des Tribus et Des Zones de Steppe En Syrie: L'exemple de La Djézireh. In: Bocco R, Jaubert R, Métral F (eds) Steppes d'Arabies: États, Pasteurs, Agriculteurs et Commerçants: Le Devenir Des Zones Sèches. Genève, Cahiers de l'IUED, pp 68–98

**Louis Le Douarin** is a third-year Ph.D. researcher at the History and Civilization Department of the European University Institute in Florence. He specializes in studies of the Middle East, historical geography and history of geography. His doctoral research focuses on the production and circulation of geographical and cartographical knowledge on Syria and Lebanon before and during the French mandate, and on the role this knowledge played in the transformation of territories in that region.

# Surveying Empires: Archaeologies of Colonial Cartography and the Great Trigonometrical Survey of India

Keith D. Lilley

**Abstract** This paper demonstrates the important contribution material culture makes to histories of cartography. Focusing on the tangible and intangible heritage of the Great Trigonometrical Survey (GTS) of India, and West Bengal in particular, the material landscape legacies of the GTS are analysed and interpreted. This reveals new insights into how surveys of the GTS were undertaken in the nineteenth century, under George Everest, and the infrastructure that was created to underpin the British mapping of India. The trigonometrical stations built by the GTS were of different designs and construction, adapted in response to local conditions and circumstances. Today, this 'survey heritage' is at risk, yet provides a basis for understanding more deeply the materiality of mapping and survey practices used in mapping empires. The paper connects the three-dimensional 'spaces of survey' with the two-dimensional 'space of the map', and concludes by arguing for greater consideration of the *verticality* of mapping.

## 1 Introduction

Over recent years, the Great Trigonometrical Survey (GTS) of India has received both scholarly and popular attention, not least through its associations with George Everest (Keay 2000; Sen 2002). The GTS represents the apogee of European imperial cartography and 'mapping empires', a huge project in terms of scale and ambition, and one which left its mark on the Indian national psyche, as well as the landscapes of the sub-continent (Edney 1997; Mukherjee 2011; Crosbie 2012). However, despite its national and indeed international importance as an example of nineteenth-century colonial cartography and scientific survey, the infrastructure created at the time, much of which still remains visible in the landscapes of India today, is neglected and to date yet to receive any systematic archaeological study.

K. D. Lilley (✉)
Queen's University Belfast, Belfast, Northern Ireland, UK
e-mail: k.lilley@qub.ac.uk

© Springer Nature Switzerland AG 2020
A. J. Kent et al. (eds.), *Mapping Empires: Colonial Cartographies of Land and Sea*,
Lecture Notes in Geoinformation and Cartography,
https://doi.org/10.1007/978-3-030-23447-8_6

101

In this regard, an assessment of 'survey heritage' of the GTS is overdue and it is but one example of Europe's legacies of overseas survey and mapping.

Indeed, despite the recognized significance of mapping in the formation of the modern world, particularly in supporting the global activities and actions of colonizing powers of the West over the past three centuries, the material cultures of European colonial cartography in former overseas territories and dominions around the globe remain undervalued and overlooked by historians of cartography and geography. By focusing on 'archaeologies of colonial cartography', this paper redresses this through new research assessing the significance of the physical structures built for the purposes of the Great Trigonometrical Survey of India.

## 2 *Surveying Empires* and 'Materialities of Mapping'

Undertaken for the subcontinent as a whole, between 1802 and 1871, by British and Indian surveyors, the GTS ultimately was 'to impact a greater coherency and cohesion to [its] mapped spaces than could be obtained by the older, astronomic methods for cartographic control' (Edney 1997: 336). Like other similar large-scale state-led geodetic surveys of the period, the GTS depended on establishing a network of trigonometrical stations in an occupied landscape, to enable observations to be made using large and cumbersome instruments in the field. While some stations were sited on mountain tops and structures such as temples and pagodas, others required purpose-built towers. It is these 'tower stations', constructed for the GTS in India, that form the subject for this paper, and are used here to exemplify and explore the 'materialities of mapping' and the three-dimensional spaces of survey.

Monuments of past surveying practices that were fundamental to colonial and imperial mapmaking are still manifest and evident in the archaeological landscape. To initiate a field-based study of the archaeology of the GTS a team of historical archaeologists and geographers from the UK and India collaborated in establishing a research project called 'Surveying Empires: archaeologies of colonial cartography in West Bengal'. The project received funding through the British Academy's International Partnership and Mobilities scheme in 2016 to explore the archaeological potential of the GTS through combining field-survey of surviving structures and monuments in the landscape constructed by GTS surveyors, with ethnographic surveys of localities where the GTS tower stations stood, interviewing local people about the sites in 2017.[1] The principal aim of the 'Surveying Empires' project concerned the identification of GTS tower stations in the field and the development of a methodological framework for future archaeological and architectural study of

[1]British Academy award PM150229. The 'Surveying Empires' project team comprised Keith Lilley, Satish Kumar and Siobhán McDermott from Queen's University Belfast, and Bishnupriya Basak, Rajat Sanyal and Sharmista Chatterjee from the University of Calcutta. The author wishes to thank the whole project team for their support and guidance in the field in West Bengal. For more project information, see surveyingempires.org.

the surviving infrastructure of the GTS in India more broadly, by focusing in one particular region, West Bengal, where the GTS had a long presence through its geodetic survey work. Calcutta, now Kolkata, was the centre for GTS operations in east India from the 1820s onwards, and Everest himself was based there in the 1830s (Edney 1997: 209, 263). The city is also the base of the project's Indian collaborators, working at the University of Calcutta, with essential local knowledge and expertise so critical to the project's fieldwork.

'Surveying Empires' coalesces around two particular 'archaeologies' of cartography, one of which involves 'excavating the map', metaphorically, to look 'beneath' the lines and features shown by maps of empire, such as historic Survey of India maps and contemporary records of geodetic observations compiled by the GTS through the survey operations during the nineteenth century; the second involves field-work, in West Bengal, combining buildings and landscape archaeology methods to survey the GTS trigonometrical station sites and monuments in situ, and undertake an analysis of their fabric and form. The idea of combining these approaches is to be able to explore both the locales of survey, under the GTS, as well as their wider connections, within India and with the British Empire more broadly, looking at the post-medieval 'material cultures' of mapping and the grounded practices of imperial cartography—geodetic and topographical surveying—through which surveying in the field actively shaped the landscape as well as the map (Lilley 2017).

To this end, in focusing on the material cultures of survey and cartography connecting Britain and India in the nineteenth century, 'Surveying Empires' has a contribution to make to ongoing debates on 'landscapes of empire' and their (post-)colonial archaeologies, in particular in terms of questioning further 'core-periphery' models of colonial relations, as revealed by the material practices of the GTS on the ground, and also in terms of highlighting issues raised by these legacies of empire in the context of a 'decolonized' India.

## 3 Being in the Field—GTS Operations in West Bengal

The operational focus of the GTS on Calcutta occurred in 1818 when the GTS, under superintendent William Lambton, was placed under the control of the Bengal government. It was also here that Everest returned in 1830 following his sojourn in England, which began in 1825 when he left India following illness incurred while surveying the Great Meridional Arc on the Deccan plateau (Edney 1997: 211, 242). Everest had himself been appointed superintendent of the GTS following Lambton's death in January 1823, and it was under Everest, as surveyor general of India (1830–1843), that the trigonometrical network series that Lambton had begun in Madras were extended across the entire sub-continent, ultimately spanning the whole of India to form 'The "skeleton" upon which the detailed geography of India might be hung' (Edney 1997: 333). By 1870 these series of trigonometrical chains were complete, providing the geodetic framework for the British mapping of India, and four of the series that formed part of this wider network of surveyed triangles converged on Calcutta (Fig. 1).

From the west, the Calcutta Longitudinal Series extended from the Great Arc at Sironj to Calcutta, a distance of 684 miles, which took just six years to complete, between 1825 and 1831, surveyed by a team including Joseph Olliver, William Rossenrode, Murray Torrick and John Peyton using a Cary 18-inch Alt-Azimuth theodolite instrument (Walker 1880a: iv). This work was undertaken while Everest was away in England, though Olliver and Rossenrode were both experienced in trigonometrical work through their earlier employment on the Great Arc under Everest. By the time the Calcutta Longitudinal had reached the flat plains of Bengal it was December 1829, and 'this entailed the building of very lofty towers and the clearing of long rays at a considerable expenditure of time and money' (Walker 1880a: vi). According to Phillimore (1958: 81), the Survey of India's historian, thanks to 'careful reconnaissance by Rossenrode a number of towers were put up in rough fashion that would just serve the purpose', but it was on Everest's return to India in 1830 that saw the series advance the final 60 miles to reach Kolkata, necessitating further discussion on the construction of the survey towers.

It is these 'lofty towers' that formed a focus for the 'Surveying Empires' project, including the GTS survey towers constructed in Kolkata itself, to stand at the two ends of the Calcutta Base line, another vital aspect of trigonometrical survey which required the very precise measurement of a distance that then forms the basis of trigonometrical calculations using the angle measurements observed through triangulation between the survey stations set up by the GTS. The Calcutta Base-line was set out on the Barrackpore Trunk Road in northern Calcutta, in 1830–1831, and here Everest and his survey teams undertook careful measurements between the two station towers using instruments Everest had witnessed being used in Ireland in 1828 under the direction of Thomas Colby, who was then leading the Ordnance Survey six-inch survey of the island (Edney 1997: 247–248; Crosbie 2012: 113–117). James Prinsep's 1832 drawing of the base-line measuring at Calcutta records this event, and also the apparatus employed by Everest, which compares closely to the instruments and system depicted being used in Ireland by Colby in measuring the Lough Foyle Base-line there (Prinsep 1832; Yolland 1847) (Fig. 2).

In 1831, following the measurement of the Calcutta Base-line and completion of the Calcutta Longitudinal Series, attention switched to creating three new trigonometrical series, to the north, south and east of Calcutta, to broaden the GTS network across eastern India and around the Bay of Bengal. First, to the north of Kolkata, the Calcutta Meridional Series was initiated in 1843, under Andrew Scott Waugh, and extended northwards from Kolkata to the foothills of the Himalaya at Darjeeling. It formed one chain linking a group of similar north-south meridional series—described as a 'grid-iron'—and designed by Everest to 'tighten' the overall GTS network and connect the Calcutta Longitudinal in the south with the North-East Longitudinal to the north (Hennessey 1883a: iii–iv). The Calcutta Longitudinal and Meridional Series both connected just north of Kolkata, where too Peyton was deployed as First Principal Sub-Assistant, the series covering 270 miles in the end and completed by 1848 using a superior Troughton and Simms 18-inch

**(a)**

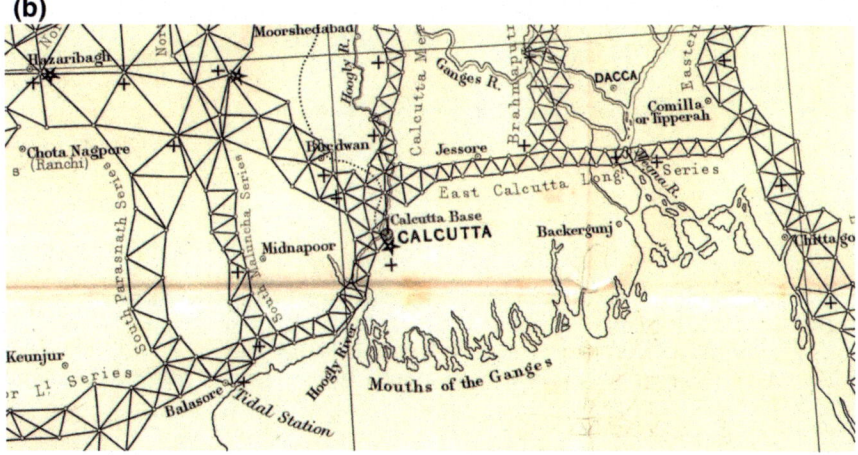

**(b)**

Fig. 1 Index chart to the Great Trigonometrical Survey of India (1870) (**a**), with an extract showing the Calcutta (Kolkata) GTS series (**b**). Creative Commons

CALCUTTA BASE LINE

**Fig. 2** Two base-lines compared, the Lough Foyle Base-line (Co. Derry/Londonderry, Ireland) of 1827–1828, and Calcutta Base-line (West Bengal, India) of 1832. From Yolland W (1847) An account of the measurement of the Lough Foyle Base in Ireland. Palmer and Clayton, London. Prinsep J (1832) Progress of the Indian Trigonometrical Survey. Journal of the Asiatic Society of Bengal 1:71–72

theodolite (Hennessey 1883a: iii–iv, vii). Secondly, and while the Calcutta Meridional Series was progressing northwards from Kolkata, the East Coast Series was initiated in 1844, working southwards along the coast from Kolkata under the authority of Captain Thorald Hill, whose previous experience was with surveying in the Madras Topographical Survey (Walker 1880b: iii–iv). The East Coast series was completed by 1860, having covered a distance of 466 miles of difficult terrain and weather conditions over fifteen years (Walker 1880b: xxxviii). Next and finally came the East Calcutta Longitudinal Series, which extended eastwards across Bengal, begun in 1862 under Lieutenant Henry Thuillier of the Royal Engineers, using a large Troughton and Simms 24-inch theodolite (Hennessey 1883b: iii–iv). Traversing the low-lying Bengal delta region of the Ganges, with its myriad of river channels, the East Calcutta Longitudinal Series eventually covered 238 miles by March 1867, connecting with the Eastern Frontier Series (Hennessey 1883b: vi). With the completion of this fourth series, the GTS project as a whole was itself drawing to a conclusion, in 1871, under James Walker, the then superintendent of the GTS who had succeeded Waugh in 1861.

The operations of the GTS in West Bengal, conducted over five decades and involving a range of personnel working on four different trigonometrical series, provides a basis for comparing GTS survey practices and their traces on the ground through the construction of survey stations in the vicinity of Kolkata. Within a 50 km (30 mile) radius of the city, the four series provide a temporal 'cross-section' of GTS activities and their associated infrastructural works, including the 'lofty towers' necessitated by the lower lying lands of the plains of western Bengal around Kolkata. To examine the differences and similarities of the GTS's survey practices that underpinned the execution of the trigonometrical series, the 'Surveying Empires' project focused on this region of West Bengal, identifying, locating and recording the surviving remains of the survey stations built and used by the GTS between the 1820s and 1860s.

## 4   Sites of Survey: The Evolution of GTS Stations

To characterize the fabric and form – as well as evaluate the current condition – of the stations associated with these four series which converged on Kolkata, required a programme of field-survey, visiting the station sites on the ground and assessing any standing remains of the structures as a prelude to formulating a typology of GTS survey stations and comparing the construction techniques and materials between the four trigonometrical series. This survey work was carried out between November 2016 and March 2017 by the 'Surveying Empires' project (surveyingempires.org/), starting with an inventory of GTS triangulation station sites compiled from 1920s Survey of India (SoI) 1:126,720-scale topographical maps (kindly provided by the Bodleian Library, Oxford) which were scanned and rasters imported into ESRI ArcGIS to provide geographical coordinates for 21 mapped GTS stations around Kolkata (all marked on SoI sheets 79B NW/SW/NE).

(For mapped locations of the station sites see the 'Surveying Empires' ArcGIS Online resource at https://arcg.is/1rLnCD.) With estimated site locations gained from the georectified historic SoI sheets, the identification of station sites on the ground involved navigating using handheld GNSS, and also much guidance gained by local enquiry. Having identified sites on the ground, survey work in the field involved digitally-recording the remains of stations using a DSLR camera, with scope for creating 3D visualizations from Structure from Motion (SfM) methods. As well as a photographic record of the sites, key dimensions of surviving structures were taken, including measurements of bricks used in construction of the stations (where they survived), while interviews with local people were audio-recorded on site to provide a basis for an ethnography of the GTS stations and to capture their meanings and significance for local communities today.

On the basis of the project's survey evidence gathered from the field, combined with historical accounts of the surveying of the GTS trigonometrical series compiled by the GTS in volumes published in the early 1880s, drawn from SoI records, the characteristics of the stations used in the four series converging on Kolkata can be pieced together. From this assemblage of sites of survey, what emerges is a typo-chronology of GTS stations and a cross-section of GTS surveying practices in West Bengal during the period of Everest's role as superintendent and his two successors. The particular types identified here owe much to the demands and local conditions met by the surveyors on the ground during the course of their field-survey work, and far from there being conformity in station construction under the overall organization of the GTS the pattern of station design and building varies spatially and temporally between series and survey teams. Yet, the purpose of the stations was always the same: to ensure accurate and reliable observations between them, to derive measurements that would provide the necessary geodetic data to map India accurately and consistently. This meant being able to rely on the stability of the structures, on which to rest the theodolite to take the instrument's readings, and having a clear line-of-site between stations, requiring the removal of vegetation, and sometimes existing buildings, by the surveyors, all of which entailed, first, the selection of sites for positioning trigonometrical stations, and often, second, construction of purpose-built structures for the survey itself. It is these structures that repay closer inspection, both as understudied evidence of the GTS trigonometrical operations on the ground, and as legacies of a neglected colonial heritage.

When Everest resumed his field-work for the Great Arc in 1833 it was after his time spent in Kolkata and the termination of the Calcutta Longitudinal Series and the base-line measurement there. The landscape the Great Arc traversed north of Sironj towards the Himalaya was similarly flat, and in his *Account of a Measurement of Two Sections of the Meridional Arc*, published in 1847, Everest noted the difficulties presented by this terrain and the need for towers to be built for the theodolite:

> It now remained to give instructions and drawings to the executive officers of the building department to enable them to proceed with the construction of the towers which the Government had authorized me to have erected at the different sites in the plains thus selected as the stations of principal triangles—of these I will beg to offer the following brief

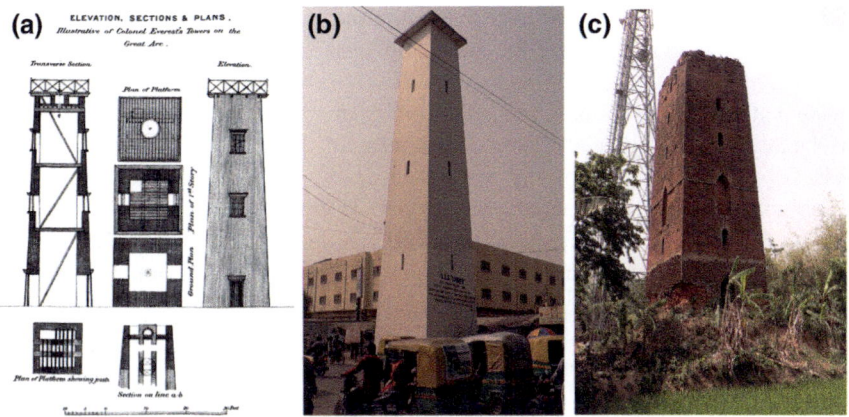

**Fig. 3** Everest-type GTS tower stations including Everest's (1847) Great Arc tower-station design (**a**), the GTS tower station at Sukchar on the Calcutta Base-line (**b**) and the GTS tower station at Āknapur (**c**)

description, which, by the aid of the drawings in Plate 25 (Fig. 3a), will probably suffice to render the subject intelligible. These edifices are of a square form at the base, and average about 50 feet in height, in some instances more and in others less. The wall at the bottom is 5 feet and at the top 2 feet in thickness, whence it appears that the interior is in mathematical language a portion of a square based prism, and the exterior a frustum of a square based pyramid. (Everest 1847: xxiii)

What Everest describes here is a design that has its earlier origins in the experiences of the surveyors working on the Calcutta Longitudinal Series at the end of 1829, when they were faced with the sixty miles of remaining land through which to complete their survey to reach Kolkata. The sequence of designs and decision-making that led to them can be traced through the historical accounts of the execution of the four Kolkata series as well as in the field. To examine how this sequence unfolded, here a 'stratigraphic' approach to the stations and their series is adopted, beginning with the earlier series and characterizing their stations' structures, and then moving through towards the later series' stations. From this cross-section of GTS stations, variations in their design and construction emerge, as do the links with those factors affecting siting and building GTS stations in West Bengal, such as local conditions and the decision-making of particular individuals and teams involved on the ground at these sites of survey. First, tower types are considered, and then second, methods of the towers' construction.

## 4.1 GTS Tower Typologies: The Calcutta Longitudinal and East Coast Series

Field-survey identified the surviving structures of GTS survey towers associated with the Calcutta Longitudinal Series, to the west of Kolkata as well as within the city itself. Broadly, two types of tower were used: a truncated pyramid structure, a rectangular tower, as evident still today at Sukchar and Rishi, in Kolkata, as well as at Bhola, near Singur, 30 km north-west (Figs. 3b and 4), and a cylindrical tower, evident at Noada, and Nibria, in the built-up environs of Kolkata and Hooghly, and at Dilākās, again situated to the north-west of Kolkata, some 40 km distant (Fig. 5a–c). The origins of the first of these two basic types are revealed by Phillimore (1958: 81–82), who describes how Everest was 'specially pleased with the two 75-foot towers built at the extremities of his base-line on the Barrackpore road by the Civil Architect, Mr Parker, but the tower built at Gopalnagar under Captain Bell [1792–1836], executive engineer, collapsed and Bell asked that the site should be shifted to better ground'. These two base-line towers are those situated at Rishi and Sukchar, both now preserved by the Kolkata Public Works Department (PWD) which has responsibility for their maintenance, and each rendered and roofed. Inscriptions on the towers commemorate their links with Everest and the GTS and are local landmarks. The dimensions and external design of the two base-line towers is identical, both having tapering façades punctuated with slender rectangular openings at three levels, and a base that measures 5 m x 5 m ($\sim 16.5$ ft$^2$). The influence of a civil architect here is itself interesting, contrasting with the station structures built by the survey team on the approaches to Kolkata, including the initial tower built at Gopalnagar. Indeed, on the basis of this early

**Fig. 4** The GTS tower station at Bhola on Singur to Tarakeswar road

**Fig. 5** Cylindrical GTS tower stations reusing earlier semaphore towers at Dilākās (**a**), and at Nibria (**b**) including a Structure from Motion (SfM) visualization of Nibria (**c**)

tower-building experience, 'Everest thought it best to entrust their design and construction to professional builders' (Phillimore 1958: 81).

As Phillimore notes, the tower at Gopalnagar built by Captain Bell had collapsed, due to the soft ground, and a new tower was constructed at Bhola nearby. The GTS tower at Bhola, on first impression, is very similar in design to the towers at Rishi and Sukchar in Kolkata designed by Parker, with a slender tapering truncated pyramid form, suggesting they were copied to create the tower at Bhola. Unlike the two base-line towers, the tower at Bhola has exposed brickwork and close inspection of the tower's exterior reveals vestiges of render, indicating that the original visual appearance of the station would have been considerably more conspicuous in the local landscape than it is today (Fig. 4). A further stylistic parallel between the Bhola and two base-line towers in Kolkata is the narrow rectangular opening at each level, all set vertically in line, offering a strong symmetrical appearance, echoing the 'mathematical language' to which Everest referred in the case of the Great Arc towers and also emulating architecturally the mathematical exactness of the GTS itself.

The tower's engineering at Bhola shows a concern for stability too, with relieving arches used above the ground-floor doorways, which because of the lost exterior rendering are clear to see. Even so, when the Bhola tower was reused at a later date for the East Coast Series survey, in the 1850s, the tower had clearly not fared well with the passage of time, and the survey team reported that it was damaged, but not as much as the nearby tower at Āknapur which was part of the Calcutta Longitudinal Series of 1825–1831 and had by 1858 collapsed entirely, perhaps as a result of the 'rough fashion' in which these earlier eastern towers had been constructed by Rossenrode? Indeed, although the GTS tower at Āknapur is a 'truncated-pyramid' form it is much less slender than the tower at Bhola, and noticeably shorter in height, reflecting the need at Āknapur for the tower's rebuilding in 1866–1867, under H. Keelan, 'Surveyor 3rd Grade', when the Calcutta Longitudinal was in the process of being re-measured due to unacceptable errors incurred in the first survey (Walker 1880a: vi–vii). At Āknapur, the tower is built on an earthen mound, and this has exposed brick foundations at the tower's corners, no doubt dating from the rebuilding, while the tower itself has a different configuration of openings compared with Bhola's, including larger windows. These, while again aligned vertically on each façade, are not rectangular in form but instead neo-Gothic, something new in GTS tower design therefore, contrasting with the earlier square-plan towers of the 1830s (Fig. 3c).

Despite subtle differences in form, the rectangular GTS towers at Bhola, Rishi and Sukchar each reveal consistent aspects in their design, and point to the continued influence of Parker's original design praised by Everest. Seemingly this model created by Parker was adopted for subsequent GTS tower construction in the Kolkata area. For example, it is evident at the GTS tower standing at Sāmālia, which formed part of the East Coast Series of the 1840s, south of Kolkata. Though here the tower is partially collapsed, its square ground-plan is clear, as are vertically arranged rectangular openings and ground floor opposing doorways with round-headed arches (Fig. 6). Not all the GTS 'tower stations' around Kolkata are of this design, however.

A second key type is a cylindrical form of tower, evident still in Kolkata at Noada, close to the Calcutta Base-line, and at Nibria, on the west bank of the Hooghly River and now subsumed into the urban fabric of Kolkata. Both the Noada

**Fig. 6** Sāmālia GTS tower station showing external façade (left) and internal view of brickwork, doorway and window opening with scale rule (10 mm gradations) (right)

and Nibria towers were in fact inherited from an earlier scheme of 1817–1828 to connect, by means of a 'visual telegraph', Diamond Harbour, to the south of Kolkata, with Benares, some 700 km (450 miles) to the north-west (Phillimore 1954: 269). By the time the Calcutta Longitudinal survey party, under Olliver and Rossenrode, had arrived in the vicinity of Kolkata, in 1830, the telegraph towers were redundant and seen to provide a suitable basis for reuse as GTS tower stations, as is made clear by correspondence sent at the time by Olliver, who observed how 'Mr. Peyton […] describes the country to be altogether alarming, […] adding it as his firm opinion of the necessity of resorting to the telegraphs' (Phillimore 1954: 263). However, 'Olliver had already noticed considerable swaying on the tops of the telegraph towers and, after making a visit to test this "with a very sensible level", Everest had the upper ten feet or so reconstructed, so that the weight of the theodolite and its pillar was carried on beams specially let into the wall' (Phillimore 1958: 81).

From exterior evidence at the tower at Nibria, the cylindrical 'semaphore' GTS towers were also rendered originally, yet have quite a different design to the square-plan towers. Round-headed arches predominate for the upper floor openings, as well as the ground-floor doorways, and small ox-eye windows are also present at regular intervals (Fig. 5b). Similar design traits between the round and square towers are efforts to strengthen the brick-courses by the use of vertical stretchers at regular intervals, as well as string-courses for the three levels of the towers. Internally, these levels were where the wooden platforms were placed, as is evident still in Noada and another cylindrical tower at Dilākās, north-west of Kolkata, and relatively close to Bhola, and likewise forming part of the Calcutta Longitudinal (Fig. 5a). The internal width of the walls at the ground floor is similar between the square and round plan towers, at around 2 m (6.5 ft), and the survival of both forms for nearly 200 years is testament to their solid construction and sound design, aiding the required stability for the theodolite, which was mounted on a platform at the top of the towers, requiring some adaptation of the telegraph towers according to GTS records. Unfortunately, in all the cases considered here, except for Rishi and Sukchar, the top-most levels of the towers have suffered most damage, and towers are now open to the sky. Even so, the site surveys of the GTS 'tower stations' around Kolkata reveal an interesting set of common design principles, despite their different shapes and origins, such as the use of rendering and the use in all cases of fired bricks; and also particularities in design, notably the stylistic differences in the form of window openings, as seen when comparing the earlier telegraph towers, the 'Parker' types of the 1830s, and the later tower as at Āknapur.

What is apparent, then, is that the GTS 'tower station' was not a fixed and uniform type, but evolving. Partly this was due to the occasional need for rebuilding. For example, under Keelan's 1866–1867 revision of the Calcutta Longitudinal, when the survey party 'proceeded with an examination of the towers employed by Mr Olliver [in 1829–1831]' and found that 'while the old telegraph towers were found to be in good preservation as when originally built […] the towers built by Mr Olliver had not proved of such good workmanship [and] Āknapur had fallen down entirely' (Walker 1880a: ix). This necessitated the rebuilding of the tower at Āknapur, hence its different form to nearby Bhola of the same series (Figs. 3c and 4) Furthermore, as well as the two main types of rectangular and cylindrical tower construction used for surveying the eastern series of the Calcutta Longitudinal, other types of trigonometrical stations were being used. Again, during the revision of this series in the 1860s, and the necessity to rebuild some of the earlier tower stations, the later GTS surveyors remarked, somewhat critically, that 'there was no uniformity' between the stations that Olliver had built, for while 'at one Madhpur a solid rectangular column of masonry 7 feet at base, 4 feet at top and 40 feet high marked the station', at Bhalki 'the station consisted of a platform having four pillars, one at each corner about 35 feet high' (Walker 1880a: ix). Such differences also contrast with the 'hollow rectangular tower' built at Bhola at around the same time, a difference no doubt owing much to Everest's intervention in 1830 and adaption of Parker's rectangular tower design for the latter which had met with Everest's approval.

## 4.2   GTS Survey Stations and Construction Methods: The East Coast Series and East Calcutta Longitudinal

Taking both the archaeological and historical evidence together, it is clear that the design of trigonometrical stations Everest later adopted for the Great Arc, in the 1840s, had an earlier gestation, with innovations occurring in West Bengal, and that contemporary influences in the construction and design of 'tower stations' came not only from the GTS surveyors working on site, such as Olliver, but also from architects and engineers advising them. The influences of those involved in constructing the GTS survey stations in the project's study area is further revealed through a closer examination of the field-evidence as well as the historical accounts of their building documented by the GTS. An acceptance to adapt designs and construction methods, according to the local conditions, both physical and cultural, that the GTS surveyors were working in can be seen by comparing the East Coast Series and East Calcutta Longitudinal Series. Both series extended the GTS triangulation network southwards and eastwards from Kolkata, and both had to be established across similarly flat and flood-prone land.

Compared with the surviving towers of the Calcutta Longitudinal Series especially, field-survey to the south and east of Kolkata undertaken in 2017 revealed scant physical remains of the GTS stations that had been constructed. In some cases, as at Bāniban, this absence of remains was due to deliberate demolition, as local inhabitants readily recounted, and in the case of nearby Mirzápur the site of the station is now a deep water-filled hole, excavated, apparently, in search for gold believed to be buried beneath it. Here, bricks visible in rubble on the site are the only tangible sign of the station's structure, some having been reused for an adjacent agricultural building. To the east of Kolkata, too, along the East Calcutta Longitudinal Series that stretched towards modern-day Bangladesh, the stations of

Berghom and Bira are no more to be seen. Sited in fields some one kilometre north from the main Hābra-Gobardānga road, on a trackway positioned in between reservoirs, the station at Berghom was reported locally to have been dismantled in the 1950s by the Survey of India (SoI), and a nearby garden wall appeared to be constructed from its former brickwork. Here too a brass plate mentioned by local people, but removed, was probably the station's marker. The nearby station at Bira has similarly disappeared and here a scatter of bricks on the site of similar dimensions to those observed at Berghom suggest they had a shared construction. While dismantling and demolition is clearly a factor in the endurance of these GTS stations, there are other factors at play, particularly the construction methods used by the GTS and the materials employed by the survey teams on the ground, especially the availability of fired bricks. This becomes especially evident in examining the stations constructed in the 1840s and 1860s for the East Coast and East Calcutta Longitudinal Series through comparing the contemporary records of the stations with the evidence gained from the site surveys in the field.

With the East Coast series, initiated in 1844 under Captain Thorald Hill, the GTS records reveal a very challenging set of circumstances that had a direct bearing on the construction of the stations along the banks of the Hooghly River, south of Kolkata. The Synopsis descriptions of the stations reveal variations in stations here, at Sāmālia 'A hollow rectangular tower 63.08 feet high defines the station, another at the bottom' (Walker 1880a: 19) whereas at Mirzápur 'The tower is hollow and 35.21 feet high, and has a mark-stone imbedded in the ground floor' (Walker 1880a: 6), and at Bāniban 'The tower is 40 feet high and has markstones at top and bottom' (Walker 1880a: 19). The lowered heights of the towers of Bāniban and Mirzápur, west of the Hooghly River, compared with the tower at Sāmālia to the east can be explained by the local conditions encountered by Hill and his men at the outset of the creation of the East Coast Series. The tall Everest/Parker-type of 'hollow' tower construction at Sāmālia, still standing though partially collapsed, suggests an effort was made initially to continue with the same construction methods as those used earlier for the Calcutta Longitudinal Series as at Bhola, a high and slender tower made of fired bricks (Figs. 4 and 6). However, the lesser heights of Bāniban and Mirzápur stations is an indication that the tall towers were perhaps too demanding in labour and materials, and that a new construction method was required as the series progressed southwards along the low-lying lands adjacent to the Hooghly, as Hill remarks:

> Of the new towers, those at Bāniban, Sāmālia and Mirzápur are unapproachable excepting by water for three-fourths of the year [...]. In the erection of these towers the usual method of making and burning bricks near the station was attempted, but not attended with success. The great price to be paid for wood fit for burning bricks, the difficulties in collecting a sufficient quantity of it, and the almost endless disputes, delays and disappointments, which occurred, and also the fact that the bricks when burned actually cost more than they could have been purchased for from the dealers, led to the abandonment of the usual system for a time. This measure was rendered almost imperative by the fact that the soil in the neighbourhood of some of the stations was so impregnated with salt... as to be totally unfit for making bricks. (Walker 1880b: viii).

The practicalities of construction clearly presented difficulties on site, and had a bearing on how the stations could be built. It was not just the suitability of local materials that were causing problems however, it was also the nature of the landscape itself. At Bāniban in particular, for example, Hill recounts the impacts of conditions he faced not just on the station and its construction but on the difficulties of surveying too, thus he wrote,

> The movement of the camp across country I have just described required a considerable degree of care, and notwithstanding all the precautions used was, in some instances, not wholly devoid of danger. The transport of the 24-inch Theodolite over rapid streams when only crazy boats were procurable, was always an operation which gave considerable anxiety. When the party were encamped at Bāniban, the bursting of a bund suddenly inundated the country; the whole camp was compelled to have recourse to boats to remove it to dry land several miles distant. A fierce North-Wester shortly afterwards blew the whole platform on the Bāniban tower out of the masonry into which it had been built, and lodged it on the ground many yards distant. Fortunately no person was on the tower at the moment. (Walker 1880b: viii).

With this in mind, it is perhaps no surprise to find so little of the station standing at Bāniban today, or indeed Mirzāpur nearby. The recourse to poorer quality construction materials coupled with wet and flood-prone sites took its toll it seems. Likewise, too, for the East Calcutta Longitudinal Series, to the east of Kolkata, begun in 1862 under Lieutenant Henry Thuillier of the Royal Engineers (Hennessey 1883b: iii–iv). Rather than the 'hollow' tower construction, encountered at Sāmālia and Bhola, for Thullier's series a different type of station construction is recorded, one that is described in the GTS Synopsis as a 'perforated pillar', standing around 30 to 35 feet in height. This is the case for the disappeared stations at Berghom and Bira, for example, and again the contemporary accounts are revealing of the reasons why. The GTS Synopsis reports,

> The design of tower usually adopted at this time [1862–1863] consisted of a central perforated pillar of burnt brick and mortar of small diameter for the instrument to rest on, surrounded by a platform of unburnt bricks and mud 14 to 16 feet square, and the whole raised to a height of from 20 to 40 feet, according to the nature of the obstacles to be overlooked. This structure had been preferred on account of its cheapness and the rapidity with which it could be constructed, and had hitherto been found well adapted to all requirements. But it appeared to be unsuited to the rainy and moist climate of Eastern Bengal, where unburnt bricks rarely have time to dry sufficiently to be safely used in raising a structure of such considerable height. (Hennessey 1883b: v).

The importance of fired brick in station construction, then, seems to be underlined, for without it the tower structures were less enduring and also lower in height, so all in all more easily removed or eroded by the passage of time in what is a region with particularly high rainfall, humidity and temperatures especially during the summer months. With just a pillar of fired brick, the stations of the East Calcutta Longitudinal could be built more quickly and cheaply. Compared with the 1830s and 1840s, under Everest, 'cheapness' and 'rapidity' are the order of the day by the 1860s, as the GTS network drew closer towards completion under Waugh and Walker. Rather than 'professional builders' being employed, as was the case in the

1830s at Bhola, Rishi and Sukchar, the GTS survey teams were later overseeing the construction of the stations on-site. Unfired bricks and mud provided enough to assist the survey teams achieve a line-of-sight above the vegetation across this relatively flat landscape of the Ganges delta. What is clear too then is a geographical and temporal variation in the construction methods used for GTS survey stations in West Bengal, both in terms of their design and form as well as their materials and fabric. Overall, there was no standard type of station design deployed by the GTS across the eastern trigonometrical series converging on Kolkata, but instead a response to the local conditions and circumstances the teams found themselves in and had to deal with in the region.

The innovations in different station forms and fabric evident in the trigono-metrical series in West Bengal are indicative of a GTS as an evolving organization responding to and very much grounded in the experience gained through 'being in the field', as survey teams grappled with their task and the materials and labour that was to hand at these geographically remote and diffused sites of survey. Here, then, as a result of examining the field-evidence and the historical sources together for GTS stations in the Kolkata region of West Bengal, the industry and tenacity of the personnel of the survey teams and their men becomes evident. It was on their activities and actions on the ground, in constructing these trigonometrical stations as well as using them for their surveys, that the overall success of the GTS—and the imperial mapping of India as a whole—depended.

# 5   Conclusion: Archaeologies of Cartography and Global Empires of Survey

Adopting a field-based approach to 'surveying the surveyors' of the GTS has wider significance for the history of science, geography and cartography, since, as Edney (2002: 432) has argued, 'Despite the centrality of fieldwork to the modern carto-graphic ideal, map historians have been happy to leave fieldwork and surveying well enough alone, unexamined and untheorized'. In response to this, this paper has set out two key dimensions of the importance of assessing 'the centrality of field-work' in furthering understanding of the GTS through its materiality; one con-cerning connecting the surveyors to their sites of survey, thus recognizing the impacts and effects this had on their survey practices in the past, and the second concerning the advantages to be gained by visiting these sites of survey in the field today, and drawing out from the landscape new evidence about these 'mapping monuments'. These 'archaeologies of cartography' represent a new field of enquiry in the history of cartography therefore, focusing on the evidence from-the-field-about-the-field, as well as highlighting the cultural value of these sites of survey and the imperative for their greater recognition and protection as 'survey heritage'.

What emerges from these archaeologies of cartography from the GTS in West Bengal is a greater appreciation of the knowledges and practices that underpinned the trigonometrical survey of India. The GTS provided the geodetic basis for the detailed imperial mapping of India, but as an institution it was not just a formidable and impressive mapmaking bureaucracy. The tangible presence of the GTS was felt just as much through the landscapes being surveyed on the ground as it was through the maps being made of those landscapes. Again, Edney (1997: 334) reflects on this process, noting how 'The British mapping of India was an exercise in discipline [...]. The disciplining of the landscape took place within the scope of the map'. The disciplining *of the map* also took place within the scope *of the landscape*. This has importance in thinking through the connections between cartography and the field, between maps and surveys, for the two-dimensional space of the map relies on the three-dimensional spaces of the survey. The contingencies of the field experiences of the GTS are thus written into the measurements and calculations that in turn fed into the Survey of India and its mapping. This, it might be argued, adds weight to the idea that 'the technological fix promised by the trigonometrical surveys was flawed', and the claim that 'The perfection of the geographic panopticon and its archive, which was promised by the use of triangulation, was accordingly subverted by negotiations and contestations between the British and the Indians' (Edney 1997: 325). However, the mapping monuments of the GTS hint at another aspect of the 'flaws' in the precision of the trigonometrical survey as well as its powers of surveillance.

One of the most impressive legacies of the GTS are its monuments in the field, the tower stations constructed to enable the trigonometrical series to stretch out across India and its landscapes. The towers are the most *visible* manifestations of the GTS, an enduring presence on those who live and work within their sights, and so it is not perhaps surprising, therefore, that local people frequently referred to them as 'lookout towers' when we conducted our enquiries. As structures standing up to 25 m (80 feet) high, their siting in the landscape in the 1830s, '40s and '50s must have seemed strikingly alien intrusions to local people at the time, a palpable sign of the 'geographic panopticon' of the GTS. Moreover, as part of the infrastructure of the GTS, the towers—with their flaws and imperfections, the result of struggles over land, materials and men—reflect too a *verticality* in mapping. There are two planes of colonial mapping, and these sites of survey forge a tangible link between the two-dimensional map and the three-dimensional spaces of survey underpinning them. Surveying the GTS stations helps us to rethink this verticality of mapping—the view from below—and the experiences of being looked down upon, surveyed, as materialized through the construction of visible monuments reinforcing surveillance, a projection of the 'geographic panopticon' instilled through the maps derived from the GTS of India.

Evaluating and recording GTS structures in West Bengal also highlights the cultural importance of these sites spanning the globe in a wider story of 'empires of survey' in the nineteenth and twentieth centuries. In India, and indeed globally, threats to this survey heritage are posed by rapid urbanization and development, and by environmental and climate change. This is especially evident in urban and metropolitan areas, such as Kolkata where currently surviving remains of GTS infrastructure and monuments in the region are at serious and imminent risk of deterioration and destruction. It is not just the loss of the station structures that poses a challenge, but also the loss of local stories about the GTS sites as a result of population change and migration. Such rising threats to 'tangible' heritage as well as 'intangible' heritage is a well-recognized concern in India today of course, but understandably much of the government focus is on the country's pre-colonial and medieval cultural heritage. This leaves colonial/post-medieval legacies—including those of the GTS—rather overlooked and thus more vulnerable in the face of India's ongoing dramatic economic development. To this end, exploring the materialities and monuments of the GTS is a basis for the two nations to recognize their shared heritage, and to use the legacies of imperial mapping as common ground for developing wider policy appreciation of surveying heritage and fostering greater public understanding of maps, both of the past and present.

# References

Crosbie B (2012) Irish imperial networks: migration, social communication and exchange in nineteenth-century India. Cambridge University Press, Cambridge

Edney M (1997) Mapping an empire: the geographical construction of British India, 1765–1843. University of Chicago Press, Chicago

Edney M (2002) Field/map. In: Nielsen KH, Harbsmeier M, Ries CJ (eds) Scientists and scholars in the field. Studies in the history of fieldwork and expeditions. Wiley, Aarhus, pp 431–456

Everest G (1847) An account of the measurement of two sections of the meridional arc of India, bounded by the parallels of 18° 3′ 15″; 24° 7′ 11″ & 29° 30′ 48″. J. & H. Cox, London

Hennessey JBN (1883a) Synopsis of the results of the operations of the Great Trigonometrical Survey of India volume XX. Descriptions and co-ordinates of the principal stations and other fixed points of the Calcutta Meridional Series or Series T and the Brahmaputra Meridional Series or Series V of North-East Quadrilateral. Survey of India, Dehra Dun

Hennessey JBN (1883b) Synopsis of the results of the operations of the Great Trigonometrical Survey of India volume XXI. Descriptions and co-ordinates of the principal stations and other fixed points of the East Calcutta Longitudinal Series or Series U and the Eastern Frontier Series, Sec. 23° to 26°, or Series W of the North-East Quadrilateral. Survey of India, Dehra Dun

Keay J (2000) The Great Arc: the dramatic tale of how India was mapped and Everest was named. Harper Collins, London

Lilley KD (2017) Mapping the nation: landscapes of survey and the material cultures of the early Ordnance Survey in Britain and Ireland. Landscapes 18(2):178–199

Mukherjee N (2011) 'A desideratum more sublime': imperialism's expansive vision and Lambton's Trigonometrical Survey of India. Postcolonial Stud. 14(4):429–447

Phillimore RH (1954) Historical records of the Survey of India, vol. III. 1815 to 1830. Survey of India, Dehra Dun

Phillimore RH (1958) Historical records of the Survey of India. Volume IV. 1830 to 1843 George Everest. Survey of India, Dehra Dun

Prinsep J (1832) Progress of the Indian trigonometrical survey. J Asiatic Soc. Bengal 1:71–72

Sen N (2002) Mapping of India and naming of Mount Everest—a bicentenary. Curr Sci 82 (7):780–782

Walker JT (1880a) Synopsis of the results of the operations of the Great Trigonometrical Survey of India volume XII. Descriptions and co-ordinates of the principal stations and other fixed points of the Calcutta Longitudinal Series or Series B of South-East Quadrilateral. Survey of India, Dehra Dun

Walker JT (1880b) Synopsis of the results of the operations of the Great Trigonometrical Survey of India volume XIII. Descriptions and co-ordinates of the principal stations and other fixed points of the East Coast Series or Series C of South-East Quadrilateral. Survey of India, Dehra Dun

Yolland W (1847) An account of the measurement of the Lough Foyle Base in Ireland. Palmer and Clayton, London

**Keith D. Lilley** is Professor in Historical Geography at Queen's University Belfast (UK). His expertise lies in exploring connections between landscape and mapping, using cross-disciplinary approaches drawn from geography, history, architecture and archaeology. His books include *Mapping Medieval Geographies: Geographical Encounters in the Latin West and Beyond, 300–1600* (Cambridge University Press, 2014). For more than twenty years he has led research projects employing digital technologies to analyse and interpret historic maps, including the Gough Map of Great Britain, and historic landscapes, including medieval and modern urban landscapes. His latest funded research projects focus on the Renaissance polymath and cartographer, Humphrey Llwyd; the archaeologies of nineteenth-century geodetic surveying in Britain, Ireland and India; and maps and landscapes of World War I.

# War Cartography in the Survey of India, 1920–1946

Oyndrila Sarkar

**Abstract** The British Empire achieved its greatest cartographical reach during the interwar period. It was therefore forced to deal with unprecedented strains, as its already bulging territories led to an increased need for securing borders. Along with anxieties about the Afghan border on the North-West Frontier of India, there were growing tensions in the virtually unmapped colonies in Africa, Mesopotamia and East Persia. A critical knowledge of the history, aims and the organization of the Survey of India, reduced to the 'Department', in the 1920s, are crucial for understanding its complex role. The SOI's mapping policy changed drastically, moving from the desire to have artistic excellence to correct, minimalistic up-to-date small-scale maps. This paper would like to show how the Empire's mapping policy laid the foundations of urban and thematic mapping in India, as well as its modern boundaries. Against the backdrop of World War II, it became imperative for the Empire to follow a constructive policy of balancing and achieving what was technically desirable, against what was economically viable. Essential geodetic foundations of topographical survey work had been laid in the past, but any future surveys would have to be dealt with differently, keeping in mind the background of these changes in conditions, needs and methods of survey.

## 1 Introduction

To understand the Survey of India's (hereafter SOI) changing role over time, one needs to consider the position in which the SOI found itself when World War II was declared, the development of the SOI into a strong base for supporting the military survey services created during World War II, and the SOI's ultimate but incomplete transition into a strong post-war reconstruction organization. Furthermore, one has to look at the years preceding World War II and what the Department stood for, right from the time of its conception. From 1767, Robert Clive appointed military officers

O. Sarkar (✉)
Department of History, Presidency University, Kolkata, India
e-mail: Oyndrila.his@presiuniv.ac.in

© Springer Nature Switzerland AG 2020                                         121
A. J. Kent et al. (eds.), *Mapping Empires: Colonial Cartographies of Land and Sea*,
Lecture Notes in Geoinformation and Cartography,
https://doi.org/10.1007/978-3-030-23447-8_7

as surveyors in India (Black 1981: 5). In all its three presidencies (Calcutta, Bombay, Madras), surveys had been carried out both with relation to military operations and under purely civil and administrative arrangements. Survey departments ranged from the Trigonometrical, to Revenue and Topographical survey operations. (Phillimore 1945: 10). These three branches were later amalgamated in 1878 under one centralized head, which came to be known as the SOI (Indian Survey Committee 1906: 10). Between 1823 and 1905, a network of triangles was established over much of the Indian subcontinent, trans-Himalayan explorations were in full swing (Montgomerie 1867: 147), maps and cartographic material grew in number at the Great Trigonometrical Survey of India (GTSI) headquarters in Dehradun, all of which would help furnish information so meticulously gathered over the years, when most needed during World War II (Newcastle Foreign Affairs Association 1860: 5; Mitchel 1883: 10).

The cartographic archive does not simply comprise records of the India Office alone, but includes records from other governmental departments, like those of the Intelligence Branch of the War Department (which came to be known as the War Office), as well as correspondence from the staff of the SOI in India and the Ordnance Office in Britain. The War Office, the precursor to the British Ministry of Defence, was crucially responsible for the training and skilling of surveyors. Most of the surveyors would then be employed in the planning and execution of military campaigns. They would explore and conduct reconnaissance surveys after which they produced line-maps and toposheets focusing on defensive features of areas in question. These areas were mostly multiple frontier areas, of particular interest to the British government in the interwar period.

Between 1905 and 1914, military authorities posted in India started questioning whether the surveys and survey information held at the offices in Dehradun were up-to-date or if there was a danger of lagging behind (Wheeler 1955: 3). The immediate reasons were that revenue surveys were developing westwards, being given more relevance than other areas, while military surveys were progressing way too slowly for comfort, and the topographical surveys were in no way modern enough for any quick reference by any survey party in a state of emergency or crisis.

As a result, 1905 saw the convening of a committee composed of high civil, administrative and military survey departments and political advisors from Britain's Ordnance Survey to put their heads together and quickly come up with a policy. Policy-making never received such importance as in 1905. The need to create a policy and programme that would then meet military needs, while already meeting civil needs, became rather urgent. Such a policy would strive to accomplish another quite ambitious feat. With the introduction of new instruments and state-of-the-art machinery, colour printing technology and techniques at the Calcutta Photography-Lithography office, the SOI would cover India within the next 25 years with 'modern' maps, coloured and contoured. Approved by the Government of India, the policy was immediately put into effect, with military surveys commencing on the western and northwestern frontiers (Pope 1905: 5).

# 2   Role of the 'Department'

## 2.1   Reorganization of the Survey of India in 1905

The modernization of Indian cartography can be studied in two distinct chrono-logical frameworks. First, through the recommendations of the 1905 Indian Survey Committee in pre-WWI British India; and second, through the post-war Retrenchment Committee recommendations of 1922. G. F. Heaney, the Surveyor General of India in 1951–56, defined the activities of the GTSI from 1905 as the product of geodesy, topographical surveys, cadastral and revenue surveys, forest and miscellaneous surveys, and map publication (Heaney 1947: 8). A departmental committee consisting of experts from the Ordnance Survey in Britain and officers of the SOI—together with representatives of the Government of India and their rec-ommendations—decided on the main tasks to be executed by the SOI. Their main instructions were to inquire and report on the state of existing maps in each pro-vince and the efforts required to bring them up-to-date, on the methods and expense of survey, on the methods of production, and on the organization of the department. Their recommendations would have to be executed through the process of survey, resurvey, revision and supplementary surveys (Indian Survey Committee 1906: 20). The main recommendations of this committee were twofold: first to have a com-pletely new contoured survey on a scale of one-inch to the mile in 25 years' time, and to be regularly revised at 25-year intervals; and second, to publish maps pro-duced from this survey in multiple colours, to bring India up-to-date with inter-national developments in cartographic printing.

For military purposes, the necessity of having accurate maps of any territory where troops might be deployed at any given time could hardly be emphasized more. The condition of almost all the existing maps was absolutely insufficient for military purposes. There were many such examples specific to the North-West frontier. Maps of the area around Peshawar were 'worse than useless' for the guidance of troops on the ground (Indian Survey Committee 1906: 46). There was a continuous pressure to keep to the standards of accuracy. General accuracy of maps provided by the Survey Department was suspiciously outdated, making any com-parison between the work done by the professional department and other agencies impossible. With the stringency in the financial budgets of the Government and the cuts in the number of British officers and Indian surveyors employed, as well as the paucity of time in hand, it became a challenge for the SOI to adopt a concrete foolproof plan. The habit of transferring officers of different parties from one part of the country to another before their assignment was finished (altering the nature of surveying in that area with new members, or sometimes even leaving it to the mercy of non-professionals), also created a complete loss of hitherto attained local knowledge and threw the organization into disarray.

Military surveying considerations had taken precedence over all other survey oper-ations. Military authorities had confined and narrowed down their focus on the tracts of the northwestern frontier, bound by the Karachi-Hyderabad-Sukkur-Bhatinda-Delhi

Railway. There were already six survey parties in operation in this area in 1904 to ensure the creation of new maps in a realistic period of 4–5 years. Operations would commence from the tracts nearest to the frontier and extend over Punjab heading towards Kashmir. The revision of maps of the southern parts of Baluchistan, the United Provinces, Rajputana and Central India would invariably be delayed for half a decade, and this sacrifice was seen as unavoidable in view of the crucial importance of British military considerations in the northwestern frontier area (Indian Survey Committee 1906: 49).

## 2.2   Retrenchment and Mapping Policy During the Interwar Period

Reorganization took place only with the return of peace after the end of World War I. The interwar role of the SOI was to be very different from what it would be after World War II. There arose an urgent need to increase the resources of India in terms of food, clothes and manpower, and the fact that the SOI was the only organized governmental department to undertake such an effort, created an unforeseen pressure. Between 1919 and 1922, recruitment of both military and civilian officers was heavy, not only to fill a vacuum created by the war, but also to make up for shortages in recruitment during the previous period. It also enabled large training programmes to be set up in an attempt to catch up on the much delayed topographical programme, with the pre-war officers of the Department supervising the work of their semi-trained counterparts in charge of survey camps and parties. It was realized that even in small wars, a potentially strong survey and map publication unit which was also trained to fight or engage in military action, would be necessary in the future. The system used in Mesopotamia, where civil units held relative military rank, was not as easily workable in the same way where larger military forces were employed (Wheeler 1955: 6).

The development of aircraft during the war and the use of aerial photography for mapping in all the war-theatres also showed the need for proper research and training in air-survey methods (Tandy 1925: 10). Such practical ventures were carried out as a first-hand experience programme in 1921–1923, for instance, in the Irrawaddy delta (Stamp 1940: 329). The military side of the Survey of India had been reorganized by the formation of military units, a large part of whose duties during peace was to find solutions for problems they might be confronted with in wartime. It was the duty of the unit officers to perfect their methods as much as possible by constant research and training, both with regard to technical work and cooperation with the army, especially with the Intelligence Department, the Royal Air Force and Royal Artillery with whose work the SOI was intimately linked. Air survey, which was introduced during the war, made the North-West frontier very easy to map through photography, where plane-tabling surveyors dared not trespass. Crucially, it helped provide suitable, accurate and updated maps for hasty operations in which strategic information could be included as and when required (Couchman 1935: 316).

Numerous training exercises were carried out between 1926 and 1930 in cooperation with the military command on the North-West frontier, with a streamlined focus on air survey training methods. War establishment and war equipment tables were worked out. Terms of service for the SOI personnel at war, along with those for the Railways and Public Works Department (which would in effect be mobilized for war), were being drawn up meticulously in 1934, only to be completely scrapped when World War II was declared in 1939, to start from scratch all over again (Tandy 1925: 8). Mobilization schemes were being prepared, based on the assumption that the SOI would supply units for a war on the North-West frontier of India, when in fact, they were not at all prepared for a full-scale world war. When the Great Depression struck India in 1929, retrenchment became the order of the day.

A Retrenchment Committee had already been set up in 1922 to make possible recommendations to the Government of India for effecting:

> all possible reductions in the expenditure of the Central Government, having regard especially to the present financial position and outlook. In so far as questions of policy are involved in the expenditure under discussion, these will be left for the exclusive consideration of the Government, but it will be open to the Committee to review the expenditure and to indicate the economies, which might be effected if particular policies be adopted, abandoned or modified (Indian Retrenchment Committee 1923: 1).

The recommendations of the 1923 Retrenchment Committee Report regarding the survey expenditure in the SOI were twofold. On the one hand, the cadre of military officers in the SOI had to be progressively reduced and all existing vacancies were to be filled by less expensive civil agency. On the other hand, survey work required for local government and local bodies had to be mandatorily undertaken on special terms, and the number of survey parties had to be immediately reduced, securing, with other economies, a suggested reduction of 709,000 rupees in the net expenditure of the SOI (Indian Retrenchment Committee 1923: 179).

Similarly, in 1932, as a consequence of the world slump in trade, there was another wave of retrenchment carried out in the SOI and the number of existing Survey Circles, survey parties and drawing offices was greatly reduced. The Central Circle (Bihar, Bengal) was completely abolished, its work left to be carried out in a subdued form by a topographical survey party working with the Geodetic Branch in Dehradun. The Southern Circle (Bangalore, Madras) and the Burma Circle were abolished, their work left to be carried out by independent survey parties working directly under the Surveyor General (Heaney 1947: 4). The acute financial stringency in which the SOI found itself made it impossible to meet the twofold standards of maintaining the existing body of small-scale maps produced and stocked earlier, and executing more surveys in order to produce maps. Little geodetic work based on the scientific principles of triangulation and rectangulation could be carried out with an equally short-staffed department. The entry of Japan and the USA into World War II also necessitated additional reorientation and reorganization in the SOI. It became difficult to obtain recruits for the Map Publication Office. By mid-1943, 'Operation Hathibarkala', the institutional move to the physical space of Dehradun, was in full swing, with a little help from the Education, Health and Land

and Finance Departments of the Government of India. The Public Works Department and the War Office supplied all necessary instruments and machinery (Wheeler 1955: 40).

# 3    Beginnings of Boundary Anxieties

## 3.1    Looking Northwest: Mapping the Northwestern Frontier of British India

British cartographical interests in the North-West were furthered by World War II. During wartime, the military survey service did not exist. Newly established Survey Circles not only needed to revise older toposheets and cartographic information, but also had full responsibility of training officers for war and training military survey teams in their respective zones, both in traditional and new air survey methods. These Survey Circles were (1) the Frontier Circle for the North-West Frontier Province, Baluchistan, Kashmir, Punjab, Delhi, Bikaner, Western Rajputana, Sind and Cutch (Headquarters, hereafter HQ at Simla), (2) the Central Circle for United Province, Central provinces, Central India Agency, Gwalior, Baroda, Ajmer, East Rajputana, Northern Division Bombay Presidency (HQ at Mussourie), (3) the Southern Circle for Bombay Presidency, Hyderabad and Mysore, Coorg, Madras Presidency (HQ at Banglaore), (4) the Eastern Circle for Bihar, Orissa, Assam, Sikkim and Bengal Presidency (HQ at Shillong), and (5) the Burma Circle for Burma, the Andaman and Nicobar Islands (HQ at Maymyo (Gunter 1926: 5).

Of these Survey Circles, the Frontier Circle was thus responsible for the northwestern front during World War II. It was created in 1925 with its head-quarters and drawing office at Simla, in close collaboration with the army head-quarters and its corresponding field units (Wheeler 1955: 5). Its different air-survey parties, e.g. 'A', 'E', 'No. 18 Air Survey Party' mostly carried out survey work in the northern and western parts of India, besides organizing and training for war. The 'E-Party', for instance, comprised standard trained British officers and a large number of Indians trained in the Indian Military Survey courses in Roorkee (Medley 1870: 30). The director of the Frontier Circle was an ex-officio map and survey advisor to the Commander-in-Chief as head of the military forces in India; the Surveyor General was his official advisor as head of the Defence Department of the Government of India. In this very heavy network of military entanglements, the northwestern frontier was kept as a top priority, with the aim of getting trained in overseeing and protecting other frontiers, like that of the North East of British India. The first exercise was the preparation of a one-inch to the mile scale topographical map (1:63,360) along with a battle map on a 1:25,000 scale by air survey alone, supported by daily triangulation data provided by the artillery to facilitate the plotting of their gun positions and targets on both these maps.

The India and Adjacent Countries Series (IAS) (Fig. 1) was a direct result of the combination of the respective map production of two aforementioned air survey parties throughout 1927–1928 at Quetta (E-Party), and at Peshawar and Murree (No. 18 Party). Using only a small portable hand press, this series was made for departmental use. The hand press also played a crucial role for producing air-survey maps during the early 'tactical exercises' in 1927 (Tandy 1928: 17). The survey party thus, over the course of WWII, managed to not only make reproductions of originals which came in from other survey units in and around the area, but also to prepare Vandyke plates for multiple reproductions at any given time, and to produce more prints to be circulated among all units and to be sent to the headquarters to keep a copy for the records.

Even with the air-survey and other specialized training of surveyors at the northwestern frontier of India, British war fronts would eventually move westwards in a matter of years (1939–1942), resulting in an increased need for updated maps and military information. The urgency for more and more maps and men was logically followed by the need for more and more reconstruction projects. Decisions needed to be taken quickly as to how the SOI could best utilize its mapping resources given the war at India's borders and British mapping policies and interests in the Middle-East. Mapping priorities were discussed at global conferences, e.g. in Cairo in 1940, and in Delhi in 1941. An important issue with regard to the northwestern frontier, was whether any of the three Indian survey services—trigonometrical, topographical and revenue—could assist or take over map publication from the others in the event of extreme destruction by 'enemy action',

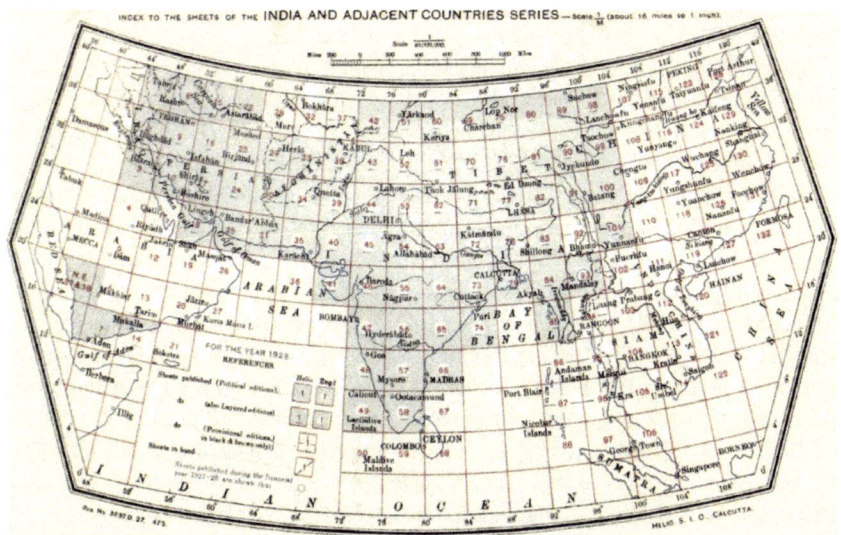

**Fig. 1** Index to the Sheets of the India and Adjacent Countries Series. Tandy (1928) (Courtesy of the Survey of India Library in Dehradun)

thereby indicating that there was an equal need to be militarily equipped and ready at all times. The SOI's commitment for building efficient military survey units grew strong, and new units were immediately formed and dispatched to Iraq in 1941, being skilled in Indian military survey methods.

## 3.2   Looking Further West: Mapping West Asia and Persia

During World War I, a major portion of the Department's military officers was diverted to military duty for regimental, non-survey employment. A few civilian officers, mostly the reserves of the officers, were called up for military duty. This resulted in the SOI being short-staffed in officers and internal survey programs slowed down considerably. The necessity for comprehensive survey and mapping work during the war and during the following period of reconstruction had to be recognized. When the war spread to large parts of Mesopotamia, which was a virtually unmapped territory, the need for both highly trained and experienced military personnel as well as for survey personnel became a matter of urgency. A number of British officers had been killed or had become key figures in the non-survey employment, so a strong survey service was formed from the SOI personnel to function in Mesopotamia and East and West Persia. There was a need to have *Hindustani*-speaking officers. The Ordnance Survey of Britain and their Dominion and other colonial counterparts had their plate full with their own set of peculiar problems in Europe and mainland Africa, and were incapable of extending their help regarding this matter. The survey material gathered in Mesopotamia and in East and West Persia during WWI proved of crucial strategic importance to the British in WWII during the occupation of Iran and Iraq.

In WWI the SOI had survey parties engaged in various parts of Iran (then Persia). Some half-inch maps (1:126,720) of these areas were created, much out-of-date by 1939. There was also a quarter-inch series (1:253,440) covering the whole country, mostly compiled from information of an explorative nature, and of course geographical maps on 1:1,000,000 (1/M) and 1:2,000,000 (1/2M) scales (Wheeler 1955: 165). Shortly before WWII, responsibility for areas west of longitude 48° East had been handed over to the War Office. In 1942, the British assumed responsibility as far east as longitude 54° (middle of Iran), and in 1943, still further east to 60°, thus taking over practically the whole of Iran (General Staff, Army Headquarters 1911: 211). Figure 2 shows a portion of the Persia-India triangulation in 1944. PA Thomas conducted this crucial survey for four major reasons. First, to assess the value of the rapid wartime triangulation of 1941–1943 by the PAI Force (Persia and Iraq Command) in terms of the 'practically errorless' trigonometrical surveys of India (de Graaff-Hunter and Gulatee 1946: 5); second, to strengthen the PAI Force triangulation; third, to extend triangulation operations and control for surveys into eastern Persia, and lastly, to contribute to geography and geodesy by completing the only missing link in the systems of triangulation extending from Europe to the Far East. Needless to say, the backstory to this survey was that it had been conceived by the mathematical advisor to the SOI, J. de Graaff Hunter (who later became Director of War Research of the SOI), only during and after the surveys of 1914–1918 (de Graaff-Hunter and Gulatee 1946: 5).

Early in World War II, an Indian force was sent to Iraq and Iran, and a number of maps of western Iran were called for. Existing fair originals were corrected and updated from any information available before being republished, interim supply coming from storage stocks. Copies of originals sent to the War Office were also obtained and corrected so far as their British information existed. When military survey units were finally dispatched to the area, they gradually took over the publication of maps as required. Up to 1943, when most of the Indian Survey units were withdrawn from the area under PAI Force control, they had carried out a considerable amount of triangulation and mapping work in western Iran as well as in Iraq. Some ground survey and air survey were also carried out in eastern Iran on the Baluchistan border by the SOI unit in cooperation with a military unit based in India, and a triangulation connection was made in 1944 between Iraq, Persia and India (Wheeler 1955: 48, 168).

Until shortly before the war, the Survey of India had been publishing maps of Iraq on a quarter-inch scale and on geographical scales. Much of the information in these maps had been obtained from surveys made by the SOI units in WWI. The

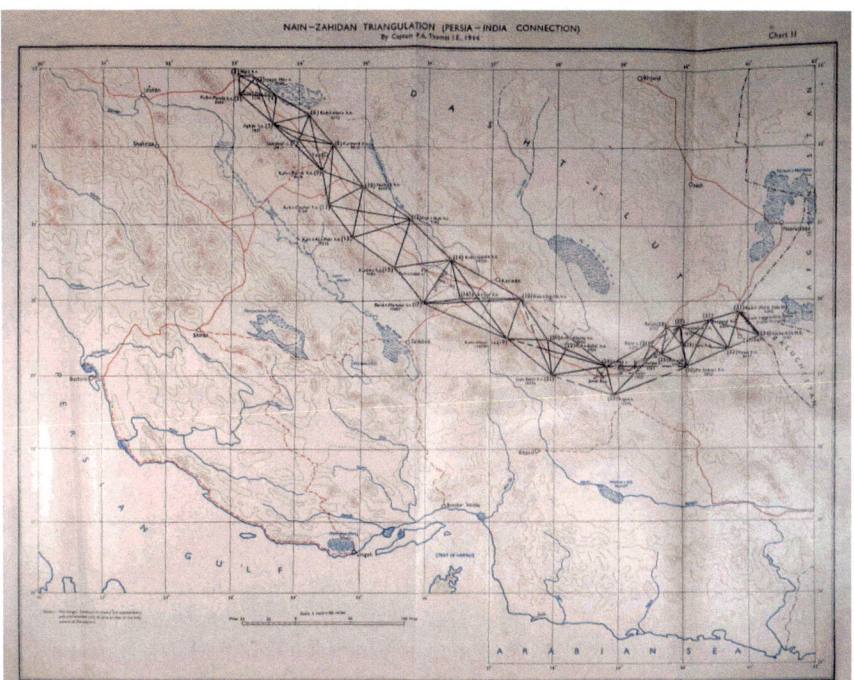

**Fig. 2** The Nain-Zahidan Triangulation (Persia-India Connection). Thomas PA (1944) SOI: War Research Series (Courtesy of Mountains of Central Asia Digital Dataset, PAHAR (People's Association for Himalaya Area Research, Nainital))

Iraq Government, which was headed by an ex-Survey of India officer, AJ Booth, had formed a survey department and some Survey of India personnel had been employed in it as instructors of survey. The SOI was in a little difficulty in providing maps for the troops first sent to Iraq from India. This was rectified by an eventual transfer of maps and cartographic material back from the War Office as time passed, and eventually the military survey units sent from India took over practically all responsibility. British Middle East units gradually replaced them, and by 1943 practically all the Indian units had been withdrawn for service in the eastern theatre of war.

At a conference in Cairo in March 1940, attended by the Director of the Frontier Circle of the SOI, British commitments in Iraq were decided upon and a map policy was framed to fit in with the Middle East. The SOI as such, had little to do with mapmaking in Iraq, though most of the mapping work up to 1943 was done by military survey units formed almost entirely from SOI surveyors. Each functional company (or survey unit) had a map reproduction section with a rotary printing machine and other cartographic equipment, demonstrating how equipped and ready the SOI surveyors were. By mid-1942, a strong survey and map publication organization was in place, which was continually being reinforced by and eventually almost completely replaced by British units from the Middle East, to enable Indian units to move eastwards as the Japanese war front developed. Priorities of mapping changed with corresponding changes in the different theatres of war.

During their time in Iraq and Iran, the Indian Survey companies surveyed or revised about 120,000 square miles in Iraq and over 100,000 in Iran, producing maps from air survey and aerial photography. To enable such surveys to be carried out, triangulation of an accuracy approximating that of the Indian topographical triangulation and supplemented in the flat desert areas by the Hunter Short Base traverse (a standard base-measurement technique) was carried out over the whole area of survey (Heaney 1947: 8).

Representatives from both the SOI and the Middle East met in Baghdad in 1940, to discuss solutions to geodetic and triangulation adjustment problems. SOI's operations in Iran during WWI were found to be disjointed and of low quality, rendering them useless for the much more accurate surveys of 1941–1942. All final adjustments of this triangulation were carried out in India and the results were then communicated to Iraq.

# 4 Priorities of Politics and Diplomacy

Against the backdrop of World War II, it became imperative for the British Empire to follow a constructive policy of balancing and achieving what was technically desirable, against what was economically viable. The stock of standard SOI maps was exhausted as a consequence of the war. Countless negatives and plates, from which reprints could have been made without the need for photography, were destroyed. There was also a reduction in the number of colours used in printing topographical maps in the post-war era. Many of the existing map sheets were restricted to be printed in only two or three colours, instead of the earlier pre-war colour scheme of eight or more. The new trends and demands for survey after the end of the war constituted a radical change, as those trends were focused heavily on the planning of numerous development projects in a newly emergent India. Proposed irrigation and hydro-electric power projects required very accurate surveys not just of the dam sites, water bodies and reservoirs, but also of their surrounding areas for laying out canals and waterlines.

Industrialization led to the urgent large-scale surveys of major cities along with wasteland reclamation for growing food crops, which in turn depended on accurately contoured survey data. With the growing availability of air-surveying instruments, new cameras and plotting equipment, it was also immediately predicted that much of the surveying work would be done cheaply, expeditiously and accurately by air survey, following close instructions as laid out by the War Office (Meade 1932: 6).

## 4.1 The Technically Desirable Versus the Economically Viable

The SOI was a civil department for which World War II was both an interruption and a digression. All normal civil activities practically ceased and all efforts of the department were directed towards war work. A large part of the staff was mobilized and formed the backbone of the Military Survey Service in Persia, Iraq and later the South East Asia Command. A few events during the war were to have a lasting effect on the Department. The first was the move of the headquarters of the Surveyor General from Calcutta to Delhi in 1940. It was followed by the move of the Directorate of Map Publication from Calcutta to new buildings in the present day Hathibarkala estate in Dehradun in 1943. These buildings were equipped by the War Office to enable the SOI to act as the base for map production for military forces in South East Asia. The formation of the Military Survey Service was to be retained as a permanent measure by the Defence Department for the post-war armed forces, relieving the SOI of some of its earlier pre-war military responsibilities. In every way the SOI had to meet very heavy demands for survey and map publication for two definite reasons: to make up for the backlog of work arising from the curtailment of normal activities before and during the war; and to provide larger-scale surveys and maps for the development of India's resources in minerals, power, agriculture and industry along with the need to meet the increased requirements of the army and in civil aviation.

In addition, the SOI had to replenish its existing stock of standard maps, incorporate all corrections in existing maps and produce new editions in pre-war style. The SOI also had to produce a totally new series of maps and charts, tailor-made for aviation purposes. Heaney in his paper on Mapping Policy (Heaney 1947: 7) highlighted the following:

(1) The completion of the modern survey of India involving 400,000 sq. miles of new surveys to be effected as soon as possible.

(2) Provisions to be made for the periodic revision of standard topographical maps for the whole of the newly independent India.

(3) Compilation of smaller-scale maps from current surveys and bringing out new editions incorporating extra-department information not to be allowed to slip into arrears.

(4) Special air navigation charts to be kept updated to meet the growing demands of civil and military aviation.

(5) A considerable area of mapping on the 1:25,000 scale to be carried out for the army and in connection with mineral development for the Geological Survey of India.

(6) Demands for accurate large-scale surveys of cities and other industrially important areas not to be refused indefinitely.

(7) Special surveys in connection with irrigation, hydroelectric power, land reclamation and other similar projects to be kept at high priority as long as development continued in India (as shown in Fig. 3).

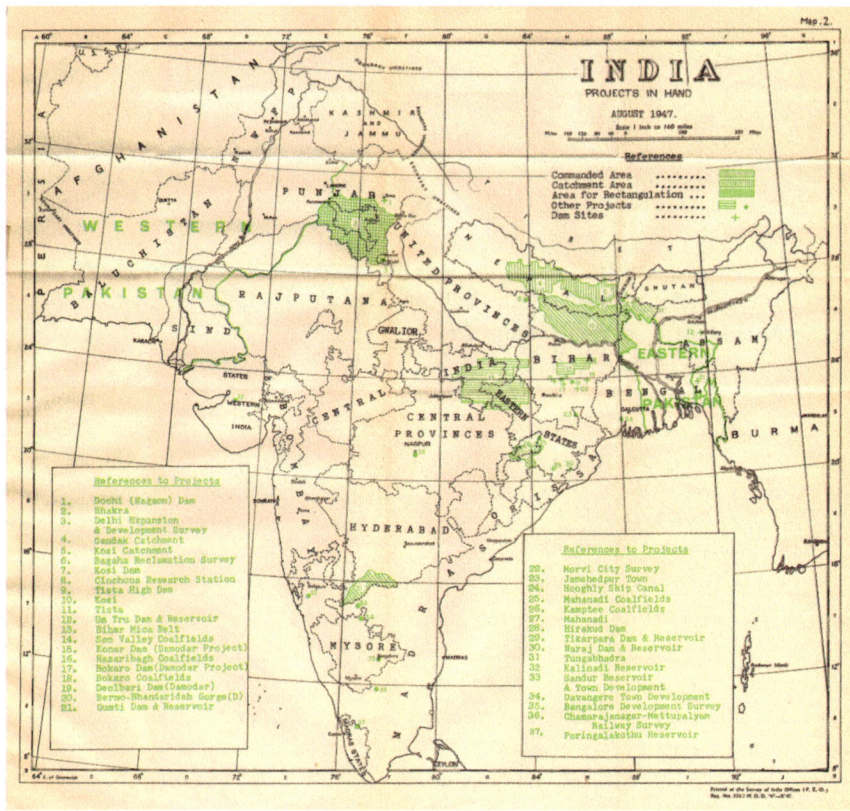

**Fig. 3** Survey of India Projects in hand in 1947 (Courtesy of Mountains of Central Asia Digital Dataset, PAHAR (People's Association for Himalaya Area Research, Nainital))

The 1905 Indian Survey Committee report had already stressed the first three points. The rest of the recommendations would follow the demands of a newly independent India. Heaney, while commenting on the need for a concrete survey and mapping policy, calls to mind the events of 1905. According to him, from 1905 until the outbreak of WWI, survey work was largely stereotyped and the bulk of the work was executed by uneducated personnel, previously trained in only one branch of work, namely whichever one the Department required of them. Specifically with the development of air survey methods, the surveyor without primary training or with only one skillset, would increasingly give way to the topographical surveyor. Heaney believed it would be wise, both for the present and the future, to increase the proportion of this type of employment in the years to come (Heaney 1947: 47).

Future policies were to be the more challenging part. In the post-war era, the survey work to be undertaken by the SOI would be more determined by what was immediately essential and by the resources available than by any broad considerations of policy. At the end of these beginning years, there would be enough resources in personnel and manpower to start working more systematically, according to some standardized policy. What such a mapping policy would entail and what personnel would be required to implement it was also something to be considered seriously. To cut down costs, air survey was the most important thing to bear in mind and be pursued. Personnel and employees would have to be trained in air survey, in aerial photography, and arrangements would have to be made to obtain a regular and routine supply of quality photographs.

The primary function of the SOI was increasingly narrowed down to the production and upkeep of topographical and small-scale maps. This new mapmaking policy with lessons learnt from the past, and having a vision for the future of cartography in India, also included several other crucial categories like a past maintenance policy, a future maintenance policy, the completion and periodical revision as suggested earlier by the 1905 report, and a focus on organizing survey parties and survey personnel for intensive training, purchase of plotting machines and air photography equipment (Heaney 1947: 18).

Before World War II, the majority of officers, both those in charge of the field survey parties as well as those heading the Drawing and Photography-Lithographic offices in Calcutta and Simla, were commissioned officers of the Royal Engineers. They constituted the Class-I Service. The hierarchical grade below them, the Class-II Service, comprised junior gazette officers recruited mostly from graduates of different survey training schools and engineering institutions (Heaney 1948: 6). Below this grade were Upper and Lower Subordinate Services of technicians. The former recruited from those with a full intermediate education, and the latter having no particular degree from an educational institute at all, viz. most of the field staff handling plane-tables and traverse instruments on the field. The newly independent Government of India changed its tactic towards the employment of regular military officers of the Corps of Engineers in the SOI.

There were talks of depreciating the employment of military officers in civil departments and shortly before the partition of India in 1947, it was decided that the composition of the Class-I service would be 37.5% military officers, 37.5% directly recruited civilian officers, and 25% officers from lower services. In the post-Partition years, the armed forces having shown an increasing interest in the SOI resulted in the Government deciding that in the future, the Surveyor General of India should always be a military officer and that 50% of the Class-I service would be recruited from military officers, at the expense of the quota of directly recruited civilians. It was also decided to keep a Military Survey Service in existence during times of peace, to be able to come up with potential solutions for any sudden geopolitical crisis (Heaney 1952: 280).

## 4.2   Fair Mapping and Thematic Mapping

'Fair mapping' is a term used in the Survey of India to indicate the drawing necessary to produce a 'fair original' from which reproduction by photographic processes can be carried out, either by reduction or by direct reproduction (Heaney 1947: 137). Though the plane-table sheets or air survey sections—or compilations, as the case may be—constitute the original field record, they are seldom suitable for direct photographic reproduction and the fair originals are also very jealously guarded, for in case of loss or damage, the whole fair-drawing process has to be done all over again. Fair originals were very carefully drawn in accordance with Departmental rules and were often used again and again for new editions of maps, being corrected as necessary from later field surveys (Wheeler 1955: 135). Fair mapping, or fair drawing, was normally carried out at 14 times the publications' scale, so as to refine drawing and minimize shakiness and errors. There usually had to be at least two fair originals, one being the contour sheet for publication in brown, the other the 'combined original' for black, blue, red and any other colours in which lines would be printed. It was usual practice to separate these colours out for printing by taking a sufficient number of photographic negatives from the combined original so that there would be one negative per colour (Couchman 1935: 14).

Preliminary town planning and urban planning started as a result of the air surveys. On the back of their success and once development projects started getting attention in India, air photography became the main ingredient for all preliminary planning, primarily because simple air photographs would take far less time to produce than line-maps. The photographs were perfectly useful for thematic mapping of urban settlements, crop investigations, forestry, and geological, archaeological and mineral surveys (Survey of India 1947: 3).

A brief summary of the basics of air photography deserves mention. Most images would be taken with the aerial camera pointing vertically downwards. Vertical air photographs resembled an ordinary line-map if the ground photographed was flat. The scale would depend on the height above the ground from which the photograph was taken, and the focal length of the camera lens used. A line-map and a photograph would differ in multiple respects (Fig. 4). A photograph usually covered a limited area. Unless the area was flat, the scale would vary. Extremely small but important features would be rendered invisible

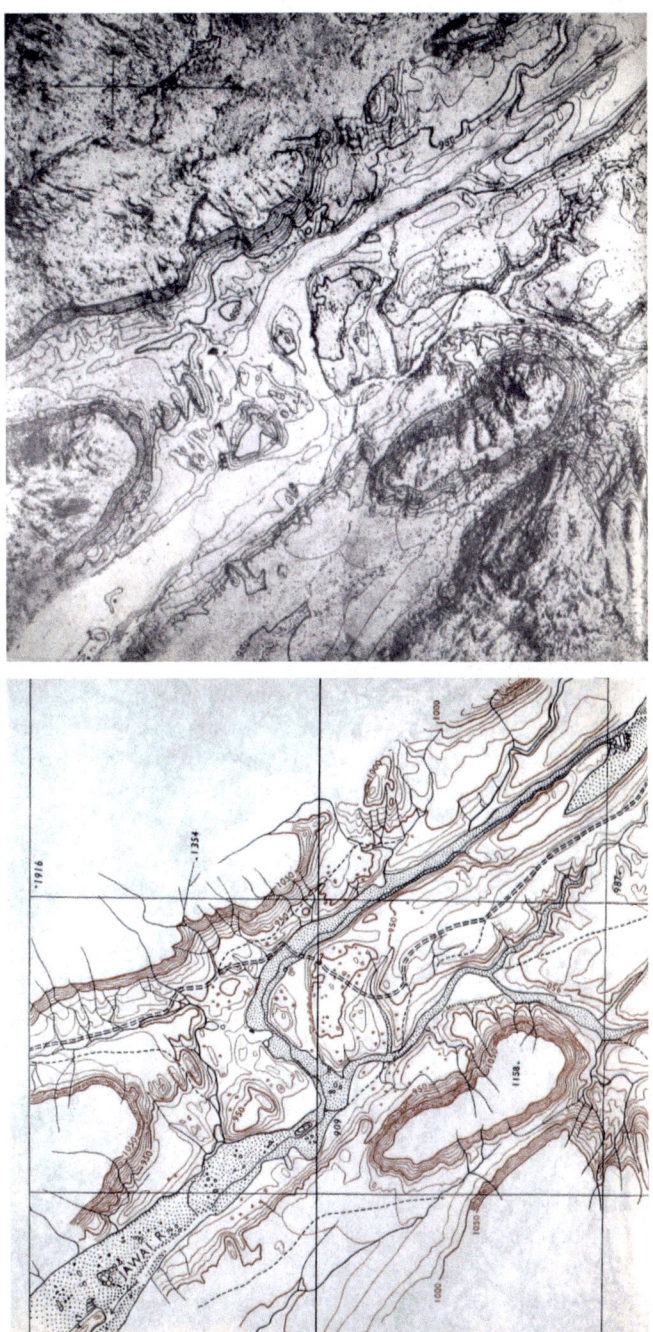

**Fig. 4** Irrigation and Hydro-Electric Projects. SOI (1947) Air Survey For Development

unless the scale was large. Often the foliage of trees would obscure features beneath them. Buildings would obscure details at their base, when they were away from the centre of the photograph. Line-maps could be readily annotated with planning notes, whereas photographs were less suitable for this purpose.

Yet, for some purposes, simple photographs were enough for preliminary reconnaissance. Photographs of flat ground would—with the aid of minimum ground measurements—be transformed to form photo-maps of sufficient accuracy for accurate measurements for revenue records and town planning. Clarity however would disappear if larger prints were required, obtained from small-scale negatives beyond 2½ times a single enlargement. Sometimes air photographs of flat ground could also be fitted together to form a mounted mosaic for use as a photo-map of larger areas than could be individually covered by a single photograph. The same was technically impossible for elevated or uneven ground or hilly tracts of land. Differing elevations led to variance in scale, which would, in turn, lead to a complicated distortion of the ground surface, which could make land seem duplicated when creating a mosaic. In such cases, the surveyor had to convert the photographs into useful maps by survey plotting. For irrigation projects it was necessary to make a rapid preliminary reconnaissance, to determine whether a project was worth pursuing. In such investigations, the probable capacity of potential reservoirs, for instance, was clearly important. This could be obtained quickly and cheaply from contoured mosaics from which capacity could be calculated to an accuracy of about 5% (Survey of India 1947: 4). Once decisions finalizing the reservoir's location had been taken, a line-map would become necessary for further precise calculations and for detailed planning and examination of what area would then be submerged. Figure 4 shows a contoured mosaic of the Erinpura Reservoir in Jodhpur state, Rajasthan, on a six-inches to the mile scale at ten feet vertical intervals. Overlying this is the line-map subsequently prepared from the photography used for the mosaic.

With the application of stereoscopy (viewing the same object from different angles on two photographs), survey maps were somewhat advantageously posited. It enabled the nature of objects to be accurately determined on the one hand, while on the other, it allowed the shape of hills and mountains to be seen, while contouring could be carried out accordingly, when height control existed on the ground. To be able to use them for stereoscopy, there should be an overlap of more than half of a photograph between the two photographs used.

Line-maps, also emerging from aerial photographs, were yet another innovative method of producing a survey map. There were two stages in line-mapping from photographs once the photographs had been taken and ground work had been completed – the first being a detailed mapping or planimetry, and the second being relief mapping or orography. For the development of urban areas, a necessary preliminary is often a map on a medium-scale of four-inches to the mile or six-inches to the mile. Planimetric details were prepared by air survey teams based on ground control, contours would be added by the ground survey on the working prints of planimetry. As a rule, every four-inch-scale map was likely to be adequate for this purpose, and each contour interval would vary depending on the type of terrain in question.

For more detailed planning than can be done on a medium-scale map, maps on a scale of sixteen-inches to the mile were likely to be required. The expenditure for the production of line-maps at this scale, before planning, was probably unjustified, since the survey would have to be completely revised after development. For this purpose, therefore, the map would have to be a photo-map. Large mosaics were mounted on Masonite boards, which were then cut into convenient and manageable sizes. In the unlikely event of a large number of such mosaics being required, copies could be printed by the lithographic halftone process in any number required. Mosaics such as those mentioned above were only recommended in flat areas, and even then, accurate measurements from them could not be expected.

Expenditure on large-scale line-maps for planning was usually unjustifiable in areas liable to much change. Extra-large-scale surveys, which may subsequently be useful as a basis for recording property rights, were desirable and justified expenditure at the preliminary planning stage. This was an innovative method combining air and ground survey for detail, while contours were done entirely by ground survey. Surveys on larger scales were used for congested and developing urban areas. Air survey was not recommended for such larger scales (Survey of India 1947: 1).

# 5 Conclusion: Asymmetry and Coherence in Modern Indian Mapmaking

The problem of the SOI Department became that of quality over time. With the Allied victory, the SOI encountered unforeseen hindrances to mapmaking and map production, which they still had to deal with: post-war reconstruction and an unofficial death count of about 2 million, along with a gory partition of India. On the other hand, the emergence of an Indian engagement with large-scale infras-tructural projects in agriculture, irrigation and hydroelectric power, made it chal-lenging for the SOI to concentrate solely on war problems. To add fuel to the fire, with the loss of Burma rice and the continuous inflow of troops into India who needed to be fed, there was an urgent need for surveys in connection with the 1943 'Grow More Food' project. It was necessary to devise an ingenious plan, which would save time and sustain manpower. Sir Edward Oliver Wheeler, during his crucial time as Director and later Surveyor General of the SOI during World War II, writes,

> To train a civilian to be a sound and efficient soldier takes a considerable time; to train a non-survey soldier to become an efficient surveyor, much longer, very much longer. (Wheeler 1955: 13)

All of these considerations elaborately question the idea of India's boundary-making and state formation by the Boundary Commission as a result of the Indian Independence Act of 1947. Historians of modern India have yet to critically theorize about the closing period of British surveying operations. With this paper, I have tried to show how the Empire's survey department had been functioning fully as both a mapmaking, map producing and a semi-military entity, in multiple geographical zones at the same time, including its immediate North-East and North-West frontiers in the Indian subcontinent. It is rather challenging to completely disregard the role of the SOI in World War II and to not have a deeper sense of military history if one needs to talk about cartography during the inter-war period. Examples of an external life, like that of the Indian Survey Company employed in the Persian Gulf, are of prime importance (Jackson 2018: 2). A surveyor had to remember that no sooner was any area of the subcontinent surveyed than the map representing that area started to become out-of-date. Future mapping was only possible due to lessons learnt during this era of war cartography and the evolution of fair-mapping and thematic mapping. Ideas of infrastructural development were, therefore, theoretically planned for execution by the heads of State in the post-colonial period with regard to the 1905 suggestions for special surveys. Proof of discussions regarding these special surveys, which would harness the hydroelectric power that was much needed for a new nation's infrastructural growth, can be found in the meetings of the older generation of British surveyors in India, and the newer generation who were taking a slow step back, after India's independence (Heaney 1952: 295). Needless to say, in the background of such an era of turmoil and political strife on the Indian subcontinent—which had a 'complex political inwardness' (Chatterji 2007: 27) making it a totally different ball game altogether—any cartographic development was asymmetrical in nature. It was, however, necessary to devise a coherent plan, which would not only save time and manpower, but also provide quick and immediate results. Newer policies of mapmaking and surveying would follow as a natural order of things, but it would take time.

# References

Black CED (1981) Memoirs of the surveys in India, c. 1875–1890. EA Black, London

Chatterji J (2007) The spoils of partition: Bengal & India: 1947–1967. Cambridge University Press, Cambridge

Couchman HJ (1935) Fair mapping. In: Handbook of topography. Photography-Lithography Office, Calcutta

De Graaff-Hunter J, Gulatee BL (1946) War research series Pamphlet No. 9. Trans-Persia Triangulation c. 1941–44. War Research Institute, Dehradun

General Staff, Army Headquarters (1911) Military Report on Persia, c. 1911. Government Monotype Press, Simla (IOR/L/MIL/17/15/5)

Gunter CP (1926) Survey of India General Report: 1925–1926. Photography-Lithography Office, Survey of India, Calcutta

Heaney GF (1947) Survey and mapping policy, Departmental Paper No. 17. Office of the Geodetic Branch, Survey of India, Dehradun

Heaney GF (1952) The Survey of India since the Second World War. Geogr. J. 118(3):280–293

Indian Retrenchment Committee (1923) Report of the Indian Retrenchment Committee, c. 1922–23. Superintendent of Government Printing Press India, New Delhi

Indian Survey Committee (1906) Report of the Indian Survey Committee, c. 1904–1905. Part I–
The Report. Government Central Printing Office, Simla

Jackson A (2018) Persian Gulf command: a history of the Second World War in Iran and Iraq.
Yale University Press, New Haven

Meade HRC (1932) Notes on air survey and map publication in England, c. 1931. Survey of India,
Calcutta

Medley JG (1870) Professional Papers on Indian Engineering Vol. 7, c. 1870. Roorkee

Mitchel JF (1883) The topographical, political and military report on the North-East Frontier of
British India. Superintendent of Government Printing, Calcutta

Montgomerie TG (1867) Report on the Trans-Himalayan explorations in connexion with the Great
Trigonometrical Survey of India, during 1865–7: Route Survey made by Pundit– from Nepal to
Lhasa, and thence through the upper valley of the Brahmaputra to its Source. In: Proceedings
of the Royal Geographical Society of London, vol 12. No. 3. Royal Geographical Society (with
the Institute of British Geographers), Wiley (1867)

Newcastle Foreign Affairs Association (1860) Falsification of diplomatic documents: The Affghan
Papers; Report and Petition of the Newcastle Foreign Affairs Association. Royal Exchange,
London

Phillimore RH (1945) Historical records of the Survey of India, vol IV. Survey of India, Dehradun

Pope TA (1905) The reproduction of maps and drawings: A handbook of instructions for the use of
government officials and others who prepare maps, plans and other subjects for reproduction in
the Photographic and Lithographic Office of the Survey of India. Photography-Lithography
Office, Survey of India, Calcutta

Stamp LD (1940) The Irrawaddy River. Geogr J 95(5):329–352

Survey of India (1947) Air survey for development. Office of the Geodetic Branch, Survey of
India, Dehradun

Tandy EA (1925) Records of the Survey of India, vol 20, The War Records, c. 1914–1920. Office
of the Geodetic Branch of Survey of India, Dehradun

Tandy EA (1928) Survey of India Map Publication and Office Work, c. 1927–1928.
Photography-Lithography Office, Calcutta

Wheeler O (1955) The Survey of India, during War and early Reconstruction, c. 1939–1946.
Office of the Geodetic and Research Branch, Survey of India, Dehradun

**Oyndrila Sarkar** is a faculty member at Presidency University, Kolkata, India. Her doctoral thesis
from the University of Heidelberg is on the *Great Trigonometrical Survey and the Mapping of
Space in Assam: 1830–1890.* She graduated from the University of Calcutta and completed her
M.Phil. from the *Centre for Historical Studies* at Jawaharlal Nehru University, New Delhi. Her
research interests include the various histories of mapping, and histories of science and technology.

# Mapping the World

# Red Star to Red Lion: The Soviet Military Mapping of Oxford

## John Davies and Alexander James Kent

**Abstract** As part of its global military mapping project, the Soviet Union produced maps of many parts of the world at several scales from 1:1,000,000 to 1:5000. These include general maps designed for military planning and terrain evaluation and highly detailed street plans of towns and cities, including Oxford. The Soviet 1:10,000 plan of the city was compiled, designed and printed in secrecy within the Soviet Union during 1972–1973. It reveals that a high level of information was collected about the location and function of buildings, from the Morris Motors and Pressed Steel Fisher factories at Cowley to Oxford Prison and the Central Post Office in the city centre. Anomalies include the omission of Marston Ferry Road (which opened in 1971) and the inclusion of the two gas holders at St Ebbe's (which were demolished in 1968). Further afield, the depiction of RAF Upper Heyford on the Soviet 1:50,000 topographic map of 1981 includes details not shown on contemporaneous Ordnance Survey maps. Focusing on the Soviet mapping of Oxford and its vicinity, this chapter provides some new insights into the global project and reflects upon the achievements, methods and supposed purpose of this unprecedented cartographic enterprise.

## 1 An Introduction to the Soviet Military Global Mapping Project

It is impossible to underestimate the challenges of undertaking a systematic and comprehensive mapping of the globe. Creating a standard symbology for portraying the wide diversity of environmental information and gathering, classifying and presenting huge amounts of topographic data require substantial human and

J. Davies
16 Charteris Road, Woodford Green, UK
e-mail: john@jomidav.com

A. J. Kent (✉)
Canterbury Christ Church University, Canterbury, UK
e-mail: alexander.kent@canterbury.ac.uk

© Springer Nature Switzerland AG 2020
A. J. Kent et al. (eds.), *Mapping Empires: Colonial Cartographies of Land and Sea*,
Lecture Notes in Geoinformation and Cartography,
https://doi.org/10.1007/978-3-030-23447-8_8

143

technological resources combined with effective management and organization. The task has eluded even the most determined attempts to map the globe through international collaboration. The idea of a 1:1,000,000-scale world map was proposed in 1891 at the Fifth International Geographical Congress in Berne, Switzerland, by Albrecht Penck, a German geomorphologist. His overall aim was to enable meaningful geographical comparisons to be made, solving supranational problems for the benefit of humanity. The resulting International Map of the World (IMW) relied on national mapping organizations to produce sheets at the fixed 'millionth' scale using a standard specification. However, the unavoidable problem of reaching international agreement on establishing cartographic conventions, the outbreak of World War I and subsequent global conflicts, and the rapid pace of industrialization in the twentieth century hindered progress and contributed to the increasing pursuit of national interests. By the early 1950s, only 400 sheets (of the 1000-plus sheets needed to cover the terrestrial surface of the globe) had been published, and after a UNESCO (United Nations Educational, Scientific and Cultural Organization) report declared the IMW no longer feasible in 1989, the United Nations stopped monitoring the project (Pearson et al. 2006).

Although the Soviet Union had withdrawn from the IMW (Russia had been involved in establishing the original specifications in 1909), the project's sheet lines and nomenclature were nevertheless adopted for the first detailed topographic mapping of the USSR (Union of Soviet Socialist Republics). Work commenced after the Bolshevik Revolution of 1917 and the first Soviet maps of Soviet territories at the scale of 1:1,000,000 were completed by 1918. In the following year, a decree from Lenin brought all mapping under the sole responsibility of the state and 1921 saw the introduction of a standard specification for military topographic maps for a range of scales (1:10,000, 1:25,000, 1:50,000, 1:100,000, 1:200,000, 1:500,000, 1:1,000,000). The nomenclature and grid lines for all these scales continued to follow the IMW model.

During World War II, greater levels of topographic detail, such as the types of forest and the widths of roads, were included on Soviet mapping to support military operations. As the focus shifted from conducting warfare to meeting domestic concerns such as advancing the national economy, Stalin issued a decree in 1945 that prioritized the production of a 1:100,000 map of the entire Soviet Union by the Military and Civil State Topographic Services. This substantial cartographic project was completed by 1954, by which time attention had been drawn to mapping the Soviet satellite states and other parts of the world (Cruickshank 2007: 24).

This new motivation gave rise to the most detailed topographic coverage that the world had yet seen. Under the authority of the Soviet General Staff, topographic maps and plans of foreign territories were produced at several scales (i.e. 1:5000, 1:10,000, 1:15,000, 1:25,000, 1:50,000, 1:100,000, 1:200,000, 1:500,000 and 1:1,000,000) according to standard specifications and by thousands of cartographers working within the USSR (Davies and Kent 2017). In addition to the topographic map series, larger scale maps and plans (mainly 1:10,000 or 1:25,000) are known to exist for over 2000 cities around the world (Davis and Kent 2017).

Prior to the collapse of the USSR in 1991, foreign access to Soviet topographic mapping was largely the preserve of the military. The US Department of the Army, for example, had been producing Soviet symbol recognition guides since at least the end of World War II. The steady flow of Soviet mapping from outlets in former Soviet republics and satellite states from the early 1990s led to Ordnance Survey issuing a statement by 1997 that declared Soviet mapping to be in breach of OS copyright and demanded an 'amnesty' of private stocks. The city plans seem to have become available since 1993, when a Latvian map dealer based in Riga began selling them at the International Cartographic Conference in Cologne, Germany. Today, digital copies of Soviet maps—including detailed city plans—may be freely accessed from the websites of several national libraries around the world, such as the US Library of Congress and the National Library of Australia (see Davis and Kent 2017 for a comprehensive listing).

Soviet military maps remain the most accurate and detailed form of topographic information available for many areas around the world, yet they have only recently become a topic for academic research (e.g. Postnikov 2002; Davies 2005a, b; Cruickshank 2008, 2015; Davies 2010; Kent and Davies 2013; Davies and Kent 2017, 2018; Kent et al. 2019). The acquisition of Soviet mapping by libraries is increasing, despite inconsistencies in how they are catalogued (Davis and Kent 2017), and there remains enormous scope for further research to contribute towards a fuller picture of Soviet military mapping. Indeed, with improving accessibility to this huge geospatial resource (e.g. downloads and prints from http://redatlasbook.com/), new applications of Soviet military mapping are being discovered and realized (e.g. Rondelli et al. 2013). In this chapter, we focus on the mapping of Oxford, including the 1:10,000-scale plan of the city that was printed in 1973. Our analysis provides some insights into the probable source material used and some reflections on the purpose of the mapping.

## 2   Soviet Map Formats

### 2.1   Topographic Maps

Topographic maps (or 'topos') were produced at seven scales, which were designated and classified as follows:

- 1:1,000,000—Small-scale/general terrain evaluation (unclassified)
- 1:500,000—Small-scale/operational (unclassified)
- 1:200,000—Medium-scale/operational-tactical and includes a description and schematic map of the surface geology on reverse of sheet ('For Official Use')
- 1:100,000—Medium-scale/tactical ('Secret' for USSR territories; 'For Official Use' if elsewhere)
- 1:50,000; 1:25,000 and 1:10,000—Large-scale/tactical ('Secret').

Also referred to as 'SK-42', Soviet topographic maps adopt a standard projection (Gauss-Krüger), datum (Pulkovo 1942), ellipsoid (Krassovsky 1940) and use similar symbology and labelling. This multi-scale approach facilitates interpretation and use between different scales. The sheets are non-rectangular; their sheet lines being defined by lines of latitude and longitude that conform to the nomenclature devised for the International Map of the World (IMW). Hence, the basic quadrangle of all Soviet topographic maps is the 1:1,000,000 sheet, which spans four degrees latitude by six degrees longitude. The quadrangles are identified by lettered bands north or south from the Equator and by numbered zones east from longitude 180°. The basic 1:1,000,000 IMW grid used in Soviet military mapping is shown in Fig. 1.

Although the IMW nomenclature adopts the Roman alphabet, Cyrillic letters are used on Soviet maps when sub-dividing sheets into larger scales. Each 1:1,000,000-scale sheet is divided into four sheets at 1:500,000, 36 sheets at 1:200,000 and 144 sheets at 1:100,000. Each 1:100,000 sheet is subsequently broken down into four 1:50,000 sheets, labelled А, Б, В, Г and each 1:50,000 sheet is divided into four 1:25,000 sheets, labelled а, б, в, г. The 1:25,000 sheets are further sub-divided into four 1:10,000 sheets, although coverage of this large-scale topographic series appears to have been limited to within the Soviet Union. The nomenclature can also be translated into a purely numerical sequence, where the letters are replaced with their position in the alphabet (e.g. the letter M becomes the number 13). Hence, Oxford appears on 1:50,000 sheets 13-30-010-3 and 13-30-010-4.

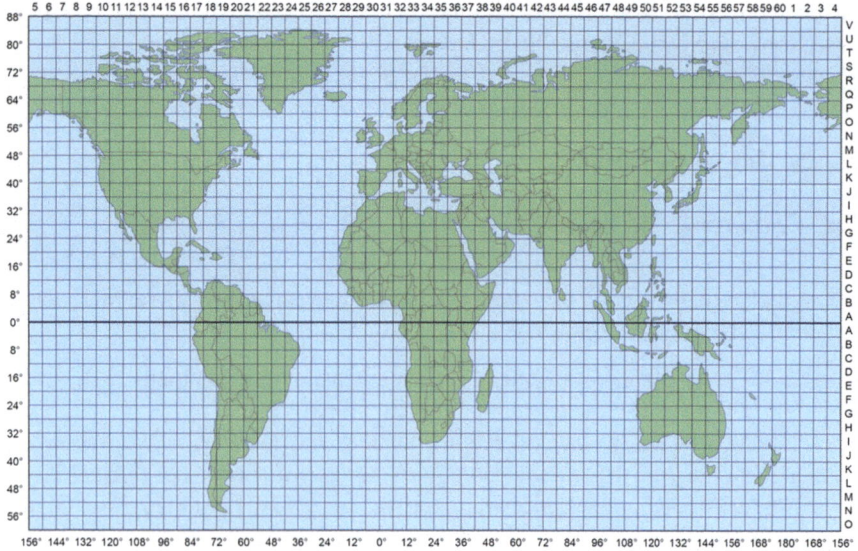

**Fig. 1** A map of the world showing the grid adopted in Soviet military mapping, based on that of the 1:1,000,000 International Map of the World or IMW (Oxford lies on sheet M-30)

The exact coverage of Soviet topographic mapping has not yet been established, although Watt (2005) estimates the production of over one million sheets, including detailed topographic maps of the USSR at several scales from 1:25,000, plus maps of the rest of the world at 1:200,000 and larger, with coverage of Europe, the Middle East, North and Central America, large areas of South America, the Indian subcontinent, south-east Asia, China, and the populated areas of Africa at 1:50,000 and/or 1:100,000 scales. Even half this level of coverage would represent the most comprehensive systematic topographic mapping of the globe yet undertaken as a singular, coordinated enterprise.

## 2.2 City Plans

Towns and cities in over 130 countries outside the USSR were mapped in detail by the Soviet General Staff, many of which are concentrated in Europe, eastern USA, western China and Japan. The list of known plans includes the world's largest cities, such as Shanghai, Tokyo, London, New York, Paris, Istanbul and Cairo, in addition to hundreds of much smaller settlements with a population of just a few thousand. Although the rationale for their selection could include strategic functions (e.g. as major railway hubs and naval bases as mentioned by Psarev 2005: 49), there remain several large cities of which no Soviet plan is currently known, such as Rio de Janeiro, Bogotá, Melbourne and Lagos (Kent et al. 2019). Toponyms are spelt phonetically in Cyrillic (e.g. the Soviet plan of Leicester is titled 'ЛЕСТЕР'), presumably to aid and standardize the Soviet pronunciation of foreign place names.

The plans typically adopt the scales of 1:25,000 or 1:10,000, though some at 1:20,000, 1:15,000 and 1:5000 are known to exist. The corresponding 1:100,000 topographic map sheet is indicated on each plan, yet these sheets were produced independently of the topographic map series described above and were classified 'Secret'. They differ in content and appearance to the topographic maps at comparable scales, mostly through their inclusion of street names and a more comprehensive symbology, especially the colour-coding of important buildings. Most city plans appear to have been compiled during the 1970s and 1980s, reflecting the cartographers' increasing dependency on Soviet satellite imagery (specifically, the Zenit satellite programme) for acquiring topographic data, and therefore their move away from a reliance on indigenous mapping—some of which is explicitly stated as source material on earlier plans (Davies and Kent 2017).

Although the sheets use the standard projection, datum and ellipsoid as per the SK-42 maps described in the previous section, they were not designed to provide continuous or systematic coverage of an area according to a global grid. Instead, each city plan is centred on a specific settlement (or conurbation), which provides its title. The number of sheets used for each plan is therefore determined by the size of the area covered by the settlement (and its immediate surroundings) to be mapped at the chosen scale. Hence, smaller towns and cities fall onto a single sheet, while several sheets are required for larger settlements. The largest Soviet city plan known to date is that of Los Angeles, which is covered by twelve sheets. Where cities are covered by multiple sheets, it is possible to remove the inner margins so that they form a unified and seamless composition.

The most visually distinctive aspect of the plans' cartography is their colour-coding of important objects and other buildings according to their function. Industrial buildings (such as factories, foundries and railway stations) are shown in black, military or communications facilities (such as barracks, radar installations and major post offices) are coloured green, and administrative or governmental buildings (e.g. town halls and law courts) are shown in purple. All other buildings are coloured brown. The important buildings and objects are numbered and correspond to a listing that is given at the edge of the plan, or, occasionally, on a separate sheet or in a booklet where they number several hundred (see Davis and Kent 2018 for an illustrated description of Soviet map sheets and a guide to understanding their metadata).

The city plans are also notable for their synthesis of rich hydrographic information with a high level of topographic detail, which surpasses that of comparable plans produced by other military mapping agencies at the time. The depth, speed and flow direction of rivers are often included and a topographic contour interval of 2.5 m is not uncommon, while annotations indicate supplementary information such as the width of streets, the construction material and carrying capacity of bridges. In addition to the numbered list of important buildings and objects, each plan also includes an alphabetical street index with corresponding alphanumeric grid squares and a 'spravka'. This is a descriptive text that summarizes the physical geography of the city (including the local topography and climate of its surroundings) and provides an overview of its economic and political functions.

In addition to the maps and plans mentioned above, other types of mapping were also produced. This includes a version of topographic mapping introduced in 1963 for civic planning and other official purposes in the Soviet republics (the 'SK-63' series omits geographical coordinates and other information) and a range of military thematic maps that use existing Soviet topographic data as base mapping (e.g. aeronautical charts, geodetic maps, transportation maps and military engineering maps). Since the Soviet military maps were compiled, edited and created in secret, details of the exact global coverage of the mapping and how the cartographic project was organized and run are yet to emerge. For the time being, it is perhaps only through the examination of individual sheets and the sharing of these observations and findings that a fuller and more accurate picture of the wider Soviet mapping project can be attempted.

# 3   Oxford

## *3.1   Topographic Mapping*

The city of Oxford appears on sheet M-30 on Soviet 1:1,000,000 scale mapping, which conforms to the standard projection, datum and ellipsoid of the SK-42 series. Grid square M-30 lies at the intersection of IMW band M (spanning latitudes 48°–52°N) and zone 30 (spanning longitudes 0°–6°W) (see Fig. 1). The earliest known M-30 sheet was printed in 1938 (and therefore predates World War II) and the latest was printed in 1985. The city also appears on Soviet topographic sheets at the scales of 1:500,000, 1:200,000, 1:100,000 and 1:50,000. At 1:50,000, the largest of these scales, the city lies between two sheets, M-30-010-B 'УИТНИ' (Witney) and M-30-010-Г 'ОКСФОРД' (Oxford). The two non-rectangular sheets abut at 1° 15′ W (east of the city centre) and together cover an area 1° 30′–1°W wide and 51° 40′–51° 50′ N tall. The print codes in the lower right-hand margin (Fig. 2) include the successive job numbers B2111 and B2112, indicating that the maps were printed consecutively, while the notation VIII-81 refers to August 1981 and the 'Д' reveals that they were printed at the Dunayev military cartographic factory in Moscow (more details on the interpretation of print codes are provided in Davies and Kent 2017). The maps are printed in five colours: black, dark blue, dark orange, light orange and green, although the range of colours used is extended to eight with the use of halftones to produce grey, light blue and light green. Regarding source material, a caption in the bottom right-hand corner of each sheet notes that the sheets were 'compiled using 1:50,000 maps dated 1974'—a clear reference to the Ordnance Survey 1:50,000 *Landranger* series that was introduced that year. The Soviet 1:50,000 topographic maps include Marston Ferry Road and bridge, which opened in 1971 and are shown on the OS mapping of that scale.

The classification and depiction of topographic features on the Soviet 1:50,000 mapping enables a very easy assessment of the characteristics of the landscape. Although hypsometric tinting (layer colouring) is not implemented at this scale, unlike the 1:1,000,000 and 1:500,000-scale mapping, the 10-m contour interval is sufficient for terrain evaluation, while the use of colour for depicting urban areas is

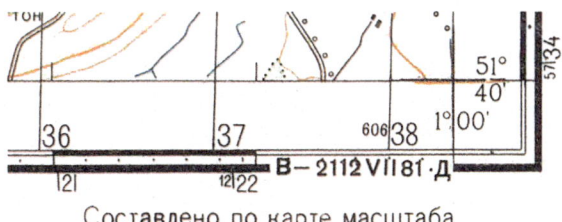

**Fig. 2** The print code on 1:50,000 Soviet topographic map sheet M-30-010-Г 'ОКСФОРД' (Oxford). The caption below the margin indicates the source mapping used, i.e. Ordnance Survey 1:50,000 maps from 1974 (reproduced from a private collection)

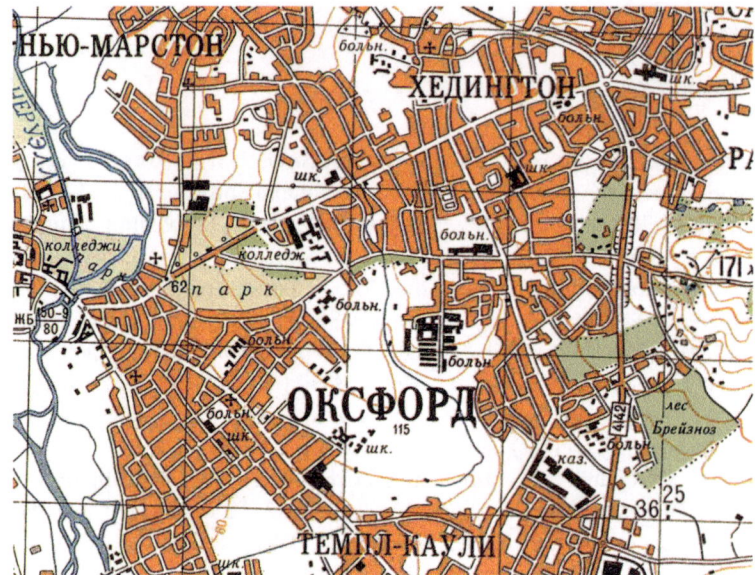

**Fig. 3** Detail from 1:50,000 Soviet topographic map sheet M-30-010-Г 'ОКСФОРД' (Oxford) indicating the eastern half of the city, which includes the Morris car factories at Cowley to the southeast. Magdalen Bridge is annotated with details of its dimensions and carrying capacity (reproduced from a private collection)

especially useful. This allows a general distinction to be made between settlements with a population over 50,000 (in orange) and below (in grey), but also for significant industrial buildings (in black) to stand out, such as the car factories at Cowley to the southeast of the city (Fig. 3). In addition, the use of two shades of green allows vegetation to be distinguished between woodland (green) and parkland (light green), as per Headington Hill and South Park respectively.

The depiction of Oxford on these two 1:50,000 sheets is typical of the Soviet mapping of the UK at this scale. However, the eastern sheet includes a rare annotation beside Magdalen Bridge that indicates its construction material ('ЖБ', which denotes reinforced concrete), dimensions (150 m long by 9 m wide), and load capacity (80 metric tonnes) (Fig. 3). There are few other examples of these annotations on Soviet topographic mapping of the UK and given that the data are not readily found on OS maps or by reconnaissance imagery, it is possible that such information was derived from eyewitness accounts on the ground.

Further afield, sheet M-30-010-3 'ВУДСТОК' (Woodstock) includes the US Air Force Base at Upper Heyford in Oxfordshire, showing the runways and assorted buildings that were omitted as 'security deletions' from OS maps at the time. However, on a road adjacent to the perimeter of the airfield, the Soviet map includes an arrow symbol that is not part of Soviet symbology but indicates a steep hill on OS maps (Fig. 4). This suggests that the Soviet cartographers had copied Ordnance Survey mapping to the extent of introducing symbols where their meaning was

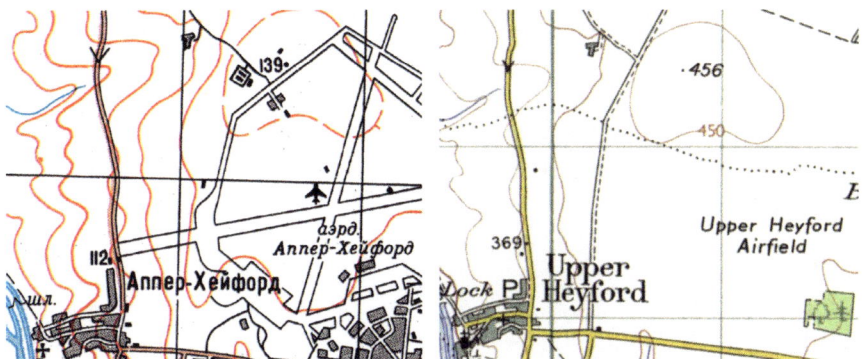

**Fig. 4** Detail from 1:50,000 Soviet topographic map sheet M-30-010-B 'ВУДСТОК' (Woodstock) (left) showing Upper Heyford USAF Base and the area depicted on Ordnance Survey 1:63,360 'One-inch' sheet 145 (right). The arrow symbol at the top left of the OS extract on the steep road to Upper Heyford has been copied directly onto the Soviet sheet, yet it is not part of Soviet symbology (reproduced from a private collection)

unknown, as well as adding other features (such as military airfields) that had been omitted from OS maps. If this was not uncommon practice, it is likely that some of the characteristics of this and other indigenous topographic mapping from around the world, such as the cartographic generalization of urban areas, will have found their way onto Soviet topographic maps.

## 3.2   The 1:10,000 City Plan (1973)

The city plan of Oxford comprises two 1:10,000 sheets, which are divided into eastern and western halves abutting approximately 1° 14′ east of the city centre. These were compiled in 1972 and printed in Moscow in March 1973 using 10 colours (black, brown, tan, green, light green, blue, light blue, orange, purple and grey, plus a lighter shade of green as a halftone). The contour interval is 2 m and forty-one strategically important buildings and objects are included in the listing at the top right-hand corner of the eastern sheet. Removing the inner margins allows the whole composition of the plan to be appreciated (Fig. 5).

The most visually prominent feature on the Soviet plan is the huge industrial complex at Cowley, whose individual factories are correctly labelled in the accompanying listing as 'Morris Motors' and 'Pressed Steel Fisher'. These are rendered in black, as strategically important industrial objects. Their numerous buildings and warehouses are shown in detail (Fig. 6), together with the adjacent railway and Morris Cowley station on Garsington Road. Publishers are also iden-tified and represented, with Alden Press on Binsey Lane and the historic buildings of Oxford University Press on Walton Street. Curiously, the premises of the former are classified as a governmental/administrative site (shown in purple) and the latter

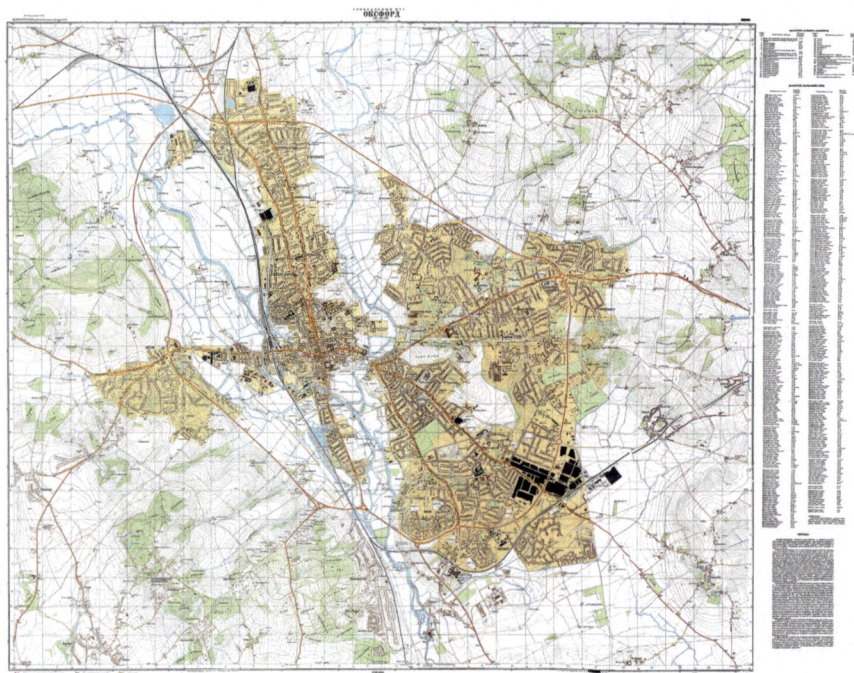

**Fig. 5** The Soviet city plan of Oxford (1:10,000) that was compiled in 1972 and printed in 1973, shown here as a single composition by joining together the eastern and western sheets (reproduced from a private collection)

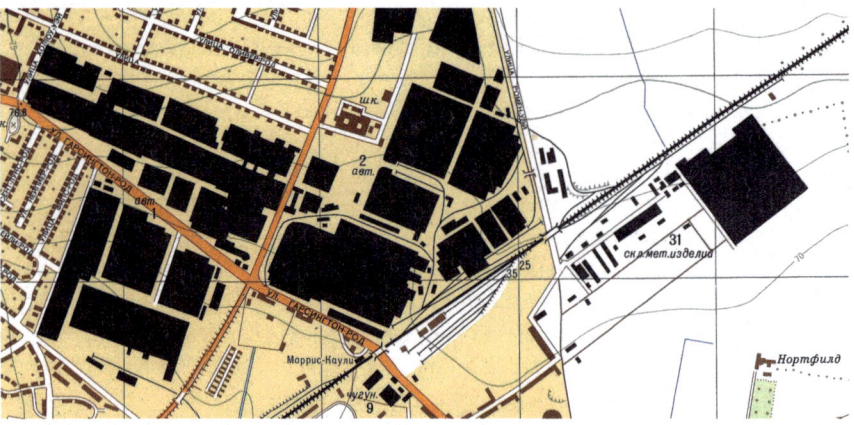

**Fig. 6** Detail from the Soviet 1:10,000 city plan of Oxford showing the industrial complex at Cowley, with the Morris and Pressed Steel Fisher factories coloured black and labelled 1 and 2 respectively (reproduced from a private collection)

as industrial, possibly because of the more substantial printing presses and machinery in use at Walton Street. The buildings of Oxford Prison are depicted clearly, as are the Police Station, Town Hall, and the main Post Office on St Aldate's.

The plan also includes some anomalies. Marston Ferry Road, which has linked Marston to Summertown since it opened in 1971 (and is depicted on the Soviet topographic maps mentioned in the previous section), does not appear on the plan despite its compilation date of 1972. The Marston ferry across the River Cherwell is shown (that was superseded by the bridge), implying that older maps were used as source material. The plan also shows the gasworks at St Ebbe's, which were closed in 1960, and the two gas holders that were demolished in 1968. As for the University, although described in the accompanying spravka as comprising sixteen departments and twenty-five autonomous colleges, the only site to be identified and labelled as 'The University' is University College (indicated by a single building, number 39 and coloured purple). This interpretation was clearly borne out of confusion, but one which would make perfect sense to those unfamiliar with the University's multi-institutional structure. Several other colleges (but by no means all) are generically labelled колледж (college), while very few are specifically named (such as Merton, St Hilda's, and Worcester) (Fig. 7).

Railways, stations and their associated infrastructure such as sidings and turntables, are shown in detail on the plan. This detail includes, for example, the location of station buildings relative to the platforms. Some stations which closed long before the compilation date of the plan (1972) are included and named, for example Abingdon Road, which was closed in 1915. Similarly, Kennington Junction, which is named and appears on the plan as a station, had only ever been a signal box (Fig. 8). As the city plan also includes areas surrounding the city, it incorporates rural areas and villages several kilometres to the east and west of Oxford. At Boars Hill, five kilometres to the west, an area of land owned and managed by The Oxford Preservation Trust is labelled 'ЗАПОВЕДНИК ОКСФОРД' (Oxford Nature Reserve). This demonstrates that this Soviet city plan (and perhaps all Soviet military mapping) was principally an endeavour to collect and present geospatial data for a wide range of potential applications, rather than for meeting a specific military objective. That the plan depicts disused and misclassified infrastructure, such as the examples connected with railways as mentioned above, supports this view.

In the right-hand margin, the plan includes a spravka of almost 1000 words. This describes the topography, economy, industry, population, civil infrastructure, utilities, transport and communications of the city and its surroundings. Much of this information would have been derived from non-cartographic sources, such as directories and gazetteers. Translated from the Russian, it reads as follows:

*GENERAL INFORMATION. Oxford is a county town and the administrative centre of Oxfordshire in the UK, a significant industrial centre, a famous university centre and a transport junction (6 railways and 9 highways), located on the river Thames,*

**Fig. 7** Detail from the Soviet 1:10,000 city plan of Oxford showing the city centre and North Oxford. University College is coloured purple and labelled '39', and is described erroneously in the numbered list as 'The University' (reproduced from a private collection)

*80 km north-west of London. In 1969, there were 109.7 thousand inhabitants in Oxford; area of the city approx. 35 square km.*

*SURROUNDINGS OF THE CITY The city lies on a hilly plain, with the bottoms of the valleys of the river Thames and its tributary Cherwell having an almost flat surface, intersected by a network of rivers and drainage canals and ditches. Hills (height 70–170 m) have rounded peaks and predominantly gentle (less than 10°) slopes. Soils in the valleys, as a rule, are sandy, in the rest of the region they are sandy loamy, sometimes clayey. The main part of the locality consists of meadows*

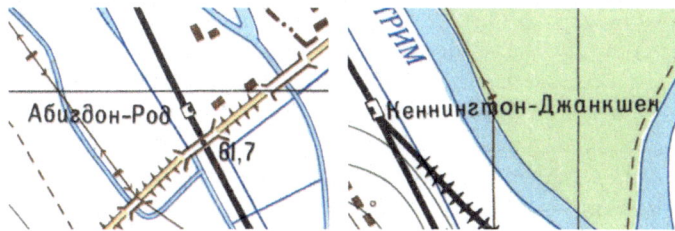

**Fig. 8** Details from the Soviet 1:10,000 city plan of Oxford showing Abingdon Road (left), a railway station that closed in 1915, and Kennington Junction (right), a signal box that is depicted as a railway station (reproduced from a private collection)

*and arable land; woody vegetation is found only in the form of small sections of forest saplings (park type), alongside roads, rivers, canals, and also as hedges along the boundaries of land. The largest water barrier is the river Thames (below the city, available for ships with a draft of 1.2 m; above, 0.9 m); its width is 20–60 m, the depth is 1.6–2 m, the current velocity is 0.8 m/s. The banks are dominated by low, shallow slopes. The other rivers are up to 20 m wide, up to 1 m deep. The largest canal, Oxford Canal, (accessible for vessels up to 21 m in length and 4.3 m deep) has a width of 13.3 m; it connects the tidal Thames with the Birmingham canal system. Highways are surfaced with asphalt and concrete. Oxford is partially visible from its surrounding heights. From the air, it is identified by its location at the confluence of the rivers of the Thames and Cherwell.*

*CITY TERRITORY. Oxford does not have a single system of planning. Its western part, located in the valley of the river Thames, is the historical core of the city. Here, the wide straight streets are combined with narrow and curved ones, the building is solid or dense, stone houses, 3–5-storey buildings; many ancient buildings of the Gothic style with numerous towers and spires. In this part of the city there are a number of administrative institutions, including the town hall (object 28), the post office (object 26), and the famous Oxford University (object 39), which unites 16 faculties and 25 autonomous colleges. At the university there are a number of scientific institutes, laboratories, museums, observatories, a botanical garden, a large library, sports grounds, etc. The eastern, newer part of the city is located mainly on the right slope of the valley of the Thames and the hills adjoining the valley. The streets here are predominantly wide, mostly straight. The building in the central part (adjacent to the Cherwell River near the Magdalen Bridge) is dense, the houses are 3–5 storeys. In the rest of the territory, as well as on the northern and southern outskirts of the old part of the city, the building is predominantly sparse 1–3-storey houses. In the eastern part of the countryside are the residential quarters of Oxford. Industrial enterprises are concentrated mainly in the southeastern and western suburbs. The city is well landscaped, there is a significant number (especially in the eastern part) of parks, gardens, squares.*

*INDUSTRIAL AND TRANSPORT OBJECTS. The leading industries of the city are machine building (including automobile, aviation and electrical engineering) and metalworking. The most important military-industrial objects are automobile plants (objects 1, 2) and iron foundry (object 9). The Oxford railway network includes several railway stations and passenger platforms, including the Oxford goods and passenger station (object 35) with well-developed track and storage facilities, a depot and a railway station.*

*UTILITIES AND MEDICAL AND SANITARY INSTITUTIONS. Oxford receives electricity from a local thermal power station (object 41), which is included in the country's integrated energy system. The city has a gas supply; there are three gas plants operating (objects 3, 4, 5). There is water supply and sewage. Oxford is provided with all kinds of modern communications. Within the city, transport is by bus. The city has 16 hospitals and a number of other health facilities.*

The spravka is remarkably succinct, with a deft combination of a wide variety of information (e.g. from the city's general location and its major functions to the predominant style of architecture) with specific details (e.g. from the flow velocity of the River Thames to the number of hospitals). Some of the more practical information—especially relating to the soils and slopes of the surrounding terrain—would clearly be of use to military planners, but most of the text provides an almost charming account of the city that would not seem out of place in a tourist's guidebook. Indeed, it is plausible that at least one guidebook from the early 1970s was used as a source that perhaps mentioned the 1969 population as 109,700 ahead of the next census of 1971. It is possible that earlier sources were also used, including maps. For example, evidence of the use of Ordnance Survey mapping is provided by spot heights on the plan, which appear to have been copied from the OS 1:10,560 'Six-inch County Series' of 1922. The OS benchmarks on the northern bank of the Thames of 197.7, 196.4, 195 and 195.8 feet correspond exactly to spot heights at the same locations on the Soviet plan when converted to the respective metric values of 60.3, 59.9, 59.4 and 59.7 m. This perhaps becomes more surprising when considering that additional details that do appear on Soviet topographic mapping, such as the annotation associated with Magdalen Bridge on the 1:50,000 sheet, are not included on the larger-scale city plan.

# 4  Conclusion

The Soviet global military mapping project was the most comprehensive cartographic endeavour of the twentieth century. Like other towns and cities around the world, Oxford appears on Soviet topographic maps of a range of scales from 1:1,000,000 to 1:50,000 and was also the subject of a city plan at the larger scale of 1:10,000. Observations from a comparative analysis based on these sheets suggests the following:

- Although the topographic maps and city plans were produced by the Soviet General Staff and used the same projection, datum, ellipsoid and similar symbology, there is little evidence to suggest that the sources or resulting geospatial or topographic data were shared in the compilation and production of the maps;
- Comparisons with contemporaneous Ordnance Survey mapping of similar scales suggests that OS mapping was used as source material, yet a number of striking anomalies (particularly the absence or misclassification of features) suggest that this was far from systematic; and
- The intended purpose of the maps, and particularly the city plans, is ambiguous. The inclusion of a wide range of topographic information that extends to the function of buildings suggests that the primary objective was to collect and portray a comprehensive level of geospatial intelligence with clarity, therefore lending to the maps a breadth and versatility of application as opposed to a specific and more limited functionality.

Clearly, further research and investigation into the comparative quality and quantity of information included and presented on Soviet mapping is required in order to draw wider and more meaningful conclusions regarding its strategic value. Nevertheless, as availability of this substantial resource increases, it is likely that more studies will be conducted and therefore more possible that this goal will be realized in future.

# References

Cruickshank JL (2008) Виды из Москвы—views from Moscow. Sheetlines 82:37–49

Cruickshank JL (2015) Military mapping by Russia and the Soviet Union. In: Monmonier M (ed) The history of cartography, vol. 16: cartography in the twentieth century. University of Chicago Press, Chicago, pp 932–942

Davies J (2005a) Uncle Joe knew where you lived: Soviet mapping of Britain (Part I). Sheetlines 72:26–38

Davies J (2005b) Uncle Joe knew where you lived: Soviet mapping of Britain (Part II). Sheetlines 73:6–20

Davies J (2010) Soviet military city plans of British Isles. Sheetlines 89:23–24

Davies J, Kent AJ (2017) The Red Atlas: how the Soviet Union secretly mapped the World. University of Chicago Press, Chicago

Davis M, Kent AJ (2017) Improving user access to Soviet military mapping: current issues in libraries and collections around the globe. J Map Geogr Libraries 13(2):246–260

Davis M, Kent AJ (2018) Identifying metadata on Soviet military maps: an illustrated guide. In: Altić M, Demhardt I, Vervust S (eds) Dissemination of cartographic knowledge. Springer, New York, pp 301–313

Kent AJ, Davies J (2013) Hot geospatial intelligence from a Cold War: the Soviet military mapping of towns and cities. Cartography Geogr Inform Sci 40(3):248–253

Kent AJ, Davis M, Davies J (2019) The Soviet mapping of Poland: a brief overview. Miscellanea Geographica 23(1):1–11

Pearson A, Taylor DRF, Kline KD, Heffernan M (2006) Cartographic ideals and geopolitical realities: international maps of the world from the 1890s to the present. Can Geographer 50 (2):149–176

Postnikov AV (2002) Maps for ordinary consumers versus maps for the military: double standards of
    map accuracy in Soviet cartography, 1917–1991. Cartography Geogr Inform Sci 29(3):243–260
Psarev AA (2005) Russian military mapping: a guide to using the most comprehensive source of
    global geospatial intelligence (ed. Filatov VN and trans. Gallagher P) East View Cartographic,
    Minneapolis
Rondelli B, Stride S, García-Granero JJ (2013) Soviet military maps and archaeological survey in
    the Samarkand region. J Cultural Heritage 14:270–276
Watt D (2005) Soviet military mapping. Sheetlines 74:9–12

**John Davies** is a life-long map collector and enthusiast. He is editor of *Sheetlines*, the journal of
The Charles Close Society for the Study of Ordnance Survey Maps.

**Alexander James Kent** is Reader in Cartography and Geographical Information Science at
Canterbury Christ Church University and Immediate Past President of the British Cartographic
Society. He is also Editor of *The Cartographic Journal* and Chair of the International Cartographic
Association (ICA) Commission on Topographic Mapping. John Davies and Alexander James Kent
are co-authors of *The Red Atlas: how the Soviet Union secretly mapped the World* (University of
Chicago Press, 2017).

# Maps Against Imperialism: Frank Horrabin and Alexander Radó's Atlases in the Interwar Period

Gilles Palsky

**Abstract** Though radical or critical cartographies are recent trends, maps that question the political order were conceived much earlier. This article deals with several atlases which were made between the two world wars, by Frank Horrabin, a British socialist, and Alexander Radó, a Hungarian communist. Both shared a conception of maps as tools to denounce and combat bourgeois imperialism. They presented their works as new projects, different from ordinary atlases: they wanted them to be rooted in current affairs and to provide dynamic approaches. They favoured small-scale representations and demonstrated a clear awareness of the interdependence of phenomena on the globe's surface. Their cartography, which may be described as 'persuasive', left much room to thematic maps, mainly on economic and geopolitical topics. In particular, their atlases outlined all forms of domination and imperial control. Horrabin and Radó used a variety of graphic means (arrows, colours, typography, and layout) to reinforce their message, which gave their maps a definite connection with those drawn by German geopoliticians in the 1920s and 1930s. The general picture that emerged from this cartography was that the world was a battlefield of imperialistic rivalries, fraught with threats and friction points. Founded on Marxist ideology, this cartography placed special emphasis on Soviet Union, presented as a peaceful State, encircled and threatened by imperialist blocs.

## 1 A Left-Wing Philosophy of Mapping

As is well known, maps were, throughout history, tools that served power. In the contemporary period, they were particularly involved in the building of nation-states and in colonial control. Over the past decades, the inseparable link between maps and the established powers has been called into question by several branches

G. Palsky (✉)
University of Paris1 Panthéon-Sorbonne, Paris, France
e-mail: gilles.palsky@univ-paris1.fr

© Springer Nature Switzerland AG 2020                                    159
A. J. Kent et al. (eds.), *Mapping Empires: Colonial Cartographies of Land and Sea*,
Lecture Notes in Geoinformation and Cartography,
https://doi.org/10.1007/978-3-030-23447-8_9

of cartography, derived from the critical thinking initiated by Brian Harley (Harley 1988, 1989). Alternative maps have flourished, as part of movements such as radical cartography, counter-mapping or indigenous cartography.

However, it is possible to identify an early map production that anticipated the practices of critical cartography. In this article I examine the case of several atlases which were made between the two world wars, by Frank Horrabin, a British socialist, and Alexander Radó, a Hungarian communist. There is no indication of any contact between these two personalities. They just shared, for three of their atlases, the same English publisher: Victor Gollanz, the editor of George Orwell, close to the pacifist and socialist ideas. It is of interest, though, to bring together a few atlases made by Horrabin and Radó, to highlight their common threads: proximity, formal and ideological. Horrabin as well as Radó advocated for inter-nationalism and against imperialism, and their atlases challenged the usual image publicized in commercial and official cartographies. Published over a short period of 12 years (between 1926 and 1938), they are to be understood within the context of a struggle between capitalist states and the Soviet Union, which was then the only state to follow the socialist model.

Therefore, the following questions are raised: what identifies these atlases, in terms of content and topics, structure, graphic design? How does this 'socialist cartography' relate to other forms of persuasive cartography produced over the same period, in particular by German or Hungarian extreme right-wing elements? More generally it seems important to rescue from oblivion an original discourse, much less common than the one designed to reinforce the traditional view of the Nation or the Empire.

James Francis Horrabin (1884–1962), also known as Frank Horrabin, was a British socialist, a representative of the left wing of the Labour Party. In 1921 he joined the newly founded Communist Party of Great Britain, but left it fairly quickly, to go back to the Labour. In the interwar period, Horrabin was especially involved in several socialist societies, notably the *Plebs League*, which sought to promote a class education of workers, based on Marxism. Horrabin was one of the key players in several institutions set up by the *League*, such as the Central Labour College, that worked until 1929. Horrabin was an autodidact, with no training in geography, but he had been trained in design, at the Sheffield Art School. He turned to cartography during his career as a journalist: he first illustrated articles and books with maps and diagrams, and later produced a large number of atlases, notably on World War II, the USSR, or Africa. Horrabin's graphical production has been mentioned in several recent books, but has only been the subject of a single general study (Bithell 1984a, b). Horrabin's work has also been addressed from the perspective of his contribution to a 'socialist geography' in England (Hepple 1999).

Alexander (or Sándor) Radó (1899–1981) shared with Horrabin a left-wing commitment, but his life was much more eventful, reflecting the political upheavals of his time. He was a Hungarian Jew who joined the Hungarian Communist Party when it was founded, in November 1918. He participated in the Bela Kun revolution, and after its failure, he took refuge in Austria, and later in Germany. During this time, he continued to be involved in left-wing politics, but also completed his studies in geography and cartography, first at the University of Vienna, and in 1922–1925 at the universities of Leipzig and Jena. Radó made a number of trips to Moscow and became a Soviet spy, in probably as early as 1919. Based in Berlin after 1925, he pursued his intelligence work, under the cover of cartographic activities for German editors. After 1933, Radó fled to France, and later to Switzerland, where he founded a news agency, which provided maps to reviews and newspapers. To be complete, it can be said that Radó is considered as one of the greatest spies of the period, and that the intelligence activity of his network allowed him to alert Stalin of the impending Barbarossa operation, among other valuable information. Radó's life trajectory is known to us from a CIA report (Thomas 1968) and Radó's own memoirs (Radó 1971, uncensored version in 2006). His work as a spy-cartographer has been addressed in numerous recent studies (Schlögel 2003; Schneider 2006; Barth 2010; Bourgeois 2011; Heffernan and Győri 2014; Boria 2014; Rivière 2016).

Horrabin and Radó's cartographic activities were motivated by their left-wing ideology. Regarding Horrabin, he claimed to produce maps specifically for the worker's education. He began to publish them, together with charts, in socialist newspapers such as the *Daily Herald* or the *Lansbury's Labour Weekly*. He also used them for booklets and cheap textbooks, or to illustrate his lectures at the Central Labour College. In a 1923 geography textbook, Horrabin presented the map as an element of pleasure and recreation for the reader and an attractive way to educate and to deliver messages. He did not define a clear political project for his atlases, but they were mainly targeted at those who studied imperialism and they strongly opposed capitalist empires compared to Soviet Russia, a federation of worker's republics. In the short texts he wrote as introductions, he clearly placed his atlases under a Marxist perspective.

Radó, for his part, connected his cartographic approach with his stay in Moscow, in 1921, at the meeting of the Third International. The young Hungarian cartographer, then looking for Russian maps for his documentation, met Lenin in a corridor of the Kremlin: 'When he learned that my scientific interest focused on geography and cartography', wrote Radó, 'he explained to me in a few words that special methods of cartographic representation had become necessary, as a result of the problems with imperialism' (Radó 1972: 48). He stated further: 'I have tried to get that done, in line with the spirit of Lenin, in my first cartographic work, the *Atlas of imperialism* (published in Berlin in 1929 and in Tokyo in 1930)' (Radó 1972: 49). In any event, this anecdote about a discussion with Lenin should not hide the fact that Radó was likely influenced by forms of persuasive cartography developed by German geopoliticians or Hungarian revisionists, in the interwar period (Herb 1997: 152).

## 2  To Show Today's World

Our corpus of analysis is made up of two atlases from Horrabin: the first he published, *The Plebs Atlas*, in 1926, and *An Atlas of Empire*, dated 1937. *The Plebs atlas* was made up of '58 maps for worker-students', reproducing illustrations made by Horrabin for the 'Geographical Footnotes' of *The Plebs* and The *Lansbury's Labour Weekly*. *An Atlas of Empire*, published in paperback format by Gollancz, contained 70 maps, each one with a facing text page. A third work, *An Atlas of Current Affairs* (Horrabin 1934) will also be mentioned, though it is less clearly ideological: its introduction sounded more neutral, as if the atlas was trying to reach a larger audience.

With regard to Radó, his first atlas is rather well known. The *Atlas für Politik, Wirtschaft, Arbeiterbewegung* [Atlas for Politics, Economy, and Labour Movement] was published in Vienna and Berlin in 1930, according to its copyright. Its subtitle, '1. Der Imperialismus', flagged it as a first volume. Radó stated his intention to issue a second volume on the workers' movement and a third one on the Soviet Union—but the rise of the Nazism in Germany prevented him from realizing this project (Radó 1980). The other work of interest is Radó's second atlas, *Atlas of To-Day and To-Morrow*, published in London in 1938. It was said to be an entirely revised edition of the previous one, and it is indeed a completely different work, despite a similar political message.

Our two cartographers presented their atlases as new projects that stood apart from existing publications. 'It is not its aim […]', said Radó about his second publication, '[…] to improve on existing geographical atlases with their accurate and multi-coloured drawings of cities, rivers and mountains' (Radó 1938: IX). Horrabin made a comparable remark: '[This atlas] is not intended to take the place of an ordinary reference atlas […]. The maps on the following pages make no attempt at crowding in all the names and facts possible' (Horrabin 1926: 3).

Consequently, they highlighted specific characteristics. First, they claimed that they wanted to focus on 'current events'. This is obvious first of all in some of the titles that were chosen: 'Current Affairs', 'To-Day and To-Morrow'. Moreover, both Horrabin and Radó insisted on this in their introductory remarks. Horrabin wanted to call 'the student's attention to essential points of present-day world geography' (Horrabin 1926: 3) or 'to illustrate current happenings' (Horrabin 1926: 4). He further suggested that his maps should be used for reading a newspaper intelligently: they could then be corrected or commented upon in the margins with a pencil, when the newspaper gave additional information. Radó stated a similar intention in his second atlas, expressing the wish to seize the evolving world of his time: 'The purpose of this atlas is to provide, so to speak, a snapshot photograph of our rapidly changing world' (Radó 1938: IX). He also shared the concern to help the reader, through the 'comprehensive picture' he proposed, 'to understand the age in which we live' (Radó 1938: IX). Of course, this common orientation was linked to our cartographers' experiences in journalism. Both prepared news maps for print media. It was even the essential cover activity of the successive news agencies that

Radó founded in Berlin (Pressgeo), Paris (Inpress) then Geneva (Geopress S.A.). It is noticeable too, that Horrabin drew maps on up-to-the-minute information for a BBC television talk, 'News Map', from 1937 onwards. With their atlases dealing with current issues, Horrabin and Radó are among the precursors of a type that primarily flourished after 1980, with for instance *The State of the World Atlas* (Kidron and Segal 1981) or the *Atlas stratégique* (Chaliand and Rageau 1983).

From this first characteristic, a second one follows: emphasis was given to dynamics, as opposed to static views of traditional geography. In the statements made by Radó or by Theodore Rothstein, the Russian Bolshevik who prefaced the 1930 atlas, one can detect a touch of irony towards a descriptive geography that recorded endless towns, mountains, rivers and so on. This orientation led to a static view and was a keynote of bourgeois knowledge. Horrabin, too, distinguished his approach from that of an 'ordinary reference atlas' (Horrabin 1934: 5). The remarks he made in a 1923 textbook for workers, *An Outline of Economic Geography*, enlightens us on that point (Hepple 1999: 86): Here Horrabin opposed a pure geography, mainly descriptive and physical, to the geography 'from the working class point of view' (Horrabin 1923: 9). Pure geography, produced by the bour- geoisie, collected 'a mass of facts' (Horrabin 1923: 9), not subjected to interpre- tation. Working class geography was 'studied in relations with history and economics' (Horrabin 1923: 9). It was deliberately selective, aiming at a simplifi- cation of information and focusing on problems that world's workers would someday have to solve. In this way, the atlases examined here were sorts of inventories of 'hot spots' of the time, of problems that carried with them the seeds of crises or conflicts. They were simple guides, as Horrabin put it, 'to key facts and key places of the world of to-day' (Horrabin 1934: 5).

Last, Horrabin's and Radó's atlases demonstrated a common concern for what we would today call globalization. It has been asserted (Henrikson 1975; Schulten 2001; Capdepuy 2011) that the sense of a globalized world emerged in cartography during World War II, as if the global war led to imagine a global cartography. However, Horrabin and Radó fully understood the interrelations between the dif- ferent parts of the world. 'I hope', said Horrabin in his first atlas, 'that readers will be able to study an area not only by itself, but in relation to its larger world-setting' (Horrabin 1926: 5). He later added, regarding his *Atlas of Current Affairs*: 'The maps have been grouped in seven main divisions—Europe, the Mediterranean area, the Americas, the Far East, and so on. But the world today is interdependent; and various cross-references will indicate the impossibility of studying any one problem *in vacuo*' (Horrabin 1934: V). For his part, Radó seemed well aware of the inter- nationalization of markets and of the global competition, whether political or economic. In his atlases, he favoured maps on a small scale, depicting the world, the oceans or continental groupings. Finally, it is important to underline that in their atlases both Horrabin and Radó emphasized the relations and flows, as we will see below.

## 3   A Prominent Theme: Economic Phenomena

If we take a closer look at the atlases, a first point of interest is their thematic orientation. It may not surprise us that Horrabin and Radó, steeped in Marxism, gave prominence to economic maps. They were both experts in the field of economic geography. Horrabin had given a central role to the subject, which he introduced and taught at the Labour College from 1918 onwards. To him, it was one of the pillars of workers' education. He had also published a popular textbook in this field, as mentioned above. Similarly, Radó conceived his study of imperialism through the prism of economic geography: raw materials distribution, activities, and international trade. He had the same experience as the British journalist, teaching the subject when he was in Berlin, at the Marxistische Arbeiterschule. Furthermore, in his preface to the 1930 atlas, Theodor Rothstein distinguished Radó's work from geopolitics by pointing out that the latter forgot the economic forces. Yet 'A truly scientific geography will necessarily be built only in the closest context to the study of economics, and that means that it will only develop into a social science and will bear fruit as a real scientific discipline if it is Marxist, that is, if historical materialism is treated' (Rothstein in Radó 1930: 5).

When Horrabin drew political maps, he regularly enriched them with indications on resources and commercial routes, as we can see for example on a map of the Near East (Horrabin 1934: 64), with the mention of the oil fields and pipelines, or a map of Morocco (Horrabin 1926: 56), which displayed the major trade routes at the same time.

This economic trend is much more systematic in Radó's atlases. In the last part of the first one, he alternated a map of nationalities and an economic map for each country. In both editions he also provided economic maps on a world-wide scale, product by product, in a part entitled: 'Targets of imperialism (resources and markets)'. Radó mixed several distributions on a single map, not least to highlight a mismatch: producers and consumers, resources and industry, as for instance a world map of rubber resources and automobile production (Radó 1930: 59). In the 1938 edition, he systematically added a flow map, showing the trade relation, for each country case.

However, most interesting in Radó's cartography is his representation of financial capitalism. This topic appeared in his first atlas, with a choropleth map that showed the impact of World War I on the value of currencies. It was much more developed in the second atlas, in particular with spectacular maps of the flows of capital (Radó 1938: 7 and 11) (Fig. 1). These were probably the first maps that had ever been made on financial flows. Only one earlier example on a similar topic is known: a representation of international credits between the different parts of the world, by the Austrian philosopher, sociologist, and economist Neurath (1936: 61). It is not insignificant to mention here the central character of the Viennese circle, inventor of the ISOTYPE (International System Of TYpographic Picture Education), a pictographic language he created with Gerd Arntz in 1920. We will see further that Radó was likely influenced by Neurath's method of data representation.

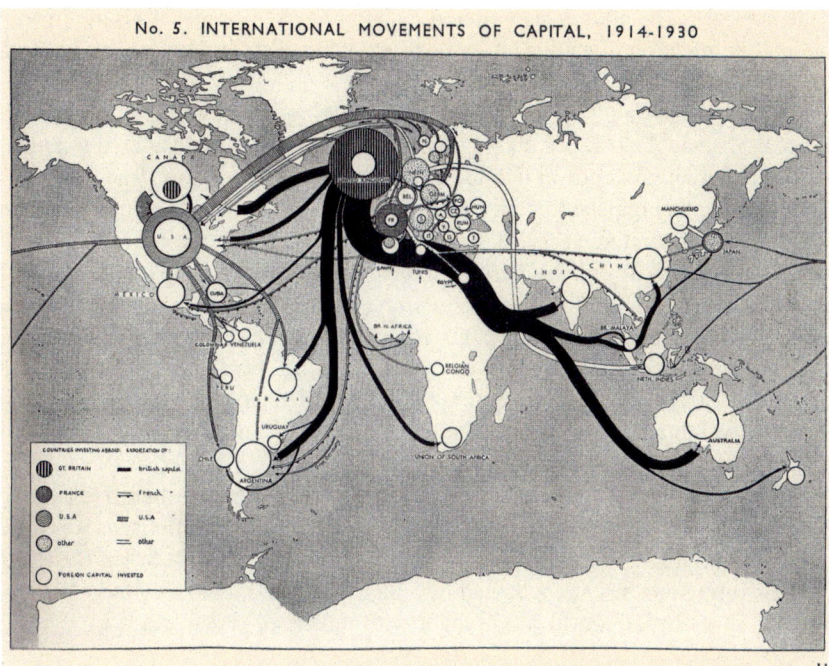

**Fig. 1** 'International movements of capital, 1914–1930', in: Rado S, *Atlas of To-Day and To-Morrow*, 1938, p. 11

In some ways, Horrabin's cartography was more traditional than that of Radó. He used statistical representations with parsimony and never drew any flow map, simply indicating routes or networks (railways, pipelines). His only reference to financial capitalism was on the dust jacket of *An Atlas of Empire* (Horrabin 1937), illustrated with a globe surrounded by a chain whose links were formed of symbols of the main currencies. Finally, it can be seen that most of the maps addressed economics in a confrontational manner: the words *struggle*, *war*, or *Kampf* recurred in titles and legends, reflecting the general idea of imperialist rivalries.

# 4   The Geopolitical Vision

Geopolitics was the second prominent theme developed in the atlases. They depicted a world divided in blocks, wholly owned and controlled by the capitalist powers. This world was the theatre of a constant fight for space, routes and resources, and the maps reviewed all the friction points or places of competing ambitions. From the authors' view, only the Soviet Union remained on the sidelines of the struggle. The atlases also offered a comprehensive study of all the forms of political domination, whether it was over colonized people or over national minorities.

With Radó, the fight took place between five colonial empires and several 'Little Colonial States' (Radó 1930: 40–41). The Soviet Union was considered separately, as the proletarian great power and the only non-imperialist power (Radó 1930: 42–43). Horrabin distinguished larger ensembles: five 'Great World Groups' (Horrabin 1926: 10) corresponding to five dominant powers. However, the overall plan of the atlases did not strictly follow these divisions. It was mostly thematic in Radó, and more classical in Horrabin, based on continents. Within these frameworks, each major power was observed through its spatial extension, often starting with decentred views (Fig. 2) which were unusual in the cartography of the period, dominated by the Eurocentric outlook. It concluded an analysis of geographic locations, which was reminiscent of the German school of geopolitics, but also of more classical geographers, like Carl Ritter.

The capitalist blocks were once again addressed according to their rivalries. Sets of arrows showed lines of penetration, strategic or political ambitions, as we can see on a map from Radó's first atlas, which depicted the opposition between British and American imperialisms, strengthened by high contrast of colours (Radó 1930: 65) (Fig. 3). This divided world was fraught with threats of crises and wars.

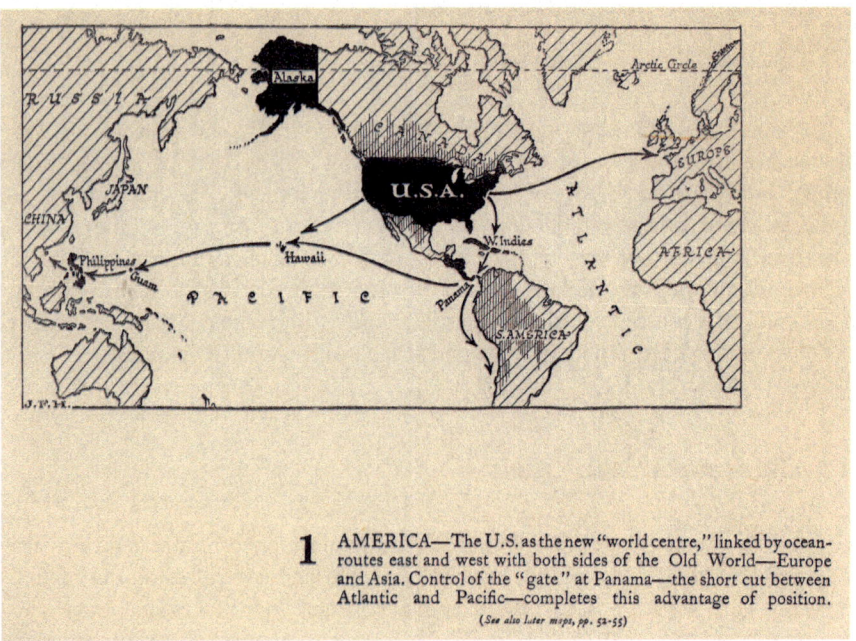

1 AMERICA—The U.S. as the new "world centre," linked by ocean-routes east and west with both sides of the Old World—Europe and Asia. Control of the "gate" at Panama—the short cut between Atlantic and Pacific—completes this advantage of position.
(*See also later maps, pp. 52-55*)

**Fig. 2** 'America', in: Horrabin JF, *The Plebs Atlas*, 1926, p. 11

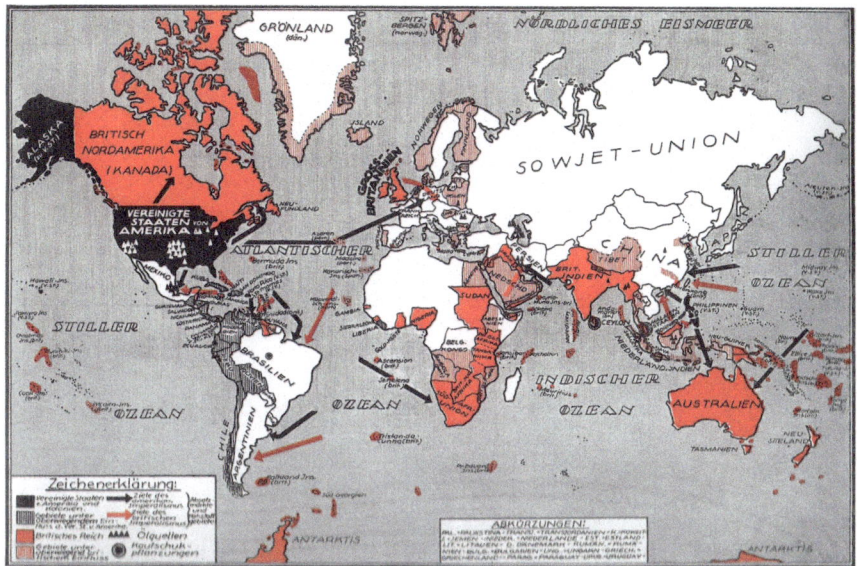

**Fig. 3** 'Der Kampf um die Weltherrschaft' (the fight for world domination), in: Radó S, *Atlas für Politik, Wirtschaft, Arbeiterbewegung*, 1930, p. 65

A second main geopolitical topic was the examination of all forms of domination, of colonized people or minorities. Horrabin emphasized the colonial control with striking maps, such as one of Africa (Horrabin 1937: 24) (Fig. 4) where he grouped all the colonies under a single pattern, and left in white the only two independent states, Liberia and Egypt. However, the common practice was rather to represent space as a political or ethnical patchwork. The patchwork primarily resulted from colonial partitions, as we can see in the case of India, with recurring maps that showed the inextricable tangle of British possessions and feudatory States under protection (Horrabin 1934: 128, 1937: 78; Radó 1930: 134–135, 1938: 37).

Most of the time, the territorial mosaic was tied to multinational States. Horrabin drew several maps on the subject in *An Atlas of Current Affairs*, primarily concerning Eastern and Central Europe, but also Spain and Belgium. He added a map of 'Nationalities in South America' (Horrabin 1934: 90), which merely identified the countries in which native Indians, blacks and mulattos formed the majority of the population. At last, Horrabin drew a unique map of religious diversity, in the case of India (Horrabin 1934: 130), as it was 'another of the complications' (Horrabin 1934: 131) that had to be solved in the sub-continent.

Radó's concern for minorities was even greater, perhaps because of his own Hungarian origin. He gave in his first atlas no less than thirteen maps on the topic and he provided the reader in the 1938 edition with a detailed cartography of all situations: one State with several nationalities or one nationality divided between several States. For him, the problem was not purely European. The notion of

**Fig. 4** 'Colonial possessions in Africa', in: Horrabin JF, *An Atlas of Empire*, 1937, pp. 24–25

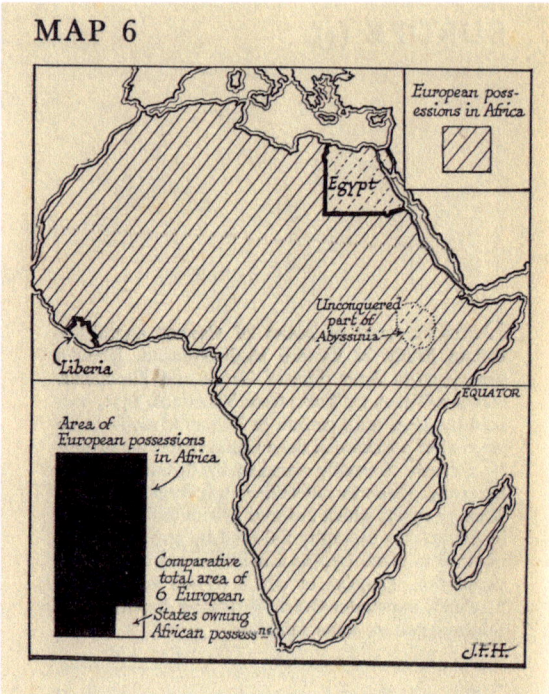

internal divisions was extended to other categories, notably the 'race', which allowed him to show a patchwork of the same type for the United States, with 'white', 'negro', 'Indian' and Asiatic population (Radó 1930: 145).

The atlases thus brought the attention to areas of potential conflicts. Several maps clearly identified immediate threats: the Germans in Central Europe, the Hungarian irredentism, Macedonia, South Tyrol, and so on. This left-wing geopolitical vision was that of a world, at the highest and ultimate stage of capitalism, filled with divisions and oppression, and pushed towards a new war. To build this image, the authors conflated very different political situations. Radó even collated all forms of domination in a single world map of 'oppressed peoples' (Radó 1930: 163) (Fig. 5).

As can be seen from this map, the Soviet Union was granted special treatment in the atlases. It was considered as one of the 'Great Groups', but neither Horrabin nor Radó marked it as an 'Empire'. It was described as a federation, whose domination on parts of Asia was non-colonial in nature. As a consequence, Russia was absent from Horrabin's *Atlas of Empire*. In the four other atlases, the vision on the country had two main dimensions. First, the Soviet Union was depicted as a victim, stripped of the territories which formed the 'cordon sanitaire'. Several maps in different atlases dealt with the same subject. Horrabin commented on it in the *Plebs Atlas*: 'Russia was robbed her Baltic coastline' for keeping it 'out of Europe' (Horrabin 1926: 34).

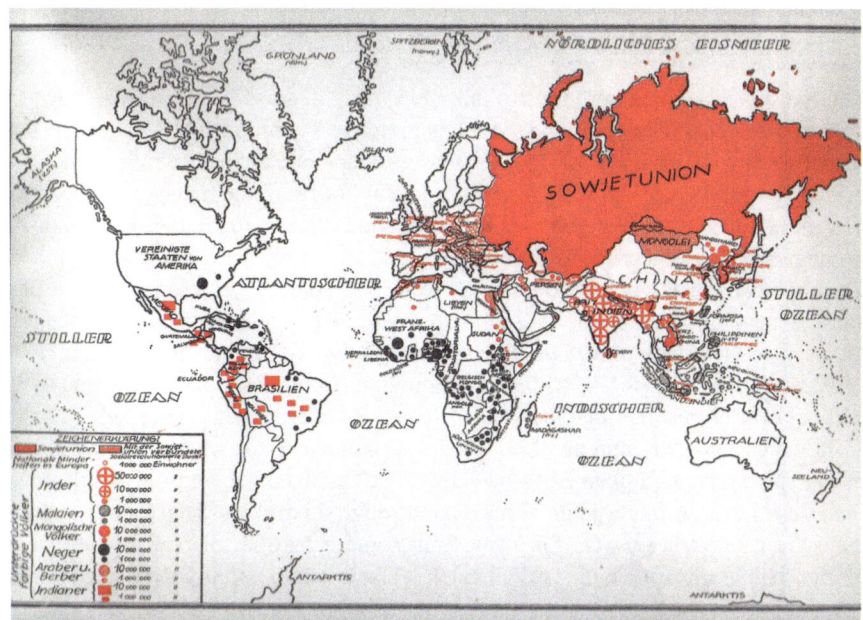

**Fig. 5** 'Die unterdrückten Völker des Welt' (Oppressed People of the World), in: Radó S, *Atlas für Politik, Wirtschaft, Arbeiterbewegung*, 1930, p. 163

The second issue was to present Russia as a peaceful State, threatened by the imperialist powers that surrounded it. With Radó for instance, a world map of weapons depicted over armed neighbours, when Russian military forces were expressed in contrast by a few pictograms (Radó 1930: 29). Thus, the Soviet Union appeared as a citadel under siege. Another map developed a thematic of encirclement by imperialist Great Britain, with a set of arrows, or axis of deployment (*Aufmarschlinien*) against Soviet Union (Radó 1930: 91). However, this view of Soviet Union was partly offset by another one: that of an emerging great power. Several maps stressed the vastness of the country, when many comments were nothing less than pure propaganda, celebrating the demographic potential, the economic development and the successful planning.

Faced with Empires often shown as composite, Russia appeared a homogeneous territory, where the problem of nationalities had been solved. In Radó's chapter on nationalities and economic issues, the Soviet Union was only depicted with administrative maps, including a general one, entitled: 'The solution of the national question in Soviet Union' (*Die Lösung der Nationalen Frage in der Sowjetunion*, Radó 1930: 157). The federal structure and the existence of autonomous towns and republics, 'a political experiment at a gigantic scale' (Horrabin 1934: 113), seemed to address all of the issues. In Radó's first atlas, Russia was always displayed with a solid red, and the problems of oppression and minorities seemed to stop short at its borders (Fig. 5).

## 5   The Language of Persuasive Cartography

The maps we examine here mostly fall under the category of 'persuasive maps', a now widely accepted expression, which is preferred to 'propaganda' or 'suggestive maps' (Tyner 1974, 1982, 2015). Persuasive maps are defined as maps intended primarily to influence opinions or beliefs, to send or reinforce messages rather than to communicate objective geographic information (Tyner 2015: 1087). Obviously, no map is solely objective, as it always depends on choices made by an author, of geographical objects or phenomena, as well as graphic signs. Conversely, a persuasive map is not without a descriptive, neutral or 'scientific' content. Today, rather than a strict opposition objective/subjective or science/propaganda, it is considered that maps fall within a continuum from descriptive to persuasive, without ever corresponding purely to one category or the other (Tyner 1982, 2015).

In the case of Horrabin and Radó, we are obviously close of the persuasive end of the spectrum: both claimed to be selective and subjective, and did not disguise their desire to transmit a political message through their maps. That being said, the maps from our atlases were far away from some categories of persuasive cartography, such as allegorical or satirical maps. Their external appearance was classical: they included title, legend and usually (but not systematically) a scale. They made use of abstract symbols or pictograms, and never included any anthropomorphic or zoomorphic figure. What, then, were their 'persuasive' characteristics, both in content and form, when compared with usual geopolitical maps of the period (Boria 2008; Muehlenhaus 2013; Herb 2015)?

First, the persuasive nature was reflected in the bias introduced by the comment texts, the titles, the legends of the maps. Most of the time, the map reading was guided towards a specific meaning, by mostly interpretative texts, that clearly reflected the author's ideological positions. Thus, Radó recalled with some pride a commentary made by Harold Nicholson in the *Daily Telegraph* about his second atlas: 'The author does not say a word about his political opinions, but everything he says bears witness to his ideological convictions' (quoted in Radó 1972: 75).We have already mentioned, for instance, the warlike tone of the vocabulary. Examples could be multiplied, such as the world map of weapons entitled 'Preparing for next war' (Radó 1930: 28–29) or a map of maritime routes in the Mediterranean interpreted as showing 'conflicting interests' (Horrabin 1934: 52).

On the content level, one can also point out simplifications that are similar to manipulations of information. In the *Plebs Atlas* for example, as pinpointed by Jeremy Black (Black 2015: 154–155), Horrabin drew *The New Map of Europe* to show 'the workings of imperialism' which were said to be 'as clearly traceable in Europe as in the other continents' (Horrabin 1926: 23). The map depicted in black the 'BRITISH possessions and "colonies"' (Horrabin 1926: 23, quotation marks from Horrabin), a category that included Portugal, Greece, Denmark, Holland, Norway, Finland and the Baltic States. In his 1930 atlas, Radó proposed a very similar map, *Europa 1929* (Radó 1930: 21), replacing the 'possessions and colonies' with a more skillful 'Influence and interests areas' (*Einfluss und Interessen Gebiete*).

In terms of graphism, the analysis requires us to distinguish between the atlases of the corpus. Radó produced two atlases very different of style. Regarding the first one, the *Atlas für Politik* [...], the maps were prepared in cooperation with Karl Metzler, a graphic artist who was a member of the German communist party (KPD). This atlas was the only one printed in colour. Actually, forty-five maps out of 120 were in black and white, seventy used red in addition, and only five red and blue. This atlas was the closest work to a propaganda visual discourse. It made use of violent contrasts, notably to oppose the Soviet Union, in red, to other powers, or to enhance Empires and their colonies. The graphic design 'screamed' at the reader, with large swatches of deep black, heavy-faced letters, thick lines or symbols. The visual manipulation blatantly appeared with the use of a Mercator projection to exaggerate the surface of Russia on several world maps. A few years later, a geographer close to the National Socialists, Max Eckert-Greifendorff, denounced this attempt to make the Soviet territory seem even more powerful, 'and so on to externally demonstrate and document the overwhelming power of Bolshevism on earth' (Eckert-Greifendorff 1939: 340, quoted in Schlögel 2003).

Radó's second atlas looked very different from the previous. It benefitted from the collaboration of a professional cartographer, Marta Rajchman, a young woman who had been trained at the School of Cartography of the Paris Sorbonne University, in the same year as Jacques Bertin, the famous French theoretician of semiology of graphics. The subjective or ideological dimension lay above all in the structure of the book, its themes, comment texts and titles, with a focus remaining on confrontations and problems. However, the tone had become more sober and composed. The presentation returned to a more neutral and scientific type, with dense comments on each map, enriched by statistical tables. The map design still relied on contrasts, for example to depict the spatial extension of empires, but it was generally speaking fine and accurate. It used in a unique modern way all the possibilities of the black and white, in particular for statistical maps that combined variations of size with different patches or patterns. At last, there were a few examples, in this atlas and the previous one, of maps that have probably been influenced by Neurath's ISOTYPE method, with statistical data expressed by a range of geometrical or pictorial symbols.

Horrabin's atlases appeared much more homogeneous with regard to their design. The British journalist was the unique author and designer of his maps, most of which he signed in the lower corner with his monogram, JFH. His style can be described as sketchy or diagrammatic. His maps, all in black and white, were very simplified and clearly legible. Horrabin probably took into account the constraint of being published in a small size, in newspapers or atlases in paperback format. He also claimed this simplicity, in a way which can recall modern discourses on map communication: 'I have aimed at leaving out everything non-essential to the illustration of a particular point. I am a firm believer in the theory that a map should be designed to make some one point clear—and other points be left to other maps. Not only elementary students, but older readers, are befogged by the wealth of detail, all of it emphasized equally, in an ordinary map' (Horrabin 1926: 3–4). He further explained that he had to draw several maps of China, in order to avoid compressing facts into a single map.

Horrabin used solid black with parsimony. He preferred patterns of etchings or points, or various types of dashes to symbolize linear objects. His maps were more qualitative than those by Radó. Economic or demographic data were often limited to locations: 'coalfield', 'Wool area within circle', 'Areas of densest population' and so on. Horrabin never drew a flow map in any of the atlases we examine here, and only two choropleth maps, both in the 1934 atlas, with a limited number of classes, three to show the levels of densities in Belgium (Horrabin 1934: 50), and five for the percentage of black population in the United States (Horrabin 1934: 74).

Furthermore, Horrabin's cartographic style was characterized by the singular use of various graphic devices to assist the reading that we would now call didactic aids. One of them was the use of arrows pointing at specific locations, whose names were inscribed in a rectangular box. Important places, lines or areas are thus highlighted as hot spots of the map. It may be tempting to establish a link between this practice and the experience of Horrabin as a cartoonist: he drew from 1919 to 1951 a strip for British newspapers, 'The adventures of the Noah family', later entitled 'Japhet and Happy'. In his atlases, the maps seemed to speak out with 'balloons', like a comic strip character. Another interesting feature was the use of double arrow lines with indications of distances, rather than showing a scale bar. It allowed bringing out phenomena of remoteness or proximity from a strategic location, such as a strait. Lastly, Horrabin strove to facilitate comparisons, whether demographic or spatial. In his analysis of empires, he often enriched the maps with histograms (bar charts) to compare population values. He also multiplied inset maps, showing at the same scale a home country and a colony, and in a more dramatic way, he sometimes superimposed their outlines (Fig. 6).

**Fig. 6** 'India (5)', in: Horrabin JF, *An Atlas of Empire*, 1937, p 86

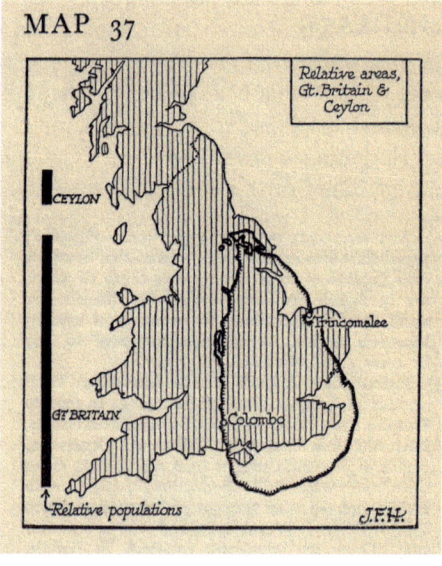

The link between our atlases and other forms of persuasive maps produced in the interwar period, notably by German geopoliticians, has to be qualified. Obviously, Horrabin and Radó's atlases cannot be understood independently of this context. Geopolitics was referred to explicitly in Rothstein's introduction to the *Atlas für Politik* [...], and Radó knew well the literature in the field, as can be inferred from his memoirs, which mentioned Karl Haushofer and his review, *Zeitschrift für Geopolitik* (Radó 1972: 75). Regarding Horrabin, the similarity of his maps with those made in Germany was underlined in his time (Crone 1934: 273). Horrabin and Radó shared this philosophy to simplify information and design, in order to convey a clear and tendentious message. However, they did not go as far as the German geopoliticians in their design practices. They almost never used the map to project a prospective scenario, as a surface of experimentation, even if highlighting the hot spots could appear as a 'conjugation in a future tense' (Raffestin et al 1995: 250) of space. In the same way, the superimposition of geometric shapes (circles, triangles) and the shift towards a form of spatial modelling are not found in their maps. Horrabin and Radó leaned toward graphic simplification, but they never developed the geometrization of space which was a hallmark of German geopolitics. If there is a lowest common denominator of these different mapping practices, it is the dynamic view and the use of the arrow as an interpretative sign, charged with a range of meanings: sticking point and danger area, aims, pressure, territorial ambition, axis of expansion, axis of deployment.

# 6  Conclusion: An Unorthodox Marxism

Horrabin's and Radó's atlases are a rarity and a curiosity in interwar persuasive cartography, largely dominated by right or extreme-right maps, attempting to spread nationalist and revisionist ideas. Their approach broke with the usual promotion of the interests of a particular country: it was more comprehensive, tinged with universalistic claims. In its own way, each atlas brought out global discrepancies, between ethnic dividing lines and post-war borders, or between economic ties and the world political divisions. Admittedly, it is true that this leaning toward internationalism went along with an undisguised support for the Soviet Union, the great proletarian power.

In their comments, the authors lifted the banner of Marxism-Leninism, and in fact we have shown that they put forward subjects linked with this political philosophy, giving for instance a key-role to economics, or placing imperialism at the heart of their analysis of international relations. However, our corpus of atlases conveyed a much-biased version of Marxism. Above all, maps show spatial objects, when the keystone of Marxism is the materialist conception of history. History had only a small presence in the atlases. Horrabin only tackled a very immediate history, as the consequences of the Treaty of Versailles, or chronological steps of some colonial conquests. Radó proved slightly more orthodox, as he made a few historical world maps that depicted the steps toward building imperialism from the late nineteenth century. Even so, the approach he generally took was to emphasize geopolitical news, far from a long-term analysis. Another notable absence in the atlases was modes of production and social issues. They put forward the struggle between great powers but neglected the struggle of classes. China was the only case which allowed evoking a form of social revolution. Horrabin drew a map of areas controlled by peasant soviets in China (Horrabin 1934: 104), but his commentary showed his uncertainty as to their communist inspiration. When Radó provided a map on the same topic (Radó 1938: 78), the China Soviet government had been dissolved and the Chinese red army had been placed under the authority of the nationalist government of Nanking.

The atlases we examined here are often considered as precursors in the history of critical cartography. They indeed responded to Lenin's wish to arouse a form of counter-cartography in front of 'bourgeois' representations. However, if they had some Marxist features, their authors were not theoreticians and they stood away from an orthodox Marxist analysis. This is consistent with remarks made by the German sociologist Karl Wittfogel in 1929, on Horrabin's economic manual (Hepple 1999). The cartographic representation invariably leads to give primacy to geographical dimension: positions, extent, distance and communications. Finally, the originality of these atlases probably lies in the connection they made between politics and economics, and in the way they anticipated a globalized world. One can ask whether a social cartography would have been possible at the time, but when it existed, it was made at a much larger scale, primarily for cities, and aggregate statistics were missing. However, these atlases may be associated with several other attempts, at the same period, to introduce a spatial dimension which was lacking in Marxism (Bassin 1996).

**Acknowledgements** The author wishes to thank Elvira Tjepkema and Michael Friendly for their assistance with the final English editing of the article.

# References

Barth BR (2010) Egy térképész illegalitásban: tények és legendák nyugati és keleti titkosszolgálati archivumokból (=A cartographer in illegitimacy: facts and legends from Western and Eastern secret service archives). In: Hegedüs Á, Suba J (eds) Tanulmányok Radó Sándorról. A Budapesten 2009. nov. 4-5-én rendezet konferencia elöadásainak szerkesztett anyaga (=Studies on Alexander Radó. Edited versions of the lectures at the scientific conference held in Budapest on 4–5 November 2009). HM Hadtörténeti Intézet és Múzeum, Budapest
Bassin M (1996) Nature, geopolitics and Marxism: ecological contestations in Weimar Germany. Trans Inst Br Geogr 21(2):315–341
Bithell A (1984a) The work of J.F. Horrabin (1884–1962) in the field of cartography and diagram. Dissertation, University of Reading
Bithell A (1984b) The maps and diagrams of J.F. Horrabin. Bull Soc Univ Cartogr 18(2):85–91
Black J (2015) Geopolitics and the quest for dominance. Indiana University Press, Bloomington
Boria E (2008) Geopolitical maps: a sketch history of a neglected trend in cartography. Geopolitics 13(2):278–308

Boria E (2014) Storia avventurosa di Alexander Radó, cartografo e spia. Gnosis. Rivista italiana di intelligence 4:142–151

Bourgeois G (2011) Sándor Radó, géographe et agent de renseignement. Hérodote 140:9–29

Capdepuy V (2011) La guerre globale enseigne la cartographie globale. In: Histoire globale. Le blog. http://blogs.histoireglobale.com/la-guerre-globale-enseigne-la-cartographie-globale_1188. Accessed 25 October 2018

Chaliand G, Rageau JP (1983) Atlas stratégique. Géopolitique des rapports de forces dans le monde, Fayard, Paris

Crone GR (1934) Reviews. An atlas of current affairs by J.F. Horrabin. Geogr J 84(3):272–273

Eckert-Greifendorff M (1939) Kartographie. Ihre Aufgaben und Bedeutung für die Kultur der Gegenwart, Walter de Gruyter, Berlin

Harley JB (1988) Maps, knowledge and power. In: Cosgrove D, Daniels S (eds) The iconography of landscape: essays on the symbolic representation, design and use of past environments. Cambridge University Press, Cambridge, pp 277–312

Harley JB (1989) Deconstructing the map. Cartographica 26(2):1–20

Heffernan M, Győri R (2014) Sandor Radó (1899–1981). In: Withers C, Lorimer H (eds) Geographers: bio-bibliographical Studies 33. Bloomsbury Academic, London and New York, pp 167–202

Henrikson A (1975) The map as an 'idea': the role of cartographic imagery during the Second World War. Am Cartogr 2:19–53

Hepple L (1999) Socialist geography in England: J.F. Horrabin and a worker's economic and political geography. Antipode 31(1):80–109

Herb G (1997) Under the map of Germany: nationalism and propaganda, 1918–1945. Routledge, London and New York

Herb G (2015) Geopolitics and cartography. In: Monmonier M (ed) The History of Cartography, vol 6(1): cartography in the twentieth century. University of Chicago Press, Chicago, pp 539–548

Horrabin JF (1923) An outline of economic geography. The Plebs League, London

Horrabin JF (1926) The Plebs atlas. The Plebs League, London

Horrabin JF (1934) An atlas of current affairs. Victor Gollancz, London

Horrabin JF (1937) An atlas of empire. Victor Gollancz, London

Kidron M, Segal R (1981) The state of the world atlas. Heinemann, London

Muehlenhaus I (2013) The design and composition of persuasive maps. Cartogr Geogr Inf Sci 40 (5):401–414

Neurath O (1936) International picture language. The Orthological Institute, London

Radó A (1930) Atlas für Politik, Wirtschaft, Arbeiterbewegung. 1. Der Imperialismus. Verlag für Literatur und Politik, Wien and Berlin

Radó A (1938) Atlas of to-day and to-morrow. Victor Gollancz, London

Radó S (1971) Dóra jelenti. Kossuth Könyvkiadó, Budapest

Radó S (1972) Sous le pseudonyme Dora. Julliard, Paris

Radó S (1980) Atlas für Politik, Wirtschaft, Arbeiterbewegung: 1. Der Imperialismus. VEB H, Haack, Gotha

Raffestin C, Lopreno D, Pasteur Y (1995) Géopolitique et histoire. Payot, Lausanne

Rivière P (2016) L'atlas d'Alex Radó, cartographe, Léniniste et maître-espion. In: Visionscarto. https://visionscarto.net/alex-Radó-cartographe-et-espion. Last accessed 6 November 2018

Schlögel K (2003) Im Raume lesen wir die Zeit. Carl Hanser Verlag, München, Über Zivilisationsgeschichte und Geopolitik

Schneider U (2006) Kartographie als Imperial Raumgestaltung. Alexander (S.) Rados Karten und Atlanten. In: Zeithistorische Forschungen/Studies in contemporary history, vol 3, no 1. http://www.zeithistorische-forschungen.de/1-2006/id=4687. Accessed 15 October 2018

Schulten S (2001) The geographical imagination in America, 1880–1950. University of Chicago Press, Chicago

Thomas L (1968) Alexander Radó. Stud Intell 12(3):41–61. https://www.cia.gov/library/center-for-the-study-of-intelligence/kent-csi/vol12i3. Last accessed 13 February 2019

Tyner J (1974) Persuasive cartography: an examination of the map as a subjective tool of communication. Dissertation, University of California, Los Angeles

Tyner J (1982) Persuasive cartography. J Geogr 81:140–144

Tyner J (2015) Persuasive cartography. In: Monmonier M (Ed.) The history of cartography, vol. 6(2): cartography in the twentieth century. University of Chicago Press, Chicago, pp 1087–1095

**Gilles Palsky** is Professor of Geography at Paris 1 Panthéon-Sorbonne University and he is a member of the research unit *Epistemology and History of Geography*. He chaired the History of Cartography Commission of the French Committee of Cartography (1999–2007) and was a trustee of the *International Society for the History of the Map* (2013–2017). His research revolves around the role of images in the building of geographical knowledge and the development of thematic mapping, nineteenth–twenty-first century. He is also engaged in theoretical issues in cartography: visualization of spatial dynamics, participative mapping, and semiology of graphics.

# Empire as Spectacle: *Harmsworth's Atlas of the World and Pictorial Gazetteer with an Atlas of the Great War*

Peter Vujakovic

**Abstract** *Harmsworth's Atlas of the World and Pictorial Gazetteer with an Atlas of the Great War*, published c.1920, is an extravagant and extremely detailed work of geography and cartography. The world is laid before its readership as a spectacle involving four hundred and eighty-five coloured maps, and over three and a half thousand photographs of peoples and places. It celebrates the British Empire, but also explores the significant penetration of competitor powers world-wide. The atlas is examined from a world-systems perspective, which recognizes inter-regional and transnational division of the world into core, periphery and semi-periphery. The information contained in the atlas is consistent with the understanding needed for restructuring of the British economy following the end of the 'Edwardian Boom' and the devastation of the 'Great War', prior to which the two main lead regions in contention with Great Britain were Germany and the USA. Germany's pre-war 'peaceful penetration' of areas of British interest are dissected forensically in the section on the 'Great War'. The world-wide development of communications infrastructure, as well as industries, is covered in great detail in the atlas; which could be described as a 'road map' for reassertion of British hegemony. The paper draws cultural geography to explore the way in which the atlas interpreted the world for its readership. It examines the visual and textual means by which an 'imperial gaze' is constructed as a set of representations that reinforce the assumed superiority of the Anglo-Saxon world. Distant places and peoples are subordinated to this gaze, and their geographies constructed according to a grand imperial vision.

## 1 Introduction: Girdle the Earth

Puck: 'I'll put a girdle round about the earth in forty minutes'

Act II, Scene I, Shakespeare's *A Midsummer-night's Dream*

P. Vujakovic (✉)
Canterbury Christ Church University, Canterbury, UK
e-mail: p.vujakovic@canterbury.ac.uk

© Springer Nature Switzerland AG 2020
A. J. Kent et al. (eds.), *Mapping Empires: Colonial Cartographies of Land and Sea*,
Lecture Notes in Geoinformation and Cartography,
https://doi.org/10.1007/978-3-030-23447-8_10

177

To '*girdle the earth*' and see its wonders is the ultimate dream for many people. To travel to exotic places has been the definitive spectacle from well before the age of global air travel. Today many people are able to follow the dream; although the majority remains earth-bound—bound by poverty and lack of opportunity. The latter, sadly, have been, and remain part of the spectacle itself, the object of the 'tourist gaze' (Urry 2002). Tourists visiting sites from Mexico City to the banks of the Ganges gaze on the 'other'—a form of neo-imperialism? For those with 'limited means', access to the web provides a vicarious solution. The world is your oyster, to be 'shucked', even if only via a computer screen. This chapter examines an example of vicarious travel in the age of rail and steam liners, through the printed word, maps and photographs of an atlas. It explores the notion of the atlas as *spectacle* in relation to the role of one specific example—*Harmsworth's Atlas of the World and Pictorial Gazetteer with an Atlas of the Great War* (hereafter referred to as the *Harmsworth's Atlas*, published c.1920).[1] This examination draws on the concept of 'the map as spectacle' (see Vujakovic 2017 for a detailed discussion) in maintaining the myth of British Empire and greatness following the 'Great War'.

Today spectacle *is everywhere/everything*; Western society is saturated in spectacle. Guy Debord's notion of 'the society of spectacle', on which this work draws, probably far exceeds what even he had imagined when he promoted the concept (Debord 2009; originally published as *La société du spectacle* in 1967). Debord's basic premise is that many people now live in societies in which the majority of the population is captured/captivated by spectacle (this includes total-itarian as well as advanced capitalist societies). After completing their work as producers, most people become complaint consumers of spectacle, limiting their real engagement in social and political life:

> In societies dominated by modern conditions of production, life is presented as an immense accumulation of spectacles. Everything that was directly lived has receded into a repre-sentation. (Debord 2009: 24)

People become so wrapped up in life as spectacle that it 'monopolizes the majority of the time spent outside the production process' (Debord 2009: 25). The society of spectacle is the progeny of state and corporate power maintained through the output of the news media, official institutions, and the *educational*, advertising and entertainment industries. Central to this conception of society is the idea that compliance and alienation is a product of socio-economic organization that achieved its climax in capitalism and advanced technologies, but which can be traced to the early modern period and the massive socio-political and economic changes that were occurring at that time. Debord acknowledged the role of the Baroque as a form of control based on spectacle and an important historical turning point (Roberts 2003)—see below. Debord died in 1994, well before his concept

---

[1]The Bodleian Library's copy has two date stamps: on the reverse of the orange title page, 21. JUN.1919; and the blank page preceding the Editorial, 29.NOV.1920. The bottom of the orange cover says 'Part 2 of "Harmsworth's New Atlas" out on Friday, July 4th.' The date of 4 July fell on a Friday in 1919.

seems to be reaching its zenith in an information-technology driven world in which advanced economies appear to be populated by iGadget zombies, entrapped in their bubbles of spectacle, oblivious to their surroundings even when walking the streets. As Callens notes, the mass media, and more recently the 'personal computer', have turned from being a means of facilitating the exchange of information to become a 'self-enforcing dominant cultural environment' in which the mass of people have become immersed (Callens 2000: 291). *Harmsworth's Atlas* (along with other products of the mass-print age), is a precursor to this total immersion.

As well as the concept of spectacle, the atlas is also examined from a world-systems perspective, an approach that recognizes the inter-regional and transnational division of the world into core, periphery and semi-periphery; consistent with both imperial and neo-imperialist forms of spatial organization. The world system is dynamic, in part as a result of revolutions in technology, especially military, industrial, and transportation and communications, and individual states may change status depending on their ability to adapt. At the time of publication of the atlas, Britain was struggling to retain its place as the global hegemonic state. Prior to the Great War, Germany and the USA were potential rivals, the war put paid to German ambitions, but left the USA a major contender. Britain did retain an empire and its position within the core, but needed to assert its position both abroad and at home. Britain along with other core states continued to be the command centres of the world economy, providing high skill, capital-intensive production, while the rest of the world provided low-skill workers for primary production and extraction industries. It is this position that is 'celebrated' by *Harmsworth's Atlas*, despite the obvious need for restructuring of the economy and society following the end of the 'Edwardian Boom' and the impacts of the Great War. The world-wide development of communications infrastructure, as well as industries, is covered in great detail in the atlas; which could be described as a 'road map' for reassertion of British hegemony.

The flip-side to the society of spectacle is the spectacle of threat. While compliance can be gained through immersive consumerism, societal anxieties need to be dealt with too. Debord noted how spectacle can also be mobilized to create compliance through fear. This is particularly pertinent with regard to the so-called 'war on terror', but it should also be remembered that the publication of *Harmsworth's Atlas* immediately followed not just the Great War, but the international reign of terror from 1887 to 1914. Regarded as a reaction to failing states within the global system, anarchist and nationalist inspired terror attacks and assassinations characterized the period prior to the war, and the assassination of Archduke Ferdinand of Austria-Hungary was the specific pretext for the events that led to the Great War. As Debord states:

> [...] democracy constructs its own inconceivable foe, terrorism. Its wish is to be judged by its enemies rather than its results. The story of terrorism is written by the state [...]. The spectators must certainly never know everything about terrorism, but they must always know enough to convince them, that compared with terrorism, everything else must be acceptable or in any case more rational and democratic. (Debord 1998: 24)

Threat comes in many forms, and it is the threat from other states that informs the atlas. Germany's pre-war 'peaceful penetration' of areas of British interest and actions in the war are dissected forensically in the section in the 'Atlas of the Great War', which forms a part of the larger atlas (see discussion below). Germany is treated as the terror threat with, for example, some very graphic images of hangings in Germany's colonies exemplifying this; in strong contrast to Britain's imperial civilizing role.

In Vujakovic's (2017) argument for 'the map as spectacle', the role of the 'Baroque' as a form of control is also explored. While it is not possible to discuss this fully, the following key points are relevant. As argued by Jose Maravall, the Baroque was a phenomenon ('historical structure') that emerged to deal with the crises and contradictions of early modern Europe. In his *Culture of the Baroque* (1986; originally published as *La Cultura del Barroco*, in 1975), Maravall makes it clear that the Baroque is not simply a concept of aesthetic 'style', but is a structure that played a role in the repression of feelings of threat amongst the masses. The Baroque emerged to control populations facing crises of economic, political and social insecurity. The Baroque project was a web of control, in which the elites mobilized the technologies of *modernity* to maintain their *traditional* position. It is this latter point that is significant here, the use of technologies of modernity to *maintain* the *status quo*. It can be argued that Baroque structures and approaches continue to be mobilized, as elites continue to dominate the technologies of spectacle; print, cinema, radio and television, and now, the networked media. Most cartography is ultimately a neo-Baroque form of production; (pre)serving elite positions, by constantly feeding on technical innovation to perform acts of spectacle. Mary Kaldor puts this well in terms of her analysis of the military-industrial complex; 'Baroque technological change represents 'improvements' to successive weapons systems which can pass through phases of invention, innovation, and integration without disturbing the social organization of the users' (Kaldor 1986: 591). It would be easy to replace, in the last sentence, the words 'weapons systems' with 'cartographic modalities' to recognize cartography as an inherently Baroque process. Recent anti-progressivist narratives of history of cartography, while genuinely seeking to address the significance of pre-modern and non-Western mapping (see for example, Edney 2018), may be in danger of ignoring the fact that cartography does 'progress' in technical terms and this can serve interests. It is a small step from education to the spectacle of 'infotainment'.

## 2 The Atlas as Spectacle

*Harmsworth's Atlas* is a sumptuous production, created using state-of-the-art process-engraving and photo-etching. The editorial team made bold claims concerning the fact that the atlas was 'up-to-date' and its maps contains 'more recent information than those of any atlas that has yet appeared' since the changes wrought by the Great War. The atlas contains 485 maps in full-colour (with an estimated

250,000 place names shown) and 3540 'photographic views' in black and white. The editorial team was particularly proud of the latter:

> But perhaps the most unusual feature of HARMSWORTH'S NEW ATLAS, and that which distinguishes it at once from all other atlases, is its pictorial Gazetteer. For the first time in history of atlas-making, the camera has been used to supplement the work of the draughtsman and the pen of the gazetteer writer [...] the latest authoritative information [is] enriched with photographic views of scenes in all parts of the world (Editorial, unpaginated)

The atlas is intrinsically Baroque—a *theatrical* staging of the continuity of imperial glory using the most up-to-date technologies of cartography and publishing, and as shall been seen, celebrating the industrial advances of the core. It opens with a series of essays (pp. i–xii), starting with 'The Story of Geography—How men came to know the World' by the Associate Editor BC Wallis, and including two essays on map making and map study respectively. Of more relevance to this examination of the atlas are the essay by Sir Sidney Low 'Geography the Key to Knowledge' (pp. iii–iv), and 'The fascination of Travel' (pp. xi–xii) by Hamilton Fyfe ('Special Correspondent of the "Daily Mail" in all parts of the World' p. xi), as both dwell on the world as spectacle.

Low's opening comments are, however, worth noting at the onset. While recognizing the terrible nature of the Great War, he is aware of the growth in both geographic knowledge and curiosity it has engendered, and the significance of geography to education in general:

> It was, I think, John Bright [British statesman and promoter of free-trade] who said that the only really valuable result of war was that it taught geography. The Great War has certainly given us a little more geographical knowledge than we possessed before [...] The NEW ATLAS will be an indispensable handbook for the newspaper-reader and the political student. It will also be accepted, I hope, with satisfaction by those who are engaged in educational work [...] for, rightly used, it [geography] is the key to nearly all knowledge, the surest and soundest foundation on which to build teaching in natural science, physics, history, economics, politics, and even literature. (p. iii)

The majority of Low's essay concerns what he sees as the foundational nature of geography in the school curriculum, culminating in an appreciation of the British Empire and its global reach:

> Special attention would be paid to the situation and circumstances of the British Empire. The pupil would see it for himself – on the globe; he would notice how it lies spread and scattered over the surface of the earth, with all its parts linked up by the silent highways of the sea.

Both Low and Fyfe's essays mobilize the imperial gaze, treating others as a form of imperial education. The following sections explore in detail the ways in which the atlas offers up the empire and others as spectacle. As Edney notes:

> Cartography [is] a human endeavour, bound up with multiple spatial discourses and (other) mapping practices. The careful analysis of those discourses and practices offers the opportunity to develop a nuanced understanding of cartography that is socially aware and culturally open. (Edney 2018: 74)

The spatial narrative exposed in the atlas is one involving issues of race, gender and class that shaped the empire (McClintock 1995). The atlas resonates with a male and Eurocentric gaze, from the opening essays to the display of ethnic types and the labelling of the exotic female form.

The narrative developed in the atlas is not novel. Two decades before the *Harmsworth's Atlas* was published, Rudyard Kipling's eponymous hero Kim had entered the public imagination. Probably Kipling's most important novel, *Kim*, was published in serial form in 1900 (McClure's Magazine), and as a book in 1901 (Macmillan and Co.). The book epitomizes the imperial gaze. Through Kim's adventures along the Grand Trunk Road, and beyond, we experience the spectacle of India and its peoples. As Edward Said notes, Kipling is one of the few novelists, along with Joseph Conrad, to 'have rendered the experience of empire as the main subject of his work with such force' (Said 1993: 160). Kipling creates an image of India as a 'timeless, unchanging, and 'essential' locale, a place almost as much poetic as it is actual in geographical concreteness' (Said 1993: 162), a viewpoint that grows from belief in an immense colonial system whose entire history and structures may be regarded as 'facts of nature'. Spectacle is essential to the *status quo*. It creates in its audiences the absolute compliance needed for elites to rule with some degree of order and certitude.

After the shock of the Great War, this imperial spectacle needs to be re-mobilized, to reinstate a sense of security—'God is in his heaven and all's right with the world'. It is the ability to generate this sense of surety, '*to create a world almost as much poetic as it is actual in geographical concreteness*' that is required (to adapt Said's words) that the atlas works.

The remainder of this chapter explores three different aspects of the atlas as spectacle, also drawing on the world systems approach; first, the display of Britain as a core industrial and imperial power, including its relationship with its Dominions; second, the rest of the empire and the wider world as subject of the ethnographic gaze, and third, reflections on Germany as a rising power and competitor;

## *2.1  Spectacle I—Britain and the Dominions*

The following extract from AS Byatt's novel, *The Children's Book*, set in the Edwardian era, clearly sums up the concept of the British Empire as a spectacular and dynamic system of interlinked spaces tied to the metropole by trade and information. Ships, rail and telegraph cables carry both goods *and* knowledge, and knowledge is power:

> Geraint saw the turning globe in his mind's eye, with its vast red imperial patches, its shifting frontiers, criss-crossed by the invisible threads of the telegrams and the visible furrows of the great iron ships forging steadily through flying foam and mirror calm seas. (Byatt 2010: 281)

This *dynamism*, focused on transport and communications, is major theme in the maps and the pictorial gazetteer of the *Harmsworth's Atlas,* especially as these pertain to the British Isles and the Dominions. Sir Sidney Low in his essay for the *atlas* 'Geography as the Key to Knowledge' (pp. iii–iv) clearly sees this as part of an essential school education.

> Let them look long and often at the big school-room globe, let them make it revolve on its axis with their hands, and navigate its blue oceans with their fingers. (p. iv)

The first four pages relating to 'The British Isles' is entirely dedicated to 'Railways', and includes twenty-six photographs of stations, bridges, tracks, 'signals and points' (York, p. 35), before turning to physical geography of Britain, the usual starting point of many geographies (and this is dealt with in only four pages). Rail embodies the late Victorian and Edwardian concern with speed, as prefigured by images such as Turner's painting *Rain, Steam and Speed—The Great Western Railway*, of 1844, that was already impressing this image on the public consciousness. Speed, as epitomized by modern transport and communications systems, was a key element of imperial control. Rail and steel had fed the 'mid-Victorian boom', steam ships and electricity underpinned the 'Edwardian boom'.

In the section of the atlas on 'England and Wales', there are a massive number of photographs, one hundred and fifty-six in total. These are dominated by buildings, the majority representing heritage (e.g. cathedrals and castles), but they also celebrate Britain's industrial prowess, for example, the railway and road bridges at Middlesbrough, Widnes, Newcastle (twice), Nottingham, Blaenau Festiniog, Barmouth, Pontypridd, St. Austell, and Looe, and the celebrated suspension bridges at Conway Castle, Menai, and Clifton. The section on London starts with three pages (of only seven in total) dedicated to its railways.

In all the pages dedicated to Britain in general, and England and Wales specifically, people are simply a distant and incidental element of a few scenes; there is no anthropological focus. The first time a person is encountered 'as a person' is thirteen pages into the section on Scotland. Now exploring the Celtic fringe of the Highlands and Islands the anthropological gaze is applied to the peoples of Stornoway (*gutting the herrings*), Harris (*wool spinning*) (Fig. 1), South Uist (*bringing home peats)* and Skye (*planting potatoes*). For the most part lowland Scotland and Ireland (pre-Free State) are treated like England and Wales. While perhaps surprising today, with Scotland included in popular notions of the Celtic nations, it would not have surprised contemporary readers that there was a clear differential between the treatment of lowland Scots and their Celtic neighbours. As Kidd makes clear in his paper 'Race, Empire, and the Limits of Nineteenth-Century Scottish Identity', lowland Scots did not identify as Celts, but 'vociferously projected a Teutonic racial identity' (Kidd 2003: 873). This even went as far as advocacy of the benefits of an Anglo-Saxon racial union, to include the USA (e.g. Stewart Lygon Murray's *The Peace of the Anglo-Saxons* (1905), cited in Kidd (2003: 889). Only in the last four pages concerning Britain (of seventy-one pages in total) are any other individuals

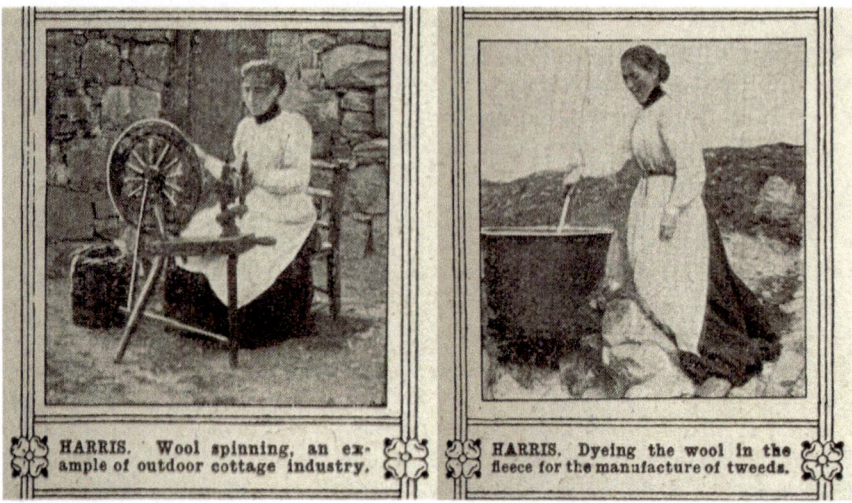

HARRIS. Wool spinning, an example of outdoor cottage industry.

HARRIS. Dyeing the wool in the
fleece for the manufacture of tweeds.

**Fig. 1** Detail from 'Scotland (Section IV)' (p. 91), showing wool-spinning and dyeing of wool for Harris Tweed

shown in close up, and these all representative of workers in key industrial trades, barring perhaps one 'folk image' of hop pickers in Kent.

Turning to the Dominions, the relationship with the metropole (the core) is clearly established; the role of the periphery as supplying primary products is recognized at the very start:

> The Australasians are chiefly interested in the great primary industries connected with the farm, the mine and the forest. Their main industrial purpose is to supply foodstuffs and raw materials to the British Empire and the rest of the world. (p. 339)

The first page consists entirely of a description (with photographs) of farming, forestry and mining. The following double page map focuses on the distribution of minerals, agricultural produce, but also railways, cables and 'steamship routes' (pp. 340–341). This relationship between core and periphery is already well established within British society, as, for example, exemplified by the well-known 'Imperial Federation Map of the World Showing the Extent of the British Empire in 1886' by Walter Crane (Fig. 2). Here, in the marginalia, the male white 'Australasian' rests on his spade, while a female colonist offers up a sheep skin (in contrast an 'aboriginal' woman holds aloft a boomerang and grasps a kangaroo by the ear!).

Of the 156 photographs dedicated to Australia, only one shows a close-up of a European settler, and one of 'aboriginals' (painted for a ceremony). With the exception of a few rural or wilderness scenes, the vast majority are industrial or urban, the latter displaying important civic or ecclesiastical buildings. This is colonial pride on display. New Zealand is treated very similarly, but now with several anthropological images of Maori people. As the reader's gaze is directed to

RATION:-MAP OF THE WORLD SHOWING THE EXTENT OF THE BRITISH EMPIRE IN 1886.

**Fig. 2** Detail from 'Imperial Federation' British Empire Map, 1886

more exotic, less Europeanized landscapes, Papua, Samoa, and British and other islands in the Pacific, the number of 'specimen' photographs of individuals (head and shoulders shots) and groups, including naked bodies, increases. The descriptions are clearly racialized ('Fine-looking Kanaka man with characteristic hair' (Fiji)). It is only in images of exotic Papua and Samoa ('Native dancer with characteristic head-dress of feathers') that the gaze is turned on sexualized images of females (p. 366).

The Union of South Africa provides a mid-way point in terms of representation. Here the scenes are a broad mixture of rural and urban-industrial. Again, civic buildings stand for colonial pride. But now, given the fact that southern Africa is a place of collision between European and African migrations and local native populations, room is given to type images, but these include no people of British 'stock'. Individuals and groups are again placed on display (e.g. 'HOTTENTOTS: Typical Male: these natives usually become house boys' p. 334; note—the male in question is obviously an adult male). The only Europeans 'displayed' are Boers (one image only).

Canada is exhibited as a blend of dramatic wilderness, modernity (e.g. 'VANCOUVER. View of water front with a background of skyscrapers', p. 395; 'TORONTO. Famous skyscraper in King Street, the C.P.R. building', p. 403), and the usual peppering of civic buildings and churches (182 photographs). There are very few images of identifiable humans, and this includes a handful of photographs that show Europeans at work. Only one image obviously 'exhibits' a specific collection of people; 'Group of Aleuts, an intelligent type of Eskimo' (p. 418).

The United States, and, to an extent, the rest of the Americas, is treated in a similar manner to the 'white Dominions'. There is in South America, like South

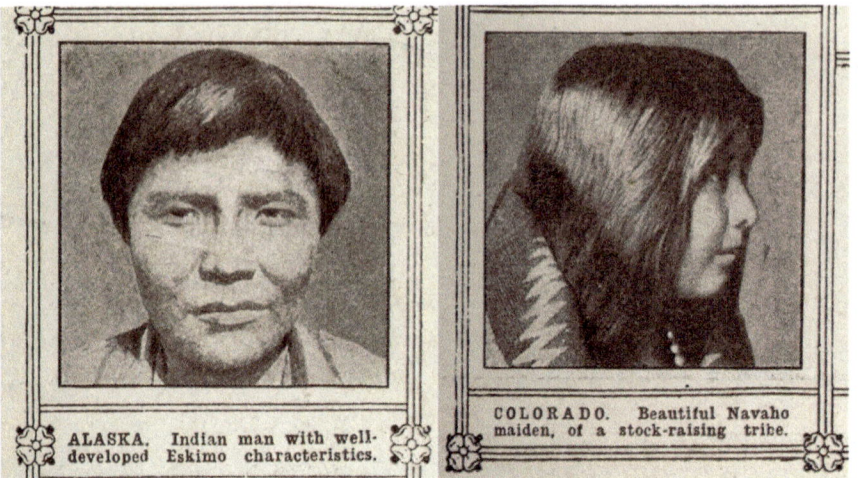

ALASKA. Indian man with well-developed Eskimo characteristics.

COLORADO. Beautiful Navaho maiden, of a stock-raising tribe.

**Fig. 3** Detail from 'America' (p. 379), 'Indian man with well-developed Eskimo characteristics' and 'Beautiful Navaho maiden'

Africa, some display of indigenous peoples. The spectacle of the peoples of the Americas is actually delivered in the margins to a two page 'Bathy-orographical' description of the continents at the very start of the section on the Americas. Twenty-two photos parade the native peoples of the continents from Alaska ('Indian man with well-developed Eskimo characteristics') to Tierra del Fuego, by way of a 'Beautiful Navaho maiden' (Fig. 3) (pp. 379 and 382).

## 2.2 Spectacle II—the Ethnographic Gaze

A large proportion of the atlas is constructed as a set of representations that reinforce the assumed superiority of the Anglo-Saxon world and the 'imperial gaze'. Distant places and peoples are subordinated and reduced to the gaze, and their geographies constructed and paraded before the reader's eyes according to a grand imperial vision. An ethnological gaze is clearly adopted in the atlas, much as it is in Kipling's *Kim*, where one of the chief protagonists, and Kim's controller within the British spy-network, Colonel Creighton, is a British Army officer *and* an ethnologist. Knowledge is power and this includes understanding the geography and peoples of empire.

In both *Kim* and *Harmsworth's Atlas* various 'native' peoples are treated as 'types' in a manner echoing the words of Sir Sidney Low in his essay in the atlas (pp. iii–iv). Low states, without any indication of irony at allowing a child to grasp death-dealing instruments:

I should let him [a child, notably male] see and handle Zulu spears and Maori clubs... I should *show him types of different races and people*, pointing out their resemblances and differences; and thus, without ever having heard of craniological measurements or brachycephalic skulls, he would have begun to acquire the elements of ethnology, and be prepared at a later stage of his studies to take in intelligent interest in Celts and Saxons, Mongolians and Caucasians. (p. iv; emphasis added)

As Said notes, anthropologists and ethnologists were important to the imperial project, as they provided advise to colonial rulers on the 'manners and mores' (Said 1993: 184) of native peoples, and, as such, are the 'handmaidens' of colonialism. The atlas invites its readership to partake of this process. Examples abound of the 'other' when the gaze is diverted from the Saxon world.

The pages dedicated to Asia, including British India, start with thirty-two 'head and shoulder' or full body displays of typical peoples. These include such 'types' as —'MONGOLIAN. Fine specimen of a hardy plateau people of N. E. Asia.' Women are typically presented as objects of the male gaze; for example, on the same page as the Mongolian 'specimen' (p. 211) two women are represented as 'a celebrated type of Eastern beauty', 'a beautiful Tamil girl' and 'CHINESE. Gorgeously clad bride of a notable Chinese statesman.' Japan is also subject to this male gaze, with one photo labelled 'JAPANESE. One of the several types of beautiful Japanese women.' (p. 275).

India consists of twelve pages of text and photographs, and six double-page maps; as befits the 'jewel of empire' India is displayed as a combination of the modern (Western buildings and railways), its ancient heritage, and a display of typical peoples. The latter includes the usual stereotypes and gendered images: 'CEYLON. Vedda man, one of the low-statured, primitive aboriginals' (p. 243, Fig. 4); 'KANDY. The beautiful daughter of a native Kandvan bootmaker' (p. 246); 'RANGOON. Sumptuous silk clothing of a Burmese lady of rank' (p. 250). Two of the twelve pages are dedicated to railways, including twenty-seven photographs, with the other pages also containing three railway photographs.

Africa is treated in a rather extraordinary manner. As with other extra-European regions, the first page is a display of types. This includes the usual ethnographic display (p. 283)—'EAST AFRICA. Masai warrior, a *specimen* of the Negro-Hamitic stock.'; 'SOUTH AFRICA. Bushman: one of the most primitive types in Africa'; 'SOUTH AFRICA. Zulu warrior, one of the most powerful of the Kaffir types'; 'CENTRAL AFRICA. Woman in mourning costume composed of grass.' The second page of text and images/photographs is, however, totally dedicated to white male adventurers who explored/discovered Africa; acknowledging three of the twelve as German.

Africa is otherwise represented as a mix of exotica and the modern. One page (p. 295) exemplifies several themes that dominate the fifty-six pages dedicated to Africa: 'native' indigenous cultures as 'other' (SOUTHERN NIGERIA. Native attendance upon queer ju-ju images'—associated with witchcraft in Africa); Britain's civilizing agenda (NORTHERN NIGERIA. Girl slaves liberated by the British Government'; and 'BAMUM. Enlightened native prince, a builder of schools'); and industrial progress ('DBOMFA RIVER. Iron bridge on the Cameroons Midland

**Fig. 4** Detail from 'India
(Section III)' (p. 243), 'Vedda
Man, one of the low-statured,
primitive aboriginals'

CEYLON. Vedda man, one of the low-statured, primitive aboriginals.

Railway'). Africa, as a newly explored world is treated as a curio, for example, a photograph of an African male wearing a bead headband, bead 'googles' and necklace, 'TANGANYIKA Territory. Native dandy bedecked with beads' (p. 307), at the same time as a resource to be exploited, 'MOCAMBIQUE. Breaking up virgin soil for the cultivation of rubber.' (p. 311). It is worth noting, however, that threat discourse is already entering representations of the African arena with the need for security and surveillance over key primary resources—'CAPE PROVINCE. Unclimbable fencing at a diamond mine, Kimberley' (p. 315).

The ethnographic gaze and the knowledge derived from this is the ultimate form of control of others. The atlas allows a wider metropolitan audience to partake in this, and thereby support and collude with the project of empire, while at the same time being captured in the web of spectacle themselves. The final section addresses the darker side of spectacle, the spectre of threat.

## *2.3  Spectacle III—Threat Discourse*

At the start of Conrad's (1899) forensic novel of imperialism, *The Heart of Darkness*, Marlow, the narrator, contemplates a map of the world, and specifically Africa, where he is bound:

> There was a vast amount of red – good to see at any time, because one knows that some real work is done in there, a deuce of a lot of blue, a little green, smears of orange, and, on the East Coast, a purple patch, to show where the jolly pioneers of progress drink the jolly lager-beer. However, I wasn't going into any of these. I was going into the yellow. Dead in the centre. And the river was there – fascinating – deadly – like a snake.

Colours denote the imperial powers listed; Conrad's readers would know from the colours and descriptive jibes at the powers involved. Colour, however, also provides subtle connotations. The red, as discussed above, connotes the health and vigour of the British imperial project. Blue, is the imperial cloth of France. Yellow, however, is reserved for negative attributions, in this case for the dark imperial project of the King of Belgium. Yellow is one nature's warning colours (Forbes 2011). Yellow, as perilous, is used to illustrate Germany's global ambitions in *Harmsworth's Atlas*. It is the only double page spread in the whole of the section dedicated to the Great War and is entitled 'The World showing Germany's Peaceful Penetration' (pp. 488–489). The use of the words 'peaceful' and 'penetration' stand in distinct contradiction as Germany's pre-war activities are clearly regarded as insidious, and far from pacific in intent. 'Penetration' can imply the acquisition of territory, but is also associated in geopolitics with sinister extension of influence. The term has generally been deployed with strong negative connotations. The British political geographer Fisher (1950) frequently deployed 'penetration' to describe Russian and Japanese expansionism pre-WWII, while during the Cold War 'the image of penetration was frequently evoked' (Ó Tuathail and Agnew 1992: 200) to characterize Soviet communism in patriarchal, sexualized and 'savage' terms (see below of other references to 'penetration' in context of Germany).

For Britain the rising power and threat to its global hegemony in the early twentieth century was Germany. And, like Sparta and Athens, these powers would plunge head-first into the so-called 'Thucydides trap' to fight a gruelling war for global hegemony, which neither ultimately achieved. *Harmsworth's Atlas*, however, seeks to address this issue from a point of strength, as the victorious power, in the ninety-one pages of '*An Atlas of the Great War*' which forms part of the main atlas (pp. 481–572). This section is an object lesson in the need to block a powerful central European power from rising again. The former German world empire is described in suitably sinister terms as a 'mushroom growth, sprung up within the last three decades' (p. 483). As noted above, German pre-war influence is shown in acid-yellow, with its strong negative association. The German empire, most notably 'German South-West Africa', 'German East Africa', 'Kamerun' and Togo, but also including their trading port of 'Kiauchan' (China) and islands in Oceania (including 'German New Guinea') are shown in green. Its global ambitions are shown in acid yellow and labelled 'What Germany Wanted'. This included most of South America and parts of the Caribbean, half of Africa, from Cape to Cairo, and influence over large swathes of Europe and the Middle East, China and south-east Asia. Areas of significant German settlement, and location of German banks and missions are shown as red symbols, and the global reach of German underwater cables and shipping routes is clearly displayed.

It is the financial and diplomatic 'penetration' of Germany that was seen as problematic. This influence was branded as 'insidious' and used to 'spy out opportunities for German aggression' (p. 490). While Germany had been defeated, it was obvious that some the threat remained in terms of German communities embedded in many parts of the world, for example, throughout Latin America (Fig. 5).

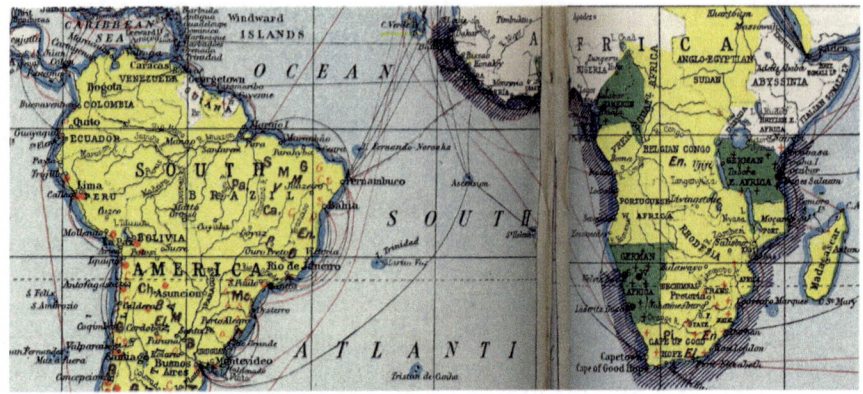

**Fig. 5** Detail from 'The World (Showing Germany's Peaceful Penetration)' (pp. 488–489)

The page of text following the map of 'peaceful penetration' makes this threat very clear (p. 490). German shipping is described as having 'actively fostered German aggression', while other forms of commercial activity are used to 'spy out the opportunities for German aggression'. German settlements across the whole globe are described as 'penetrations' and their populations as 'pioneers of Deutschtum' [Germanness], and giving, for example, 'a distinctly German colour to certain parts of South America'. The suggestion of insidious influence is fostered by the fact that of thirteen photographs surrounding the text, four show the Chilean military adopting Germanic uniforms and forms of organization (Fig. 6). In the decades after the atlas's publication, especially with the rise of Hitler, fear of German 'penetration' in this region was to grow again, and be implicated in issues related to US hemispheric defence; for example, the following quote from Shepardson writing for the Council on Foreign Relations (US) '[…] every new act of commercial or financial penetration by the Nazi regime might also mean the spread of their political influence' (Shepardson 1938: 264). Commercial airlines, rather than the merchant shipping discussed in the *Harmsworth's Atlas*, become implicated as opportunities for espionage; for instance Hall and Peck's classic paper for *Foreign Affairs* notes 'The drone of German and Italian airplanes over South America is not a new sound […] [b]ut we in the United States have been slow to recognize it as a warning of Nazi-Fascist penetration in the Western Hemisphere' (Hall and Peck 1941: 347). Note the use of the term 'penetration' in both the previous quotes. Discussing German settlement in Latin America more generally Hubert Herring (1941) describes the population as containing 'good Hitler-Germans' and danger of fifth-columnists, while the famous US journalist Gunther (1942) described in detail the threat of German 'Fifth Columnism'—propaganda, espionage and subversion.

Much of the rest of the *Atlas of the Great War* is focused on the events of the war and its outcome. Mention is made of German 'frightfulness', including attacks on hospitals, while the allies are cast in heroic mode. Of specific significance to this

**Fig. 6** Detail from 'The World (Showing Germany's Peaceful Penetration)' (p. 490) showing German military influence in Chile

analysis, however, is the clear difference in the way that the atlas' authors portray British and German colonialism in the periphery. The loss of German colonies at war's end, and their subsequent administration by Britain and France is depicted as a blessing for the inhabitants. The conditions under Germany are graphically displayed in the photographs surrounding p. 495 'The Fate of German Colonies'; one image shows 'Chained natives crammed into a very tiny compound', while two others are very graphic images of hangings in German East Africa. The title of one these images is extremely explicit concerning the nature of German rule, 'A native finally freed from his German taskmasters', while the second, showing well over a dozen hanging corpses, notes 'Natives punished by the Germans for minor offences'. The message is very clear, better part of the British and French empires. The post-war system of Mandates, under the League of Nations, is praised as a system applicable to those areas in the world were a large native population has to be mothered by the representatives of a 'more highly civilised people […] the serfs liberated from the domineering arrogance of German taskmasters will be subjected to a kindlier control' (p. 495).

At a distance of roughly a century the conceit of this statement is obvious, but as Said notes:

> Neither imperialism nor colonialism is a simple act of accumulation and acquisition. Both are supported and perhaps even impelled by impressive ideological formations that include notions that certain territories and people require and beseech domination, as well as forms of knowledge affiliated with domination: the vocabulary of classic nineteenth-century imperial culture is plentiful with such words and concepts as 'inferior' or subject races', 'subordinate peoples', 'dependency', 'expansion', and 'authority'. (Said 1993: 8; emphasis in the original)

A statement to which could be added the *Harmsworth's Atlas*'s 'more highly civilised' and 'kindlier control'.

# 3 Conclusion: The Empire as Spectacle

The *Harmsworth's Atlas* provides an exemplar of the concept of the 'map as spectacle'. The world is displayed before its British readership, acknowledging both the right to rule over much of the Earth's population and land mass, and to be regarded as moral guardian of the 'lesser' peoples of the globe. The capitalist world system of core, semi-periphery and periphery is presented as natural, while recognizing that aberrant forms of the colonial/imperialist enterprise might exist, and can be dealt with within a new world order represented by the League of Nations. This exhibition must, however, be seen for what it is, part of a (Baroque) structure that ensures compliance; achieved through both an egregious display of the core's technical accomplishments and the bounty of the colonies, and through a constant reminder that this prize is threatened and needs to be constantly secured.

The atlas is a classic example of intertextuality; both within its own pages, as text, map and photographic images support each other, and as a wider validation of the imperialist narrative. While, with the distance of a century between its publication and today, it may appear outmoded, it undoubtedly represented the cutting-edge of publication at the time, especially in its use of photography to parade the peoples of the world before its readership.

The history of cartography only really has value if it works for us now, if it helps people envisage a better future. Equitable and sustainable relations thrive when individuals and communities understand history, including the mistakes made in the past, and interweave this into their debates about contemporary issues. The map as spectacle continues to have a hold on contemporary audiences; hopefully this essay will be a step towards dismantling such hegemonic narratives.

# References

Byatt AS (2010) The children's book. Vintage Books, London (originally published by Chatto & Windus, 2009)

Callens J (2000) Diverting the integrated spectacle of war: Sam Shepard's '*States of Shock*', Text Perform Quart 20(3):290–306

Debord G (1998) Comments on the society of spectacle. Verso, London

Debord G (2009) Society of the spectacle (trans. Knabb K, originally published as La société du spectacle in 1967). Soul Bay Press, Eastbourne

Edney M (2018) Map history: discourse and process. In: Kent A, Vujakovic P (eds) The Routledge Handbook of mapping and cartography. Routledge, London, pp 68–79

Fisher CA (1950) The expansion of Japan: a study in Oriental geopolitics, part I. Continental and maritime components in Japanese expansion. Geogr J 115(1–3):1–19

Forbes P (2011) Dazzled and deceived: mimicry and camouflage. Yale University Press, New Haven

Gunther J (1942) Inside latin America. Hamish Hamilton, London

Hall M, Peck W (1941) Wings for the Trojan horse. Foreign Aff 19(2):347–369

Herring H (1941) Good neighbors: Argentina, Brazil, Chile and seventeen other countries. Yale University Press, New Haven

Kaldor M (1986) The weapons succession process. World Politics 38(4):577–595

Kidd C (2003) Race, empire, and the limits of nineteenth-century Scottish identity. Hist J 46 (4):873–892

McClintock A (1995) Imperial leather: race, gender and sexuality in the colonial contest. Routledge, London

Roberts D (2003) Towards a genealogy and typology of spectacle; some comments on Debord. Thesis Eleven 75(11):54–68

Said E (1993) Culture and imperialism. Chatto & Windus, London

Shepardson WH (1938) The United States in world affairs: an account of American foreign relations yearly review—1937. Council on Foreign Relations, New York

Tuathail GÓ, Agnew J (1992) Geopolitics and discourse: practical geopolitical reasoning in American foreign policy. Political Geogr 11:190–204

Urry J (2002) The tourist gaze, 2nd edn. Sage, London

Vujakovic P (2017) The map as spectacle. In: Kent A, Vujakovic P (eds) The Routledge handbook of mapping and cartography. Routledge, London, pp 462–474

**Peter Vujakovic** is Professor of Geography at Canterbury Christ Church University (UK). Peter's research interests in cartography span a variety of areas associated with culture and politics, from the local (disability access mapping, and 'sense of place') to the global (news media maps and geopolitics, and maps in development education). Peter is an Associate Editor (and former Editor) of *The Cartographic Journal*, co-editor of *The Routledge Handbook of Mapping and Cartography* (published in 2017) and an expert contributor to *The Times Comprehensive Atlas of the World* (14th Edition).

# Mapping Boundaries

.

# Mapping Changes in Ottoman-Austrian Borders During the Eighteenth Century

Uğur Kurtaran

**Abstract** Significant political, military and diplomatic changes took place in the Ottoman Empire during the eighteenth century. The Treaty of Karlowitz was signed in 1699, introducing territorial losses that dramatically changed the Empire's borders. Since the determination of these borders was important for international relations, particularly with Austria, boundary mapping was one of the main subjects of Ottoman cartography during the eighteenth century. This chapter examines these maps with regard to their political context, starting with the Treaty of Karlowitz and the subsequent treaties of Passarowitz (1718), Belgrade (1739) and Svishtov (1791).

## 1 Introduction

Border relations constitute one of the most important aspects of intergovernmental diplomacy. Austria was one of the most important rival powers to the west of the Ottoman Empire, as is reflected in the frequency of border changes resulting from extensive political and military developments in the eighteenth century. These began with the Treaty of Karlowitz in 1699, which introduced heavy territorial losses for the Ottoman Empire and saw the balance of power shift in favour of Austria. Throughout the eighteenth century, the Ottoman-Austrian borders were in flux with the successive treaties of Passarowitz (1718), Belgrade (1739) and Svishtov (1791), all of which built on the changes that were introduced at Karlowitz. The aim of this chapter is to examine changes in the location of these borders and to explain how they were established. The investigation begins by exploring the historical dimensions of Ottoman-Austrian relations and the general characteristics of the new borders that emerged as a result of this series of treaties.

U. Kurtaran (✉)
Karamanoğlu Mehmetbey University, Karaman, Turkey
e-mail: ugurkurtaran@gmail.com

© Springer Nature Switzerland AG 2020
A. J. Kent et al. (eds.), *Mapping Empires: Colonial Cartographies of Land and Sea*,
Lecture Notes in Geoinformation and Cartography,
https://doi.org/10.1007/978-3-030-23447-8_11

197

## 2   The Treaty of Karlowitz and Its Impact on Ottoman-Austrian Borders

The Hungarian Kingdom was eliminated in Mohaç during the fourteenth and fifteenth centuries and the Ottoman Empire gained overall superiority in its military and political relationship with Austria after the Battle of Mohaç in 1526. This situation was reflected in diplomatic relations with the inclusion of Austrian territory within the Ottoman borders. However, as a result of the wars of the Holy Alliance that began in 1683, the situation began to change, particularly with the signing of the Treaty of Karlowıtz in 1699 between the Ottoman Empire and the three European states of Austria, Poland and Venice. Indeed, the Ottoman Empire ceded territory in Poland, Ukraine and Podolya to Lehistan; it left Mora and Dalmatia to Venice; and it ceded all territory in Hungary to Austria except for Erdel and Banat (BOA. A. DVNS. DVE. Nemçelü Ahidnamesi 57/1: 21–8; BOA. KK. d. 53: 2–6). New borders were therefore created for the new Ottoman Empire in the west and border commissions were established (beginning with Austria) to delineate these new borders. The Habsburgs commissioned Ferdinando Marsigli, an engineer in the Austrian army, as a border commissioner (Kurtaran 2017: 578). In this way, traditional methods were used to reach an agreement between the two parties in border negotiations, i.e. by adopting the principle of 'alâ halihi' (where each of the parties owns the territory they have already acquired) (Kurtaran 2017: 576). Under this principle the mountainous regions were completely ignored, while new borders were delineated according to natural elements such as rivers (Molnar 1999: 477). Consequently, as a result of the negotiations between the Ottoman Empire and Austria to delineate new borders, the Salankamen region was accepted as the common borderland. The borders between the two parties were subsequently determined in three regions: Belgrade (see Kurtaran 2018b: 119–145), Bosnia (see Gökçe 2001: 75–104) and Timisoara.

These new borders were recorded in the book *Tar Hudutname Daha* (Kurtaran 2017: 584), which documents the successive changes in the Ottoman borders from the Treaty of Karlowitz. According to *Tar Hudutname*, the borders of the Ottoman Empire gradually withdrew from Salankamen to Belgrade in twenty stages. The regions of the Danube and Tisa rivers were left to Austria, while some areas of Belgrade and Timisoara were given to the Ottomans (Molnar 1999: 477). The process of delineating the new borders was carried out in accordance with the provisions of the Treaty of Karlowitz and was based on the locations of natural and artificial features. It can be concluded from related documents that natural features, such as the Tuna, Sava and Tisa rivers, and mountains and hills, were used to establish borders as well as artificial features with markers called 'humka' (border stones) (BOA. A. DVN. DVE. Venedik Hudutname Defteri, 17/5: 1–54; BOA. AE. SMST. II. 10/956).

# 3   The Treaty of Passarowitz: New Losses and New Limits

The Ottoman Empire had suffered heavy losses with the signing of the Treaty of Karlowitz in 1699. In seeking to compensate for these, the Ottomans began a war with Venice to re-capture the lost lands, which resulted in the Treaty of Istanbul in 1700 (see Kurtaran 2018a: 287). The Ottoman Empire also invaded Russia at the beginning of the eighteenth century (the Prut War), which led to the signing of the Treaty of Prut in 1711. Mora was conquered by the Ottoman forces in 1715 following the war with Venice (Silahdar Findiklili Mehmed Ağa 2001: 838–840) leading to an alliance between Austria and Venice against the Ottoman Empire (Zinkeisen 2007: 357). Austria, accusing the Ottoman administration of violating the Treaty of Karlowitz, demanded that the Ottomans return the territories they had gained from Venice. However, the Ottoman Empire did not respond to this demand and the Grand Vizier declared war against Austria at the behest of Damad Ali Pasha (Raşid Mehmed Efendi-Çelebizâde İsmail Âsım Efendi 2013: 981–983). The Ottoman army, which was commanded by the Grand Vizier himself, was heavily defeated by the Austrian armies under the command of Prince Eugen near Varadin. Banat and Timişoara were recovered by the Austrians as the Ottoman armies withdrew to Belgrade. With intervention by England and the Netherlands, the Treaty of Passarowitz was signed on 21 July 1718 (BOA. A. DVNS. DVE. Nemçelü Ahidnamesi, 57/1: 55–61). The agreements the Ottoman Empire signed with Austria, with the Austrian and Venetian States, and at the Treaty of Pasarowitz, comprised 20 items over 24 years. The treaty signed with Venice comprised 26 items (BOA. A. DVNS. DVE. Venedik Ahidname Defteri, 16/4: 99–107; Mecmuasi 2008: 170–196; Kurtaran 2018a: 288).

In the Treaty of Passarowitz that was signed with Austria, the principle of 'alâ halihi' was adopted as it had been at Karlowitz. Although this principle reduced the losses of the Ottoman Empire, the Ottoman-Austrian borders in Passarowıtz largely changed in favour of Austria. In fact, as mentioned in the first seven articles of the treaty, all of Banat and Eflak to the west of the River Olt was left to Austria. In addition to the western part, which is also called Little Wallachia, the northern part of Serbia, including Belgrade, and Northern Bosnia were also ceded to Austria (BOA. A. DVNS. DVE. Nemçeli Ahidnamesi, 57/1: 56–58; Samardzic 2011: 17; Savaş 2002: 559; Kurtaran 2018a: 290).

Commissions were formed to finalize the boundaries between the parties, as in the Treaty of Karlowitz (Özcan 2007: 180). In this case, the Austrians appointed Anshelm Franz von Fleischmann as the border commissioner to determine the borders and General Petraş to assist him. Nigbolu Mutasarrifs Vezir Ahmed Pasha and the other Selanik Mutasarrifs Mustafa Pasha were appointed to determine the Danube borders on behalf of the Ottoman Empire. The border commissions both commenced their work to delineate the borders one year after the treaty was signed. After this, new boundaries were created from natural and artificial features, as for the Treaty of Karlowitz. Accordingly, the agreed borders began from the River Olt and the Ottoman lands to the west were given to Austria. In addition, a common border was determined from where the River Olt flowed into the Danube until the River Timok. As a result of these geographical delineations, the exact boundaries between the parties were created on the condition that all the islands in the River Danube would remain in the Ottoman Empire. The new Ottoman-Austrian borders, determined by the Treaty of Passarowitz, remained valid until 1736 (Kurtaran 2018a: 291–292, 295).

## 4 The Treaty of Belgrade: Compensation for Lost Territories

Although the Treaty of Passarowitz signed on 21 July 1718 lasted for 27 years, aggressive policies by Austria led to the deterioration of the peace process in 1736. The Treaty of Passarowitz was broken in 1736, when the Ottoman Empire took the opportunity to attack Persia at that time and made an alliance with Russia (Uzunçarşılı 1988: 559). However, the war did not go as Austria had planned, and the Ottoman forces regained Belgrade (which had been lost at Passarowitz in 1718) (Köse 2002: 220). In the aftermath of the war, the two states signed the Treaty of Belgrade on 18th September 1739 under the mediation of France, as a result of the withdrawal of Nis and the defeat of the Austrian forces in Hisarcik (BOA. A. DVNS. DVE. Nemçeli Ahidnamesi, 59/3: 185–191; BOA. A. DVNS. MHM. d. 147: 29; Erim 1953: 81–94; Kurtaran 2018c: 384). Russia, upon accepting the peace conditions, also signed the Treaty of Belgrade between Russia and the Ottoman Empire (BOA. A. DVNS. DVE. Rusya Ahidname Defteri, 83/1: 82–118; BOA. HH. 1428/5845; 1428/58455). The Treaty of Belgrade signed with Austria comprised 23 articles and clauses and lasted for 27 years.

In the final treaty the Ottoman Empire signed with Austria (and with any European power), the Ottoman Empire took back many of the lands that were lost at the Treaty of Passarowitz. While Belgrade and Sabacz were retained by the Ottomans, the Danube and Sava rivers were accepted as borders between the two states (Uzunçarşılı 1988: 289–291; Karagöz 2008: 284). This situation created some changes in the borders between the two states that had been determined by Passarowitz. As a result, both parties began the task of delineating the borders just nine months after signing the treaty, with the Grand Vizier Nişanci Ahmed Pasha acting on behalf of the Ottoman Empire (Sertoğlu 2011: 2519; Kurtaran 2018c: 436). In addition, muhaddits were appointed to establish the location of the borders at Belgrade, Timisoara and Bosnia. Ibrahim Efendi, one of the judges of the Divan-i hümâyûn, was appointed to delineate the borders of Timişoara; Mevkufâti El-Hac Mehmed Efendi for Belgrade; and Mehmed Said Efendi was commissioned to establish the Bosnian borders (BOA. A. DVNS. MHM. d. 147: 29). They also participated in the meetings of Abu Sehl Numan Efendi as border mollas and reported the negotiations between the parties. Efendi remained in office for the duration of the report and his book *Tedbirat-i Pesendî* consists of three parts.

Austria employed Tümgenaral Kont Guadigny for its border demarcation and General Engelsofen was appointed head of the border delegation. There were also ten architects, engineers and painters in the Austrian delegation (Ebû Sehl Numan Efendi 1999: 22).

Negotiations began in 1740, when the border rulers appointed by the two states arrived in Belgrade. There were some problems of diplomatic protocol in contrast to those at Karlowitz and Passarowitz (Ebû Sehl Numan Efendi 1999: 57–58). However, as a result of the successful diplomatic activities of the Ottoman muhaddits, these problems were overcome and a border agreement was signed between the parties on 2 March 1741 (BOA. İE. HR. 19/1724; BOA. A. DVNS. NMH. d. 8: 12–13; Kurtaran 2018c: 438). The result of the collaboration between the parties by Code of Civil Procedure are the definite limits that were determined on 10th May 1741 (Kurtaran 2009: 174). Again, natural and artificial features were used by both sides to delineate the new bordes. The Danube and Sava rivers were accepted as common borders, while many mountains, rivers and hills were also used. As per earlier treaties, the border markers were called 'humka'.

As a result, after the Treaty of Belgrade in 1739, after the border-setting activities lasting about 10 months between the parties, 11 fortresses and arbors from Belgrade to Bosnia and more than 900 villages and towns remained in the Ottoman Empire. (Vak'anüvis Suphi Mehmed Efendi 2007: 666–667). The limits set out here are based on a number of minor amendments to the 1791 Treaty of Svishtov that were made by the end of the century (Karagöz 2008: 293; Kurtaran 2018c: 443).

# 5   The Treaty of Svishtov: Preservation of the Status Quo and New Borders

The new Ottoman-Austrian border, which was established by the Treaty of Belgrade in 1739, was re-approved by the Treaty of Istanbul in 1747 and secured a lasting peace until 1787. In the last quarter of the eighteenth century, new wars with Austria and Russia began to emerge. Catherine II of Russia signed an alliance (the Greek Plan) with Josef II of Austria to dismantle the Ottoman Empire (Kurtaran 2009: 186–187). Towards the end of the eighteenth century, the Ottoman Empire therefore faced a new Russian-Austrian war (Beydilli 2013: 467). However, the war did not go as intended and Austria was forced into peace with the Ottomans. After lengthy discussions between the Ottoman Empire and Austria, the Treaty of Svishtov was signed on 4 August 1791 (BOA. A. DVNS. DVE. 59/3:31-34; BOA. A. DVNS. NMH. d. 9: 225–229; Kurtaran 2009: 264–270; Kuzucu 2012: 257). According to the clauses of the 14-article treaty, it was agreed that the Treaty of Belgrade and the Treaty of 1739 would be respected. In fact, a *status qua ante bellum* was observed by both parties with the return of some lands and the seizure of captured lands. This complied with the principle of renewing the validity of the treaty by both parties (Beydilli 2013: 469–470). Under this framework, Austria was to follow the *status quo* in 1791 and to have friendly relations with the Ottoman Empire. According to this, the land, city, fortress and palaces that were invaded by Austria would be granted to the Ottoman Empire (BOA. A. DVNS. DVE. 59/3: 31–32). In addition, after establishing the border treaty between the two states, they signed another treaty relating to the seven-point assay (BOA. A. DVNS. DVE. 59/3: 35–36; Kurtaran 2009: 273–278). The delineation of borders in the Treaty of Svishtov also used natural and artificial features in a similar way to the earlier treaties. The most important difference between the Treaty of Svishtov and the earlier treaties is that the boundaries between the parties were shown on maps for the first time. For example, maps were prepared by the Ottoman engineer Abdurrahman Efendi on the limits determined in the Svishtov negotiations (BOA. HH. Nr. 241/13550 and BOA. HH. 242/13588; 13589) (Figs. 1 and 2).

The ongoing war between the Ottoman Empire and Austria, which had begun in the early sixteenth century, ended with the Treaty of Svishtov in 1791. With the end of the war with Austria, the Ottoman Empire, first of all, was freed from waging wars on two fronts and could focus on the Russian front. However, with the impact

**Fig. 1** The map created by Ottoman Engineer Abdurrahman Efendi, who showed the borders of Ottoman Austria in the Treaty of Svishtov (BOA. HH. Nr. 242/13588; 13589) (Courtesy Prime Ministry Archive)

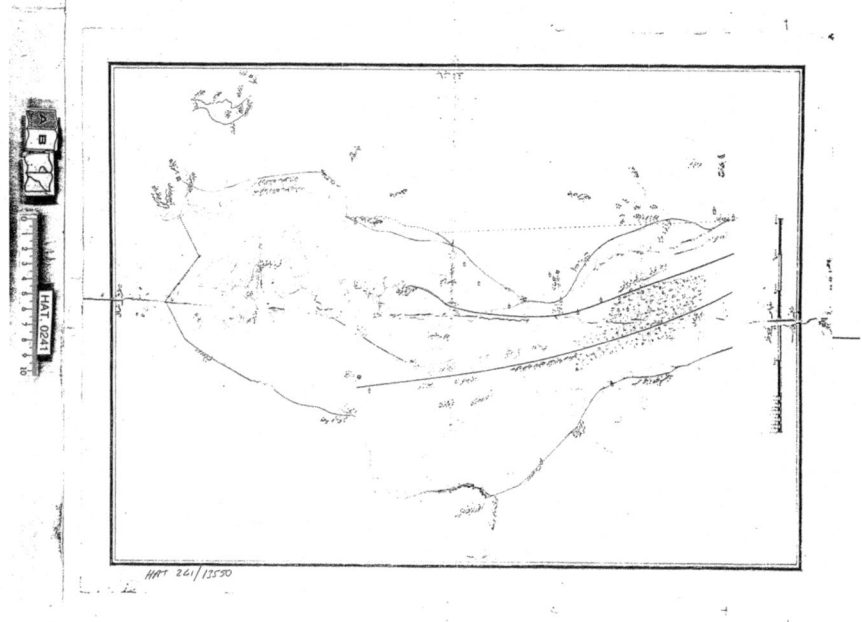

**Fig. 2** A map showing the Ottoman-Austrian borders in the Treaty of Svishtov with Austria (BOA. HH. Nr. 241/13550)

of significant political developments in Europe, a peace agreement was soon reached with Russia. The Treaty of Svishtov constitutes the last phase of Ottoman-Austrian relations, bringing the war to a close and is an important turning point in its 300-year history. After this point, Ottoman-Austrian relations began to improve and the two states formed an alliance in the nineteenth and early twentieth centuries. This period saw the reforming of Ottoman statesmen, who reinforced the joint Austrian-Ottoman policy.

# 6    Conclusion

This chapter describes how borders between the Ottoman Empire and Austria developed in parallel with political and military developments in the eighteenth century. A new process, which started with the Treaty of Karlowitz in the eighteenth century, introduced significant changes to these borders (including large territorial losses to the Ottoman Empire). This was followed in 1718 by the Treaty of Passarowitz, when the Ottoman Empire faced further territorial losses. However, with the Treaty of Belgrade of 1739, new borders were drawn that compensated most of its losses against Austria since the beginning of the century. The same treaty was extended in 1747 and continued until 1787. The war between the Ottoman Empire and Austria ended with the Treaty of Svishtov in 1791. In this treaty, the boundaries between the two states remained unchanged as of the Treaty of Belgrade and its extension in the Treaty of Istanbul in 1747.

Another finding described in this chapter is how the process for establishing borders evolved from the beginning to the end of the century. It was concluded that these procedures were put into practice by the commissions formed by both states and by adhering to the conditions set out in the treaties. In these processes, natural features were used to delineate boundaries and artificial boundary markers (such as stones) were used in regions where such natural features are absent.

Finally, it is important to note that Ottoman-Austrian borders were constantly changing throughout the eighteenth century and these changes resulted in new diplomatic relations between both parties. The results of these changes were recorded under the so-called border delimitation report or *Hudutname*. Consequently, it is possible to observe the outlines of the border changes in the eighteenth century more precisely.

**Acknowledgements** This study was supported by Karamanoğlu Mehmetbey University Scientific Research Project Coordination Unit. Project Number: 16-AG-18.

# References

## Archive References

Prime Ministry Archive (BOA)
BOA. A. DVNS. DVE. Nemçelü Ahidnamesi, 57/1; 59/3
BOA. A. DVNS. DVE. Venedik Ahidname Defteri, 16/4
BOA. A. DVNS. DVE. Rusya Ahidname Defteri, 83/1
BOA. A. DVN. DVE. Venedik Hudutname Defteri, 17/5: 1–54
BOA. A. DVNS. NMH. d. 8
BOA. KK. d. 53
BOA. AE. SMST. II. 10/956
BOA. HH. 241/13506; 241/13550; 242/13588; 13589; 1428/5845; 1428/58455
BOA. A. DVNS. MHM. d. 147
BOA. İE. HR. 19/1724

## Other References

Beydilli K (2013) Ziştovi Antlaşması. DİA C 44:467–472
Ebû Sehl Numan Efendi (1999) In: Savaş AI (ed) Tedbîrât-ı Pesendide. TTK Publications, Ankara
Erim N (1953) Devletlerarası Hukuku ve Siyasi Tarih Metinleri, CI. TTK Publications, Ankara
Gökçe T (2001) 1699–1700 Tarihli Bosna Vilâyeti Hududnâmesi. Tarih İncelemeleri Dergisi 16:75–104

Karagöz H (2008) 1737–1739 Osmanlı-Avusturya Harbi ve Belgrad'ın Geri Alınması. Unpublished Ph.D. thesis. Süleyman Demirel University, Institute of Social Sciences, Isparta

Köse O (2002) XVIII. Yüzyılda Osmanlı-Rus Esir Mübadelesi. XII. Türk Tarih Kongresi Kongreye Sunulan Bildiriler. Ankara, pp 536–549

Kurtaran U (2009) Osmanlı Avusturya Diplomatik İlişkileri (1526–1791). Ukde Publications, Kahramanmaraş

Kurtaran U (2017) Sultan II. Mustafa (1695–1703). Siyasal Publications, Ankara

Kurtaran U (2018a) Pasarofça Antlaşması'na Göre Yapılan Sınır Tahdit Çalışmaları ve Belirlenen Yeni Sınırlar Uluslararası Sosyal Araştırmalar Dergisi C. 11 February 2018, pp 285–300

Kurtaran U (2018b) XVIII. Yüzyıla Ait Hudutname Ve Sınır Tahdit Örnegi: Karlofça Antlaşması'na Göre Belgrad Sınırlarının Belirlenmesi ve Tespit Edilen Yeni Sınırlar. Osmanlı Diplomasi Tarihi Kurumları ve Tatbiki, Grafiker Publications, Ankara, pp 119–145

Kurtaran U (2018c) Sultan I. Mahmut (1730–1754). Altınordu Publications, Ankara

Kuzucu S (2012) XVIII. Yüzyıl Son Çeyreğinde Osmanlı Avusturya Siyasi İlişkileri ve Ziştovi Antlaşması (II. Josef ve II. Leopold Dönemi). History Studies, Prof. Dr. Enver Konukçu Armağanı, pp 251–261

Molnar FM (1999) In: Eren G (ed) Karlofça Antlaşması'ndan Sonra Osmanlı-Habsburg Sınırı (1699–1701). Osmanlı, C. I. Ankara, pp 472–479

Muâhedât Mecmuası (2008) C. II/III. TTK Publications, Ankara

Özcan A (2007) Pasarofça Antlaşması. DİA C 34:177–181

Râşid Mehmed Efendi-Çelebizâde İsmail Âsım Efendi (2013) Târîh-i Râşid ve Zeyli C. II (Prepared by: Abdülkadir Özcan-Yunus Uğur-Baki Çakır-Ahmet Zeki İzgöer). Klasik Publications, Istanbul

Samardzic N (2011) The Peace of Passorowitz 1718: an introduction. In; Ingro-Nikola C, Samardzic-Jovan P (eds) The peace of Passorowitz 1718. Purdue University Press, Indiana, pp 9–39

Savaş Aİ (2002) In: Güzel HC, Çiçek K, Koca S(eds) Osmanlı Devleti İle Habsburglar Arasındaki Diplomatik İlişkiler, Türkler. Ankara, pp 555–566

Sertoğlu M (2011) Resimli-Haritalı Mufassal Osmanlı Tarihi, C.V. TTK Publications, Ankara

Silahdar Fındıklılı Mehmed Ağa (2001). Nusretname (Tahlil ve Metin) (1106–1133/1695-1721) (Prepared by: Mehmet Topal). Unpublished Ph.D. thesis. Marmara University, Institute of Social Sciences, Istanbul

Uzunçarşılı İH (1988) Osmanlı Tarihi, C. III/II and IV/I. TTK Publications, Ankara

Vak'anüvis Mehmed Suphi Efendi (2007) Suphi Tarihi, (Prepared by: Mesut Aydıner), Kitabevi. TTK Publicationsİstanbul, Istanbul

Zinkeisen JW (2007) Osmanlı İmparatorluğu Tarihi (Translated by Nilüfer Epçeli and edited by Erhan Afyoncu). Yeditepe Publications, Istanbul

# Lines on the Map: International Boundaries

**Rose Mitchell**

**Abstract** This chapter discusses how maps were made and used throughout the process of boundary creation, from draft maps made by statesmen in European boardrooms to boundary demarcation in the field, and to final records attached to treaties. It aims to show how different stages tended to promote the creation of different map types, and how these maps can be important for the history of cartography. It draws upon examples from one of the largest and most important accumulations of these maps in the world. Held at The National Archives of the United Kingdom, they document British government involvement in shaping boundaries and in resolving boundary disputes over many centuries, whether as a colonial power, a neutral observer or an independent source of surveying expertise. The maps discussed not only encompassed the British Empire, they also defined neighbouring lands and other past empires. They shaped many of the countries in existence today. They can still hold authority as part of the evidence portfolio in current international boundary discussions.

## 1 The Historical and Documentary Context for the Maps

Rarely in British history have boundaries been such a hot topic as they are currently, as these islands of the United Kingdom attempt to define a new shape for themselves on the global scene. At European and international level, borders feature in news stories about immigration, customs and trade deals, intelligence, and military transgressions. With around 17% of international borders in dispute at any one time, historic maps help to inform discussions and contribute to the peaceful resolution of boundary questions.

R. Mitchell (✉)
The National Archives, Richmond, UK
e-mail: rose.mitchell@nationalarchives.gov.uk

© Crown 2020
A. J. Kent et al. (eds.), *Mapping Empires: Colonial Cartographies of Land and Sea*,
Lecture Notes in Geoinformation and Cartography,
https://doi.org/10.1007/978-3-030-23447-8_12

207

This paper uses examples from workshops I have run with Durham University's Centre for Borders Research for those investigating current boundary issues using historic records.[1] They are drawn from the holdings of The National Archives (TNA) of the United Kingdom. These maps, along with the treaties, reports and correspondence to which they relate, document the role played by British governments in the creation of international boundaries over the centuries in a surprisingly large number of areas of the world.[2]

There are a number of reasons for this wide geographical coverage. British colonial interests were carefully nurtured and jealously guarded. Strategic concerns meant that Britain was constantly on the alert for threats to or encroachments on her Empire, particularly with regard to Persia, Russia and China, so she favoured the creation of clearly defined and defended boundaries. In the nineteenth and early twentieth centuries particularly, British surveying expertise was widely respected, with the result that there was frequently a British commissioner on international boundary commissions, even when Britain had no direct interest in the territories concerned. Hence TNA holds certified copies for instance of Balkan boundary maps, and of the Argentine-Chile boundary.

## 2 Stages in the Process

### 2.1 Overview

Europeans took to the colonies their notions of statehood, which entailed the delineation of international boundaries, especially where colonies of different powers adjoined, or when a colony of a foreign power became independent but the adjacent district remained a British colony. I focus in this chapter specifically on the role of maps and the types of maps produced and used at different stages in the boundary-making process, mainly within the historical period of the conference, the century from about 1850 to about 1950. From the nineteenth century, especially in Africa, Asia and the Americas, it became common for international boundaries to be created between British colonies and neighbouring colonies of other European nations by a process of definition by treaty, demarcation on the ground by a joint Boundary Commission, and delineation on a record map to be preserved among the

---

[1]Those researching live boundary disputes are advised to consult the International Boundaries Research Unit (or IBRU) of the Centre for Borders Research at Durham University, which gives advice and training on how to conduct all angles of current research. Its website also carries news stories: https://www.dur.ac.uk/ibru/news/.

[2]TNA's online research guide on International borders records gives details of how to research among maps and textual records and section 4 lists the main geographical areas covered: http://www.nationalarchives.gov.uk/help-with-your-research/research-guides/international-boundaries-in-maps-surveys-other-records. A podcast on this subject is also on TNA's website, Mitchell (2014).

archives of both colonial powers concerned. Maps were used throughout the process of boundary-making and creating a record of it.

The little material that has been published about international boundary maps has mostly been written to inform those working on and making maps for current or recent boundary disputes. For modern maps, readers may refer to an article by Gerald Blake, founder of IBRU, which posits four phases in the process, terming the initial phase 'allocation', switching the terms 'demarcation' and 'delimitation' as I use them below, and adding a fourth, 'management' or 'maintenance' stage which requires large scale maps for patrols and repairs.[3]

My research among historic records suggests that maps were not made for every boundary, nor necessarily for each stage in every case. It is also a question of working with what remains; in some cases TNA holds the only surviving boundary maps, where other originals have been destroyed by conflict, climate or disaster. An outline of the development of boundary mapping will be given here as I have found it, through examples from the archives.

## 2.2 Early Examples

From early times, rulers and governments have been concerned to define and protect their frontiers against invasion and conquest. In Britain, early examples of physical boundaries between nations are Hadrian's Wall between England and Scotland, and Offa's Dyke on England's border with Wales. Early boundaries were often defined verbally, a practice which left much scope for confusion and dispute. The earliest map I know drawn for an international boundary dispute shows the area between England and Scotland called the 'Debatable Land', in 1552. Four straight lines show proposals by the Scottish, the English, and the French ambassador acting as mediator in the case, with the fourth and 'Fynal line' agreed, still part of the border today[4] (Fig. 1).

The practice of surveying and mapping international boundaries did not become the norm until the nineteenth century, although earlier examples can be found, perhaps showing boundaries incidentally on a map made for other matters. A map dated 1767 of land grants in the Lake Champlain area also happens to show the boundary between the American provinces of New York and Vermont, with the

---

[3]Blake (1995). For an example of an article on border maps written by a map archivist see Beech (1986).

[4]MPF 1/257 from State Papers Scotland; one of a number of examples of maps illustrating the theme of boundaries in the book Mitchell and Janes (2014). The story of this map and a reproduction are at pp. 28–29. Another example used to illustrate the talk and not here mentioned or illustrated is a chart of Antarctica showing territorial claims under consideration in 1953 pp. 218–219.

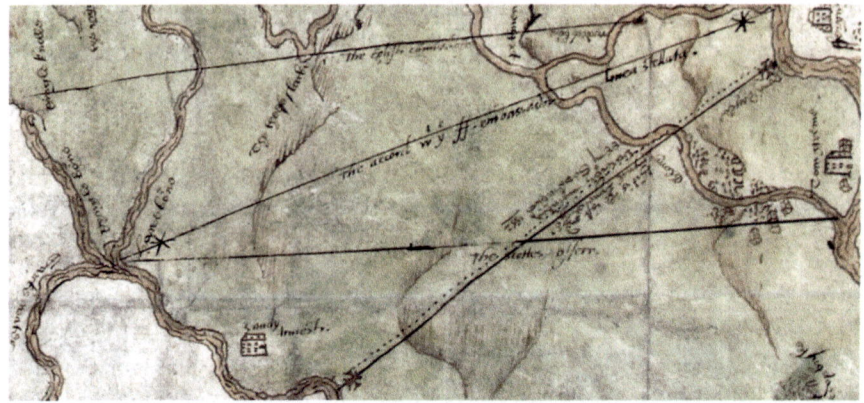

**Fig. 1** Part of a manuscript map of the 'Debatable Land', Anglo-Scottish border, 1552 (Courtesy of The National Archives)

province of Quebec in 'Canada', as it is called on the map.[5] Insight into the map's genesis is found in a letter to the Colonial Office from the Governor of New York province, which mentions a larger map which the Governor General of Canada had sent him. He had this map made, he says 'at a reduced Scale to make it more portable'. This is of interest for cartographic history and gives the source of data shown on the map. As is so often the case, records with the map are vital to its context and interpretation.

## 2.3 Maps for Treaties and Initial Border Allocation

Maps made or used for this phase tend to show a boundary as envisaged during negotiations, before demarcation on the ground. As we have seen with the example of four straight lines on the map of the Anglo-Scottish border, the process to find a preliminary agreement on the location of a boundary was often theoretical rather than based on information about practicalities on the ground. While the Debateable Land map was specially made for those boundary discussions, it was more often a question of diplomats in Europe using whatever maps were already available of overseas territories on which to draw lines. Since political advantage rather than topographic or cartographic reality was usually a driving factor at this paper boundary-making, lines drawn might be vaguely located to indicate a general area envisaged, and may be rather thick and seriously out of scale.

---

[5]Manuscript map of French and British land claims, Lake Champlain, 1767, MPG 1/367 extracted from CO 5/1098. Mitchell and Janes (2014: 180–181).

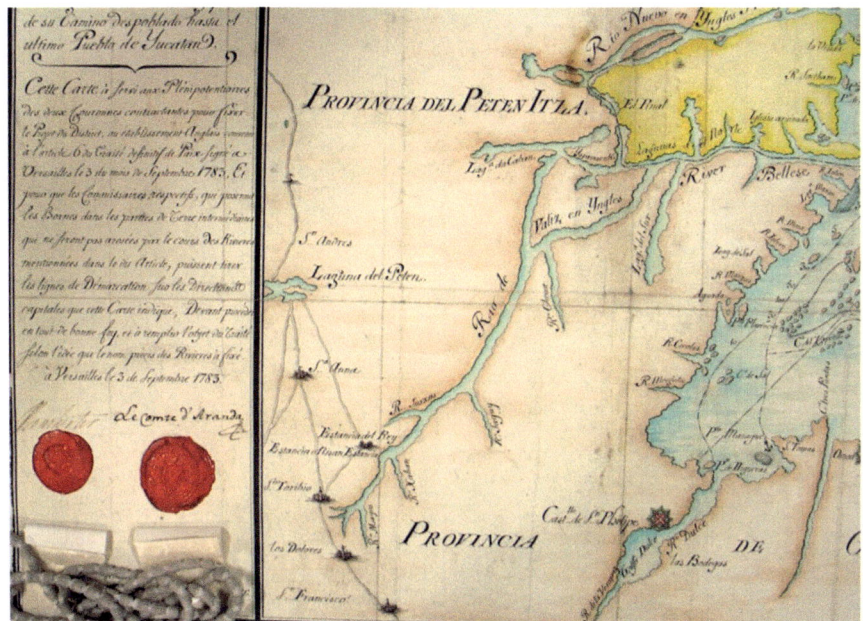

**Fig. 2** Part of South American treaty map, 1783 (Courtesy of The National Archives)

### 2.3.1 Maps in Treaties

Maps found in treaties were drawn in many styles. The examples shown here are taken from the long series of treaties to which Britain has been party, FO 93, which runs from 1695 to 2003 and is still accruing.[6]

An early example from South America was originally annexed to the Treaty of Versailles signed in 1783, and bears the typical seals and signatures of plenipotentiaries, for authenticity.[7] It shows areas now in Guatemala, Belize and Mexico, and the yellow on the part of the map shown indicates the agreed boundary of the English settlement in what was otherwise an area controlled by Spain (Fig. 2).

The treaty of 1898 by which China granted the United Kingdom a 99-year lease of an area called the New Territories[8] contains an outline map which shows in white this large extension of the existing Hong Kong colony, with a dotted line as the boundary with China, shown by pink wash. The place-names are in English and

---

[6]For information about treaties see section 9 of TNA's research guide to Foreign Office records: http://www.nationalarchives.gov.uk/help-with-your-research/research-guides/foreign-commonwealth-correspondence-and-records-from-1782/#7-treaties-1695-present.

[7]MPK 1/155 extracted from FO 93/99/2.

[8]FO 93/23/18. The story of this map and a reproduction are at Mitchell and Janes (2014: 208–209).

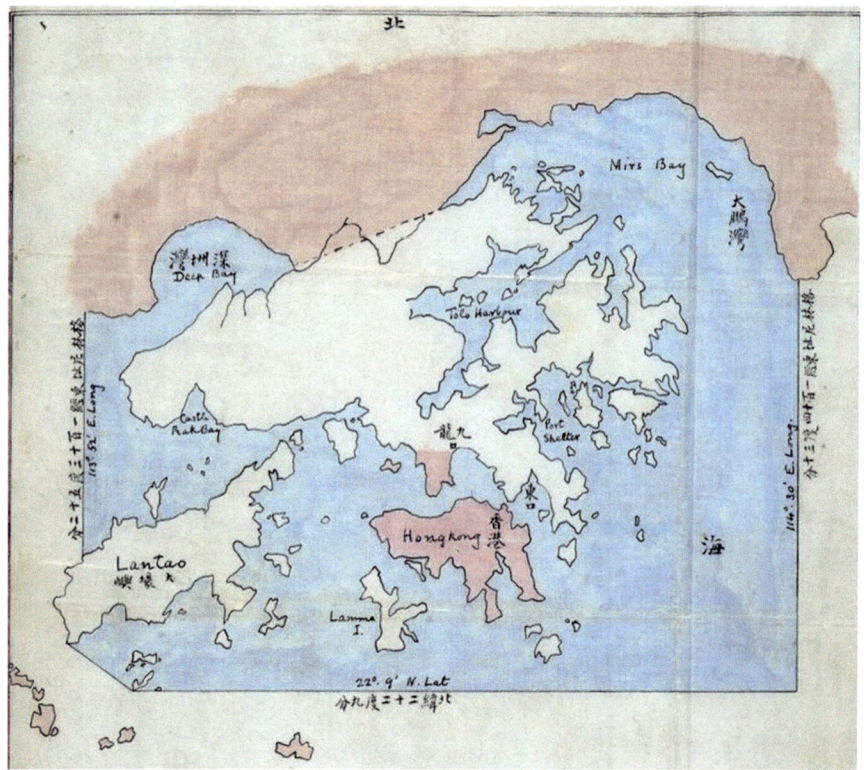

**Fig. 3** Treaty map showing New Territories of Hong Kong, 1898 (Courtesy of The National Archives)

Chinese. It is a small-scale map with not much detail, so the treaty noted that a ground survey was needed to fix it more precisely (Fig. 3).

### 2.3.2 Other Preliminary Boundary Maps

Maps to show outline allocation of boundaries are not just found with treaties. One of the most famous maps among Foreign Office records is the map signed in the lower right-hand corner by Mark Sykes and François-Georges Picot in 1916. On a printed base map was drawn a chinagraph pencil line from the Mediterranean north-eastwards across the desert to the Persian frontier, the basis of their proposed secret agreement to establish British and French spheres of influence in the Middle

**Fig. 4** Sykes-Picot Agreement map, 1916 (Courtesy of The National Archives)

East, supposing the defeat of the Ottoman Empire in the World War I—'B' for the British and 'A' for the French.[9] It illustrates the Foreign Office copy of correspondence about the agreement (Fig. 4).

Another type of preliminary boundary map just shows the outline of an area. A South African example shows part of the border between Cape Colony and Natal in 1885.[10] Despite the word 'sketch' in the title, this map was lithographed, by Dangerfield, a firm which specialized in printing large colour advertising posters. It is a simplified depiction of quite complex boundaries between Griqualand, Pondoland and Natal: lands in yellow-pink were 'Under Her Majesty's Rule', referring to Queen Victoria; those in green, of 'Her Majesty's Allies'. Locations of

---

[9]MPK 1/426 from FO 371/2777. For more on this map and its context see Barr (2011) and a blog on TNA's website, Desplat (2016).
[10]FO 925/1011.

**Fig. 5** Cape Colony and
Natal, 1885 (Courtesy of The
National Archives)

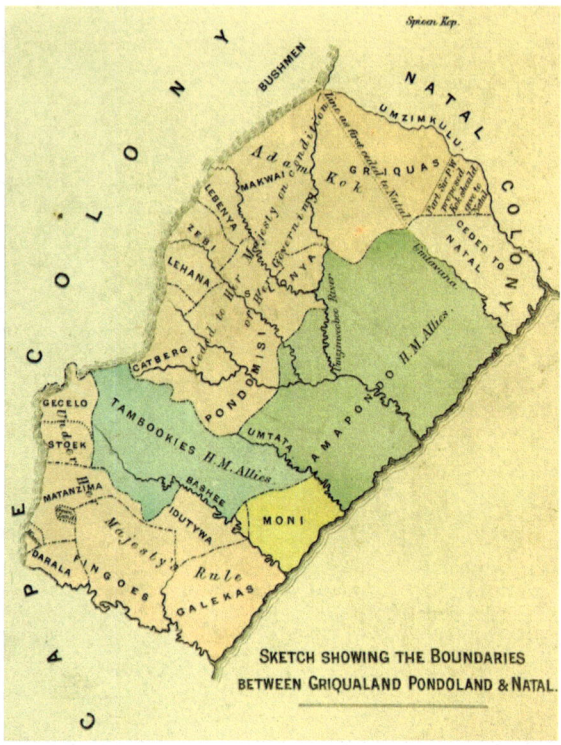

SKETCH SHOWING THE BOUNDARIES
BETWEEN GRIQUALAND PONDOLAND & NATAL.

peoples such as the Galekas, Fingoes and Pondomisi are shown within the border,
with that of the Bushmen at the top. The diagrammatic form of the map aids
presentation of the interplay of boundaries, alliances and peoples in the area, from
the British perspective (Fig. 5).

## 2.4   Demarcation on the Ground

The next stage in the process was to mark the provisional boundary line on the
ground, and to record its location on a map and often in tables of latitudes and
longitudes as well as by written description. The survey maps produced plus reports
of a survey's progress can contribute much to our understanding of the technical
process of mapmaking at specific points in time, and in field conditions varying
from desert to mountains and virgin forest. Maps produced in the earlier phase,
especially in the nineteenth century, were often, too, the first European survey of
remote areas which would otherwise only have been mapped much later, if at all.
Lands adjacent to the boundaries may be shown, and border maps often provided an
outline on which to carry out later detailed topographic survey.

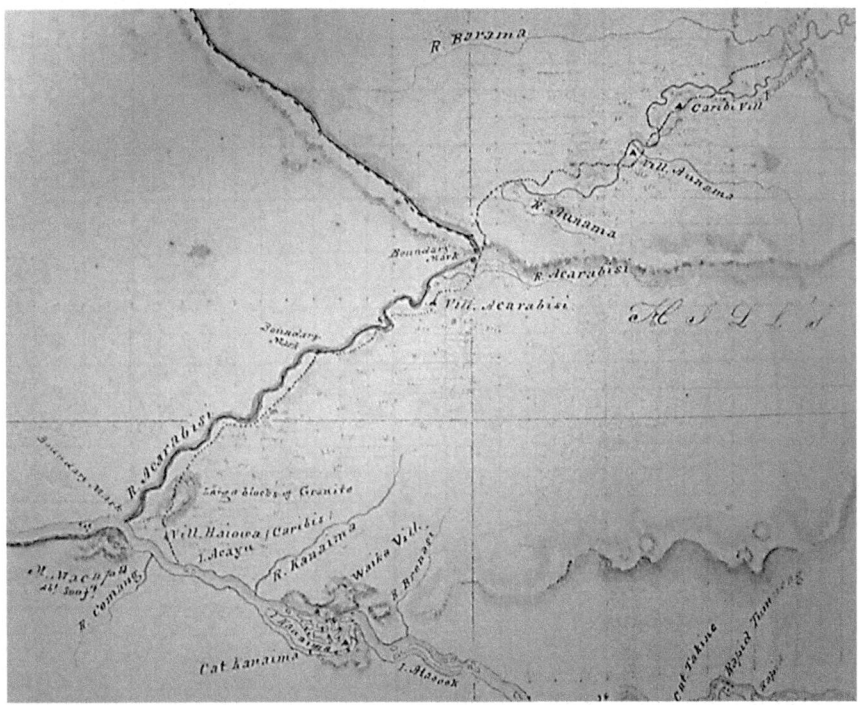

**Fig. 6** British Guiana-Venezuela boundary, part of Robert Schomburgk's manuscript map, 1841 (Courtesy of The National Archives)

Physical survey and mapmaking might also take place before agreement, as we shall see with the case of the British Guiana-Venezuela border. Robert Schomburgk was commissioned by the Colonial Office to survey British Guiana, now Guyana, a colony finally transferred from The Netherlands in 1814. His 1841 map shows the interior with some topographic detail, and its western limit as claimed by the British government.[11] Locations of boundary markers are indicated on the excerpt of this large manuscript map illustrated here. This dispute is still running, as gold and more recently oil have been found in the contested area (Fig. 6).

### 2.4.1  Boundary Commissions

Once a treaty had been ratified by the Powers concerned, generally a Boundary Commission would be established, usually with military survey officers from each country, to fix or demarcate the line on the ground and to produce definitive maps at a relatively large scale.

---

[11]CO 700/BritishGuiana22.

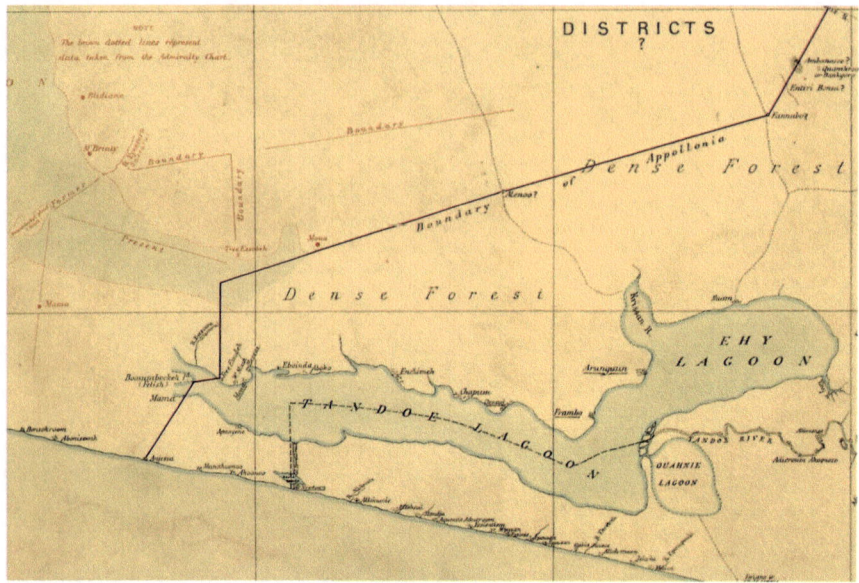

**Fig. 7** Part of Assinie Boundary Commission map showing differing lines, 1884 (Courtesy of The National Archives)

For instance, an Anglo-French Boundary Commission worked on the Gold Coast of Africa (now Ghana) during 1883–1884.[12] A map of their findings shows the boundary area on the Gulf of Guinea coast, with an angled dotted line through Tano Lagoon representing the French claim as the eastern boundary of the French protectorate, and a red line to show where the English considered to be the western boundary of the Gold Coast Colony. Broken brown lines represent what are labelled as 'former' and 'present' boundaries. The English commissioner was a Royal Naval officer, Lieutenant T F Pullen, who explored inland of the contested boundary, and the map also records his findings. This is similar to a strip route map, showing only those hills, swamps and rivers encountered and local boundaries which intersected with the roads, with notes on the passability of roads. Pullen underlined those places for which he had fixed the positions astronomically, and put a question mark against place names whose location was 'conjectured from information obtained'. The map was lithographed at the Intelligence Branch of the War Office (Int. Br. 398) (Figs. 7 and 8).

---

[12]Correspondence and 10 maps produced or used by the Commission is at CO 879/19/10. The map mentioned has the reference CO 700/GoldCoast22/1.

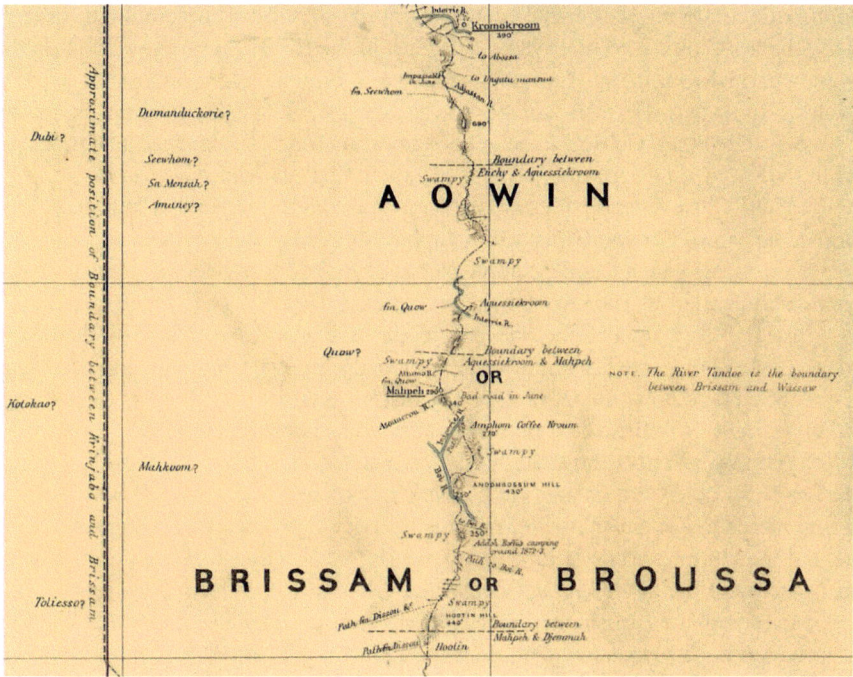

**Fig. 8** Part of Assinie Boundary Commission map showing Lieutenant Pullen's route, 1884 (Courtesy of The National Archives)

In addition to the maps themselves, reports and correspondence can give an insight into the workings of boundary commissions, as well as the work and life of individual mapmakers tasked with recording boundary demarcation. Specific international boundary material is often found among official correspondence and reports from these men back to London, where it was often issued as Confidential Print, for circulation in official circles, with another advantage that it was much easier to read than the original manuscript.

### 2.4.2 Mapmakers on Boundary Commissions

Many boundary surveyors were military men—perhaps officers of the Royal Navy such as Pullen, or more often in the army, specifically in the Royal Engineers. Notable examples of such men include Colonel Thomas Holdich, Major General Sir

John Ardagh and Field Marshal Lintorn Simmons.[13] One of these military officers was Charles Close, who in 1898 as a captain in the Royal Engineers served on the joint British-German Nyasa-Tanganyika boundary commission (now Malawi-Tanzania). He also worked on the Survey of India, and on the boundary between the Niger Coast Protectorate and Cameroon. Close was later head of Military Survey and Director General of the civilian Ordnance Survey.

In 1899 Close's report on his work on the Nyasa-Tanganyika boundary was printed as part of Foreign Office Confidential Print 7115, together with copies of the treaty protocol, general and detailed maps, description of boundary posts, and a triangulation chart.[14] The report discussed technical matters such as the question of two meridians upon which the boundary depended, and observations on terrain and local usage. Close noted that the boundary mostly followed natural features such as a watershed, and how 'the natives readily understand that if a village is on a stream flowing south, it is British: if on a stream flowing north, German'.

The report described how the work was carried out, and gave Close's opinion that he judged the German maps to be made at too large a scale (1:100,000) making it impossible to use a plane-table and thus 'full of sketchy work'. Close related how he and the German commissioner, Captain Hermann, refused to sign each other's maps. Despite this, he commented that the German maps showed no serious discrepancies with the British ones affecting the boundary, and he thought that the final map would be 'in fair agreement with ours'.

The report also gave some insight into Close's thoughts and experiences during his boundary work. His impression of the boundary zone: 'as worthless a bit of country as any in Africa, but [...] as a road into the heart of the continent it deserves [...] attention'. He recounted how one of the party made a river survey in a dinghy and was pursued by a hippopotamus, but lived to tell the tale.

These records thus can give an insight into the extent of co-operation or antagonism between the representatives of European powers working on a boundary, and can reveal their opinions on working methods and equipment, as well as the conditions and difficulties encountered in the work of boundary demarcation in the field. Bad weather, local fauna such as hippopotami, hostility of local tribes, extreme terrain, ill health and local customs (Close mentioned the dangers of blackwater fever and of falling into game pits with lethal stakes)—all might play havoc with Commissioners' schedules.

---

[13]For Holdich's work on the Russo-Afghan boundary through the Pamirs in 1895 see FO 93/81/54, and on the 1902 Argentine-Chile boundary commission in FO 925/1209B; for material on Ardagh's work early in his career on the Bulgaria and Turco-Greek frontiers see his private papers in record series PRO 30/40; Simmons' papers are in record series FO 358, including his work on the Turko-Russian border in Asia. For more on this subject see the blog on TNA's website, Mitchell (2014).

[14]FO 881/7115 Hippopotamus story p. 22; Songwe river course p. 24; comments on German maps and note on direction of stream flow p. 34. A version of Hermann's printed map is appended to the copy of the 1901 Nyasa-Tanganyika boundary agreement in FO 93/36/49.

### 2.4.3  Difficulties of Demarcation

The difficulties encountered during the process of translating preliminary board-room paper borders to the actual ground were legion. Physical features of the landscape such as mountains and rivers were often used as theoretical borders, but found not to exist or to differ wildly in size and location from those on the map drawn in Europe.[15] An example is Kiepert and Moisel's 1895 map of the area between the Rio del Rey and the Cross River, on the Gulf of Guinea.[16] It shows the Anglo-German boundary (now the boundary between southern Nigeria and southern Cameroon) as quite a thick straight red line, an approximate location rather than a sign that the boundary was really miles wide.[17] The Agreement which preceded delineation set the boundary from what it termed the 'right bank of the Rio del Rey', but the actual place of that name is an estuary of the Niger River rather than a river per se, not as specific an area as intended by those making the Agreement.

In another instance, the British and the Americans agreed in principle that the boundary between the United States and Canada would run straight through the north-western point of the Lake of the Woods, without reference to the actual ground. When the American North-West Boundary Commission carried out a survey on the ground, the anomaly they found was such that a chunk of land on the Canadian side of the straight line became an exclave of Minnesota and is still part of the United States today, as recorded by the Commission on a map.[18] These are the kind of problems that arise when transferring a boundary line drawn in a European boardroom to real places half way across the world.

Maps themselves might later cause problems, for example through inaccuracy or vagueness about where exactly a border was to run, or simply because they recorded a natural border such as a river, which changed over time. In the case of the Nyasa-Tanganyika boundary Charles Close commented that the Commission decided to make the Songwe River part of the boundary as it considered that 'it is

---

[15]Gerald Blake *op cit* p. 46 mentions the case of the Argentine-Chile agreement of 1881 which mistakenly assumed that the highest peaks and the watershed in the Andes would coincide, causing a long-running dispute settled only in 1967. He also cites Sir Claude Macdonald's description of the findings of the 1889–1890 Nigeria-Cameroon Boundary Commission, that the European-set boundary based on a map showing an 800-mile river called the Akpayaff turned out to be about three and a half miles in reality.

[16]CO 700/WestAfrica55.

[17]The understanding in this and similar cases usually was to achieve first a temporary line to separate the respective spheres of influence, which subsequently, as ground knowledge and administrative needs evolved, were to be replaced by more 'refined' borders (Professor Imre Demhardt, private correspondence).

[18]MPK 1/55 formerly FO 302/27; the map is shown and its story told in Mitchell and Janes (2014: 206–207).

**Fig. 9** Part of map showing boundaries in Lake Victoria, East Africa, 1924 (Courtesy of The National Archives)

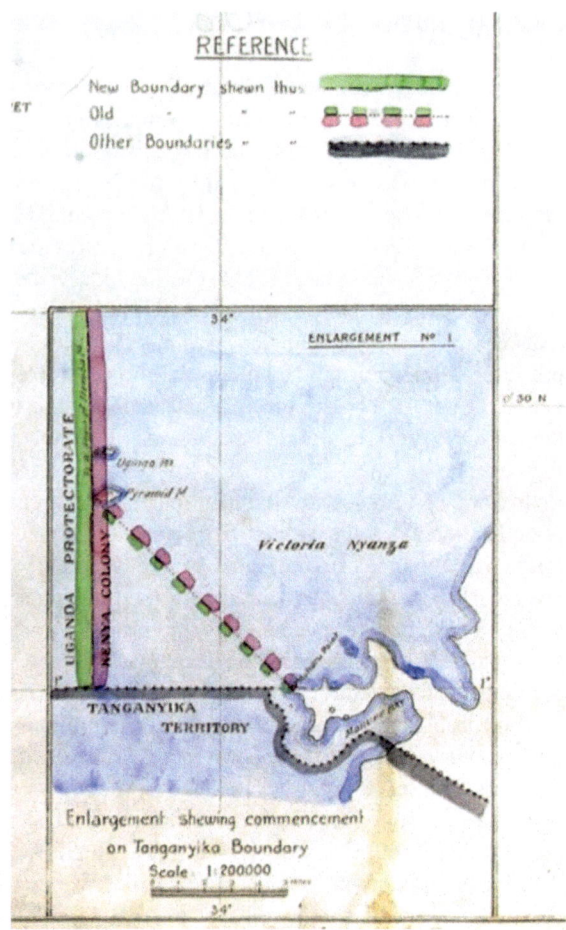

improbable that this river will change its course' based on local evidence that it had not done so within living memory; only after the boundary was signed off would the river's changeability be found to pose a real problem.[19]

Lack of clarity in defining the complex tripartite border across Lake Victoria has contributed to continuing problems in this area of Eastern Africa to modern times. Maps dated 1924 in the former Colonial Office map library (now at TNA) show proposals under the title 'Description of Boundary between Kenya Colony and the Uganda Protectorate', also showing the relevant part of Tanganyika. A main map is supplemented by enlargements of portions of boundary, one of which is illustrated here. It shows the new boundaries of Kenya and Uganda in green and pink as

---

[19]Some later problems are identified in a file on the Malawi Songwe River boundary 1980–1982, at OD 6/1826.

adjacent solid lines drawn in thick pen; to the literal-minded this makes them miles wide compared to the scale bar, when international boundaries have no width. However, this kind of map was likely to have been made to give an idea of the changes and what they would mean. The map also shows as dotted lines the previous boundary lines, and the line of latitude set at 1° south, mentioned in a relevant colonial Order in Council[20] (Fig. 9).

## 2.5  For the Record

### 2.5.1  Record Copies

A certified copy of any detailed work on a boundary, especially by a boundary commission, is likely to be deposited in the archives of all the countries concerned, for the record. This may be manuscript or printed, but in either case will bear the signatures of plenipotentiaries, and was usually attached to the relevant treaty or articles of agreement. Ideally the record will contain all materials needed for clear future interpretation of a boundary. Where Britain was involved, whether her colonies were directly affected or not, a copy of definitive records, including maps showing the agreed boundary, was usually brought back to London by the British commissioner for deposit in the archives of (usually) the Foreign Office.

The 1895 Protocol between Great Britain and Russia concerning the Russo-Afghan boundary through the Pamirs includes a manuscript map signed by both parties.[21] It locates the boundary pillars through the mountains by Roman numerals, but these are difficult to read against the brown-painted deep hachures. Appended tables of latitudes and longitudes of the boundary pillars site them more precisely. The British military surveyor was Colonel Thomas Holdich, who also worked on the Argentine-Chile boundary, one of the military men for whom this became more or less a career.

Boundary questions arise not just on land, but also at sea. A map accompanies a relatively recent Agreement of 1988, to show delimitation of the continental shelf between the UK and the Republic of Ireland.[22] Akin to a sea chart, it shows only the outline of the countries concerned with no topographical detail or place-names. The zigzag pattern of the boundary line may appear rather haphazard compared with the more usual straight line; an annex to the treaty lists the geographical co-ordinates of each of its 'corners'. Maritime boundaries can be hotly contended because of the

---

[20]CO 1047/142; a sketch map in the same portfolio shows only by noughts and crosses with no other locators the spot where the boundary across Lake Victoria was to reach the east shore, which might be rather hard to identify on the ground.

[21]Agreement with FO 93/81/54; album at CO 1069/104; 1901 treaty with map FO 93/36/49.

[22]FO 93/171/23. For more on this subject see the blog on TNA's website Janes (2013).

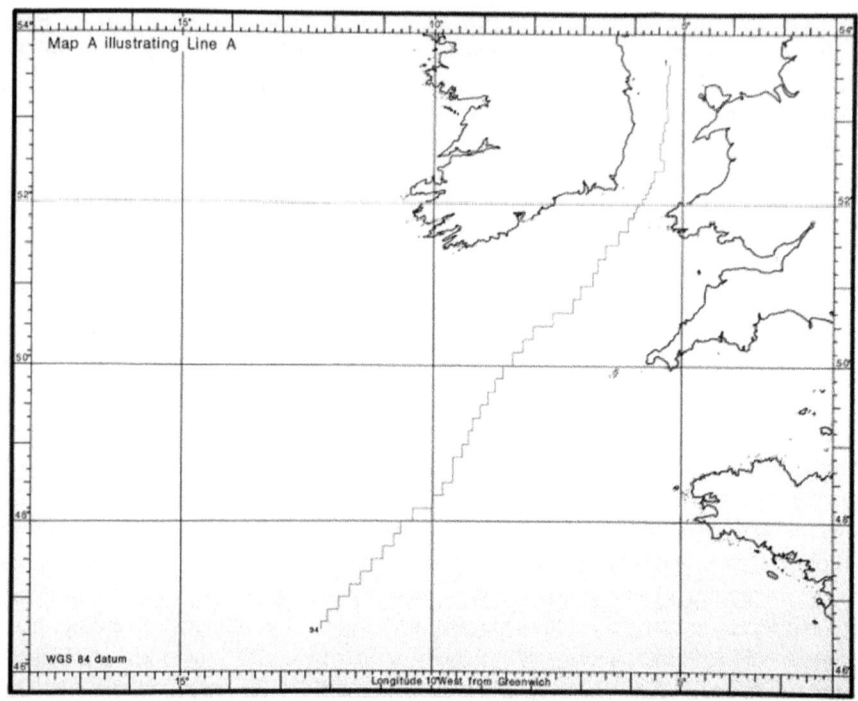

**Fig. 10** Map of the continental shelf between the UK and the Republic of Ireland, 1988 (Courtesy of The National Archives)

potential economic importance of what may lie under the surface of the sea in the form of mineral resources, such as oil and gas (Fig. 10).

Sometimes the work of a boundary commission was published. The hugely long boundary between the United States and Canada along the 49th parallel, roughly 3,500 km, was surveyed and demarcated on the ground by the Joint North-West Boundary Commission between 1872 and 1876. The Joint Commission published an atlas containing 24 photo-lithographed detailed maps of the agreed border area which were sold to the general public. Illustrated here is the title page of the atlas, from the copy in the Foreign Office map library held at TNA, illustrated with a picture of a peak called Mountain-of-the-Chief by the Blackfeet, and now in Glacier National Park, Montana.[23] Many other documents including manuscript maps, photographs and reports were produced in the course of the Commission's work which together give a fascinating insight into its task of setting boundary markers across difficult terrain, from plains to mountain tops and requiring the cutting of swathes through tall virgin forest (Fig. 11).

---

[23]The records of the North-West Joint Boundary Commission are in the record series FO 302. A copy of the atlas has the reference FO 925/1565.

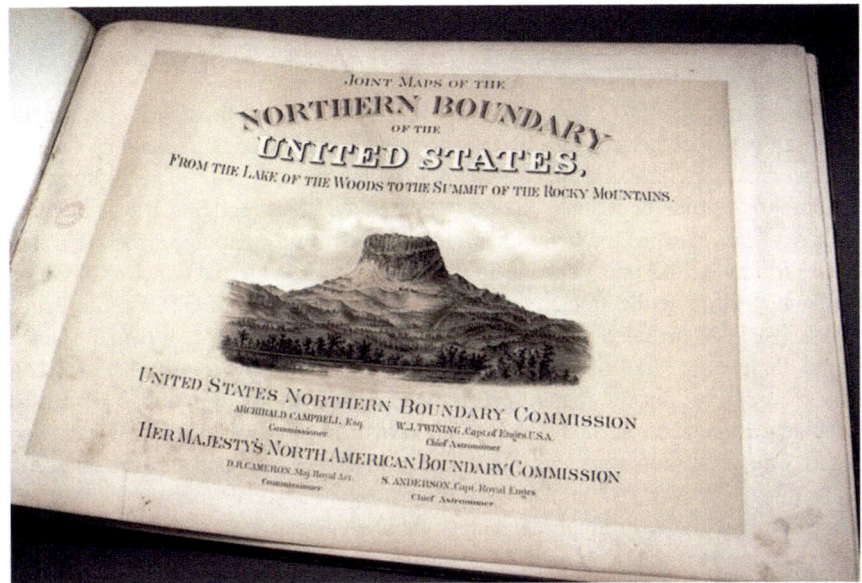

**Fig. 11** Title page of Joint North-West Boundary Commission atlas, 1878 (Courtesy of The National Archives)

These kinds of maps, then, were rarely created or kept solely as cartographic constructs. They are often found with related correspondence or reports about the boundaries concerned. These papers may relate information about the maps, their makers and the circumstances of their making. For instance, the 1898–1899 Nyasa-Tanganyika Boundary Commission archive comprises, in addition to the report of British commissioner Charles Close mentioned above, a range of materials including a photograph album of the boundary pillars and the British and German boundary parties. A map by the German commissioner, Captain Hermann, was printed and appended to the final treaty.

### 2.5.2 Other Types of Publication

A different type of boundary map is the printed compilation map made from original sources. In the late nineteenth century the Foreign Office librarian Edward

Hertslet compiled helpful publications drawing together available data and maps on European and African boundaries as defined in successive treaties.[24]

There is a modern successor to this genre covering Asia.[25]

The arbitration atlas is a specific and important type of boundary cartographic product. A type of factice atlas, a partisan form of historic atlas, it contains versions of relevant maps reduced and redrawn to support each case. The influence of Schomburgk's map of British Guiana has extended across time, as it was the basis for many later maps of the boundary, and versions of this map appear in arbitration atlases for the British and Venezuelan cases, held in the Colonial Office and Foreign Office map library collections respectively (TNA CO 700 and FO 925). The British arbitration atlas included printed versions of Ptolemy's early world map, and of Henry Popple's 1730s map of what he termed the British Americas. It also contained foreign-produced maps, which were collected by a number of British intelligence agencies, so that it would be known in advance what Venezuela proposed. The arbitration atlas was a tool of diplomacy, exchanged with the other side in the dispute. It was also an instrument of propaganda, so versions reproduced in it when compared with original maps, may show changes made as visual persuasion.

It is important when looking at a map to have some idea of the part of the process for which it was made, as its degree of accuracy may reflect its place in that process. For example, a manuscript map with sketched lines may show preliminary ideas about where to locate a border, contrasted with detailed work which may indicate a Boundary Commission working in the field, or the same map printed and signed by both parties for the archives. The fact that a map was printed was, as we have seen with the arbitration atlas, not necessarily an indication that it was officially approved, or that it was not partisan. It is useful for the researcher to be aware of the reason and viewpoint from which a map was made, which may influence what is shown – or what was there but is not on the map.

Many maps were produced in a specific administrative context: an understanding of that context and of all the documents created in conjunction with a given map is essential for a valid interpretation of the map. Maps are no more infallible than any other documents created by humans, and the use of related documents can help to identify deficiencies in the maps. Particularly useful in this context are likely to be manuscript drafts, printers' proofs and other preliminary versions of definitive maps.[26]

---

[24]Hertslet E (1875–1909). Note that while TNA holds maps and other records of European boundaries, they were outside the geographical focus of this volume.

[25]Prescott (1975).

[26]An example is an unrevised proof of a confidential military map showing the border area of the Anglo-Egyptian Sudan in 1901, IDWO 1567, which may be unique owing to manuscript additions and overprints. MR 1/1786/8 from Foreign Office Abyssinia correspondence FO 1/45.

# 3   The Importance of International Boundary Maps for Research

Historic maps are important for borders research, where their degree of clarity may either contribute to disputes or help to inform discussions and contribute to the peaceful resolution of boundary questions today. Their value as an object of study for the history of cartography is also clear on a number of grounds.

Firstly, there are huge numbers of them, which gives a wide field for study of the evolution of map types over time and comparison of processes and products of different countries. The initial need by colonial powers to define boundaries often produced outline maps that served as the basis for later detailed country topographical surveys. Where there are multiple maps of the same border area, these allow comparison not only of how the landscape has changed over time, but also the different ways in which maps can represent the same area. There is also a huge variety of terrain depicted: border maps and surveys made lines across sand, snow and ice, water, forests, plains and mountains around the globe.

Secondly, they are the products of a wide range of mapmakers, from politicians in European offices drawing lines on maps with thick-nibbed pens and scant regard for scale, to seasoned military surveyors, and colonial governors with knowledge of the ground. Officers had been trained to make maps – it was for this reason, besides experience in rough terrain, that Royal Engineer officers were often engaged in boundary surveys in which Britain was involved, because they might have to conduct their own basic surveys of the region adjacent to the boundary to give the border line context and fix key points on it in relation to prominent nearby physical features such as hills and water features.

Maps and related records such as correspondence or reports may also give insight into the degree of map awareness among policy makers, colonial governors and civil servants; for instance by showing how far these people were comfortable in reading, drawing and using maps to carry out their work. They may also give information about the maps, their makers and the circumstances of their making. Text and map together tell a broader story than either alone, and give insight into maps and their role specifically in boundary making and more widely into their social and political role in the colonial arena.

# References

Barr J (2011) A line in the sand. Simon & Schuster, UK

Beech G (1986) Sources for the History of Canadian Cartography in the Public Record Office: the Alaska Boundary. Association of Canadian Map Libraries Bulletin, 58/59

Blake G (1995) The depiction of international boundaries on topographic maps. IBRU Boundary and Security Bulletin, April, pp 44–50

Desplat J (2016). https://blog.nationalarchives.gov.uk/blog/dividing-bears-skin-bear-still-alive-1916-sykes-picot-agreement/, 16 May 2016

Hertslet E (1875–1891) The Map of Europe by Treaty, 1814 to 1891 4 volumes, and The Map of Africa by Treaty (1895–1909) 3 volumes both HMSO London

Janes A (2013) The boundary line. https://blog.nationalarchives.gov.uk/blog/boundary-line/, 7 November 2013

Mitchell R (2014) Lines on the map: records of international boundaries. http://media.nationalarchives.gov.uk/index.php/lines-map-records-international-boundaries/, 25 November 2014

Mitchell R, Janes A (2014) Maps: their untold stories, map treasures of The National Archives. Bloomsbury, London

Prescott JRV (1975) Map of mainland Asia by treaty. Melbourne University Press, Australia

**Rose Mitchell** is map specialist at The National Archives of the United Kingdom. She is the co-author of *Maps: Their Untold Stories*, for which she wrote chapters on colonial maps and sea charts, among others. Rose contributed a chapter to the published proceedings of the 2014 ICA symposium on military cartography, entitled *Contours of Conflict*. She gave papers at the ICA colonial maps symposium in Utrecht in 2006 and at the 2010 IMCoS symposium 'Britain, power and influence', for which she also provided a display of original overseas maps. Rose runs workshops on archival research with the International Boundaries Research Unit of Durham University. She is a Fellow of the Royal Geographical Society.

# Toponyms

# German Names in the Kilimanjaro Region

Wolfgang Crom

**Abstract** When the colonial powers finally divided up East Africa, in 1885 the Kilimanjaro volcano massif fell to the German Empire. At the time, the assault on the summit of Kilimanjaro was regarded as an important national task and tackled as a scientific research project. For various reasons, its first climber Hans Meyer had given numerous names to landscape elements in the higher uncultivated and uninhabited summit region. These names became part of the international nomenclature. In addition to descriptive toponyms, there are mainly names of individuals with added generic terms on his maps. Subsequent explorers have continued to apply the practices introduced by Meyer for naming prominent landscape elements. The designations have been included on official topographic maps as well as on current trekking maps.

## 1  Kilimanjaro or Kibo?

At the western end of Sanssouci Park in Potsdam, there is the New Palace, built under King Frederick II of Prussia, but in late nineteenth century it was the summer residence of the German Emperor William II. One of the four banquet halls is designed as Grotto Hall and is decorated with minerals, shells and semi-precious stones. In the centre of these decorations, there is a stone labelled to be a rock sample from the main peak of the Kilimanjaro. The first climber Hans Meyer had brought two pieces of lava rocks with him back home to Germany and had given one of them as a present to the Emperor. After the ascent on 6 October 1889, Hans Meyer described his heroic act: '[…] and in virtue of my right as its first discoverer christened this hitherto unknown and unnamed mountain peak—the loftiest spot in Africa and in the German Empire—*Kaiser Wilhelm's Peak*' (Meyer 1891: 154). The Emperor accepted the stone with pleasure and ordered that it should be made a part of the decorations in the Grotto Hall after initially he had used the stone as a

W. Crom (✉)
Staatsbibliothek zu Berlin, Berlin, Germany
e-mail: wolfgang.crom@sbb.spk-berlin.de

© Springer Nature Switzerland AG 2020                                                            229
A. J. Kent et al. (eds.), *Mapping Empires: Colonial Cartographies of Land and Sea*,
Lecture Notes in Geoinformation and Cartography,
https://doi.org/10.1007/978-3-030-23447-8_13

paperweight. In the 1950s the stone disappears and, albeit several versions about its whereabouts, remained stolen. Since then another lava stone has filled the empty space; a stone, said to have been brought from Africa by a female tourist.

This perhaps lesser known anecdote leads to a well-known fact: in May 1848, the missionary Johannes Rebmann was the first documented European to have seen the Kilimanjaro and he was also the first to give a report of the impressive sight of the snow-covered summit crater (Fig. 1). Rebmann relayed the name Mount Kilimanjaro, which stuck until today, to Europe. However, it is necessary to investigate whether or not this passed on name is correct. He ascribed the name to his first sources, the Swahili people from the coastal region around Mombasa, where he was stationed as a missionary. In his diary notes, dated 7 April 1848, that is before his first sight of the mountain, there the following explanation is given: 'Die Ableitung des Bergnamens Kilimandaro wäre also zu machen von Kilima da Aro, d. h. Berg der Größe, großer Berg' (Rebmann 1997: 13).[1] The missionary and philologist Johann Ludwig Krapf, who published not only Rebmann's but also his own travel experiences in that region, added another explanation, namely that *Kilimanjaro* has to be understood as '"mountain of caravans" (Kilima, mountain,— Jaro, caravans)' (Krapf 1860: 255).[2] This interpretation should point out that the mountain, which easily can be seen from very far away, acted as a helpful landmark for caravans. Rebmann himself soon discovered that the Swahili people from the coast were possibly misleading informants and in his report he explains: 'On my first journey my guide had misinformed me, when he said that the people of Jagga had no word for snow; but when I asked the natives of Jagga themselves, their various statements, for example, that the Kibo when put into the fire turns into water, convinced me that they not only knew it as "Kibo", but knew no less well its nature and properties' (Krapf 1860: 255). One of the various stories by the Chaga or Swahili the people inhabiting the southern flank of the volcano massif, is the tale that the mountain is inaccessible because it is a gold and silver mountain and its interior is inhabited by evil spirits. Therefore, Rebmann introduced a new name: *Kibo*, as the Chaga people call it. He translated Kibo as white, snow, ice, cold or similar, which suggests snow coverage (Rebmann 1997: 33) and found that it was synonymous with 'Beredi', the Suahili term for cold (Krapf 1860: 236). Therefore, he put 'Mt. Kilimandjaro covered with eternal snow' on his sketch map (Rebmann 1850). Rebmann did not limit the term to the main summit, however, and in his writings there is no indication of a name for the distinguished second or eastern summit, which he also described and which is clearly noticeable on an early printed map (Petermann 1859). The often opaque classification is not to the least a result of how missionaries perceived things through their European perception. This concerns the structure of indigenous languages as well as the description of landscapes.

---

[1]'The deduction of the mountain's name Kilimanjaro would be made from Kilima da Aro, i.e. mountain of greatness or great mountain' (translated by the author).

[2]For the German version see Krapf (1858, 2: 73), where he explains his addition in a footnote: '"Berg der Karawane" (Kilima = Berg, Dscharo Karawane)'.

To come to a common understanding on landscape elements, Europeans and their African informants inevitably had to fail, because there were neither terms nor toponyms for a direct word to word translations. Instead, the desired information had to be understood from the description. In this context, Voigt (2012: 34) speaks about a transformation of 'Sprachbilder in adäquate Kartenelemente'.[3]

The phenomenon of snow and ice-covered mountains in equatorial Africa, described for the first time by Rebmann and Krapf, was confirmed by the travels of Carl von der Decken, undertaken in 1861 together with geologist Richard Thornton and in 1862 accompanied by the naturalist Otto Kersten. As a result of their work it had—for the first time—become possible to describe with some precision the morphological structure of the volcano massif; however, the uncertainty regarding naming and classification still remained. The map to log von der Decken's first journey, drawn by Kiepert (1863), and the map of Bruno Hassenstein (Hassenstein 1868) are both based on the observations by von der Decken together with Thornton and Kersten. Both maps refer to the *Schneeberg* [=*snow mountain*] *Kilima-Ndscharo* either in the map title or naming it directly beside the mountain. On Hassenstein's map, the Kilima-Ndscharo is in addition named as the *Kibo der Wadschagga*, but printed in another character style. The two main peaks are already distinguished, but they do not yet have their own toponyms. Instead they are simply called *Grosser* and *Kleiner Kilima-ndjaro* or *Gipfel*. However, on a map dated 1864, which Hassenstein also created according to the travel accounts of different authors, we find yet another attribution: *Schneeberg Kilima-Ndscharo der Wateita* (Hassenstein 1864).

Rebmann had already pointed out the difficulties, which arise when different groups name the very same object. Onomasticians, the experts on geographical names, know very well this phenomenon, which is also evident on the maps of early explorers in East Africa. On a map created by Oscar Baumann in 1894, one can find, besides the term *Kilima-Njaro,* yet another toponym, which is set in parentheses: *Dónyo Ebor i.e. Weisser Berg*, which is a name used by the Maasai people (Baumann 1894). This means that three different language groups use three different terms for the highest mountain of Africa: Kilima Njaro is used by the Swahili and Wateita people, Kibo by the Chaga people and Dónyo Ebor by the Maasai people. However, the question remains, whether the toponyms refer either to the whole mountain massif or only the white covered summit area.

The spellings *Kilima Ndscharo* or *Kilima Njaro* on the first maps still show the attempt of etymological derivation and pronunciation. Until today, however, no definite clarification of this toponym existed. Simo (2002) speculates about the possible translation *Kleiner Hügel* [=*small hill*] *von Njaro* as a result of an indirect and therefore inadequate question-answer dialogue (cf. Hamann and Honold 2011: 59). Regarding the naming of the highest African mountain there is the frequent occurrance of communication barriers and misunderstanding when questioning people: What was said to Rebmann and what did he understand? It will probably no longer be possible to clarify whether *Kilimanjaro* originally was a toponym or not,

---

[3]'verbal images in adequate map elements' (translated by the author).

**Fig. 1** Richard Kiepert *Deutscher Kolonial-Atlas* (1:3,000,000) was published in the 1890s in a general scale to get an overview for use in the colonies. Sheet Aequatorial-East-Africa (amended 1895) shows the Kilimanjaro and adjacencies with the routes of the most important explorers (Courtesy Staatsbibliothek zu Berlin, Kart. C 16732-4)

or whether the inhabitants really had a name for the whole mountain massif or only its peak region. However, this name, which for European ears has a wonderful exotic sound, has become the well-established name for the impressive appearance of the volcanic mountain.

## 2  New Toponyms by Hans Meyer

When the colonial partition of East Africa happened in 1885, almost all of the Kilimanjaro massif became part of the German protectorate Deutsch-Ostafrika. Along with the creation of administrative structures, economic development and military control, the scientific exploration of the snow-covered mountains became a particular focal matter in East Africa. It was regarded as a major national goal to reach the summit and that was tackled as a research project under the direction of Hans Meyer.[4] The efforts of his pushes for the summit region are impressively

---

[4]Hans Meyer (1858–1929) was a member and at times director of the publishing house Meyer in Leipzig, the family wealth permitted the financing of his expeditions.

noted on the early maps by the names given to numerous before his ascents still unknown and therefore unnamed landscape elements in the Kilimanjaro region. In fact, Meyer produced a real flood of names and neologisms when he entered the uninhabited and uncultivated heights of the massif. Further down we will a closer look at them.[5] Already before Meyer there had been some attempts at climbing the peak, but they only reached the lower ice boundary. On the maps of these expeditions no new toponyms were recorded.[6]

On his first attempt, Meyer and an official of the Deutsch Ostafrikanische Gesellschaft (German East Africa Company), Ernst Albrecht von Eberstein, at an altitude of about 4500 m, reached the saddle plateau between Kibo and Mawenzi and even progressed further to the lower ice boundary at about 5000 m. The relevant map draft does not contain any toponyms, but only descriptive indications of distinctive landscape elements like *Verwitterte Lavahügel* [=*weathered lava hills*], *Lavaströme* [=*lava flows*], *Schneefelder* [=*snow fields*], *Gletscher* [=*glaciers*], *Asche* [=*ashes*] *and so on*. The central crater of the Kibo is represented as an ice dome (Meyer 1887).[7]

In 1889, Meyer and the Austrian mountaineer Ludwig Purtscheller, as previously mentioned in the opening paragraph, successfully ascended the summit, where they gave a name to the highest point of Africa and the German Empire. Meyer was successful in making money from his travels by publishing them as fascinating and widely read scientific adventures, furnishing them with a number of maps in different scales.[8] Map II,[9] created by Bruno Hassenstein, has an increased density of toponyms, especially on the southern side of the mountain, where it shows settlement areas, political units and the landscape names of the Chaga people. What is remarkable, is that there are no names for settlements with only a few exceptions like the Bomas (palisade fortified buildings or ruler's residences) as well as the few places of colonial administration or mission stations. Typical for this map type are descriptions along vaguely dotted lines with terms instead of the use of symbols. The northern part of the mountain contains only such designations, which have been obtained by observations from a distance or observations by other travellers without the possibility of a precise localization (*Grassland and Herbs,*

---

[5]Further information on the history of mapping the Kililmanjaro s. Brunner (1989, 2004), Demhardt (2000), Pillewizer (1941), Sriguey and Cullen (2014) and Uhlig (1909). General information on colonial cartography s. a. Crom (2003), Demhardt (2006), Eckert (1924), Finsterwalder and Hueber (1943), Hafeneder (2008), Obst (1921) and Sprigade and Moisel (1914).

[6]Compilations of first ascents and pioneers at Kilimanjaro s. Meyer (1891: 6–20) and http://kilimanjaro.bplaced.net/wiki/index.php?title=Erste_Besteigungen. Accessed 27 November 2018.

[7]Map without scale. Supplement to: Meyer (1887) Vorläufiger Bericht über meine Besteigung des Kilimandscharo, Petermanns Geographische Mitteilungen 33: 376–378).

[8]Shortly after the German edition Meyer also published an English edition of this expedition which is the basis of the following examination (Meyer 1891).

[9]A Map of Kilimanjaro. 1:250.000. In: Meyer, Hans: Across East African Glaciers, 1891. Originally published as: Originalkarte des Kilima-Ndscharo 1:250.000. In: Meyer, Hans: Ostafrikanische Gletscherfahrten, 1890. https://archive.org/details/acrosseastafrica00meye/page/n501. Accessed 27 November 2018.

*Upper* and *Lower Line of Primeval Forest, Grassy Terrace with permanent Maasai Kraals*). Map III[10] shows the first detailed cartographic representation of the landscape in the summit region, although some things could only have been guessed. Altogether the map shows only a few toponyms, they are mostly descriptive names for peculiar landmarks: *Triplets, Red Cent, Front Hill* or *Red Wall* along the route, which were a help for orientation or offered a camp and protection for climbers: *Lava Cave, Kibo Camp* and *Rock of the Four men* (or in the text Meyer 1891: 139 *Four Men's rock*).

Remarkable are the now well-legible and identifiable toponyms named after a person. They can be found in the area of the peaks of Kibo and Mawensi: beside *Highest Peak* or *South Peak,* there is a second main summit with the name *Purtscheller Peak.* At Kibo one now can read alongside *North Glacier* or *Cone of Eruption* the names *Hans Meyer Notch, Ratzel Glacier* and the already mentioned *Kaiser Wilhlem Peak.* The designations of *Purtscheller Peak, Ratzel Glacier* and *Kaiser Wilhelm Peak* are not meant as parallel German or English names of already existing endonyms, but they are new creations referring to unnamed landscape elements. Meyer also provides explanations for his inscriptions, for example: 'In memory of a friend, we called this, our first glacier on Kilimanjaro, the "Ratzel glacier"' (Meyer 1891: 145 ff.) or he honours his climber Ludwig Purtscheller, probably in order to mitigate his disappointment as he has not reached the Mawensi summit (Meyer 1891: 177). The *Hans Meyer Notch* was originally described as *Ostscharte auf dem Ringwall*; it was named after him, because in 1887 he had failed to successfully conclude his ascent of Mount Kibo (Meyer 1891: 180). It is a trench at the crater's outer rim on the northeast end of the Ratzel Glacier.

Until the next ascent of Mount Kilimanjaro by Meyer in 1898, there were no new toponyms on the maps in the holdings of the Map Department of the Staatsbibliothek zu Berlin. However, the morphological division of the mountains into in three parts, namely Kibo, Mawensi and Schira Ridge are well worked out. The now defined third area of the summit, for which there is obviously no endonym, has a name derived from the local Chaga language, whose people settled the landscapes at the southwestern slopes of the Kilimanjaro.[11] The maps created around that time are mostly compilations from already published maps or reports of various travellers. The cartographic focus is mainly on the corrections regarding the positions of objects. The respective sources are generally given in the title.[12]

---

[10]A Map of the Upper Kilimanjaro. 1:85.000. In: Meyer, Hans: Across East African Glaciers, 1891. Originally published as: Spezialkarte des oberen Kilimandscharo. 1:85.000. In: Meyer, Hans: Ostafrikanische Gletscherfahrten. 1890. https://archive.org/details/acrosseastafrica00meye/page/n175. Accessed 27 November 2018.

[11]Instead of Schira, the government used the landscape names of Kibonoto or Kibongoto of the Swahili people (Meyer 1900: 186).

[12]Cf. Heymons (1891) (after Höhnel/Meyer), Hassenstein (1893) (after Meyer/Höhnel/Baumann) and Baumann (1894) (after Fischer/Spring/Werther).

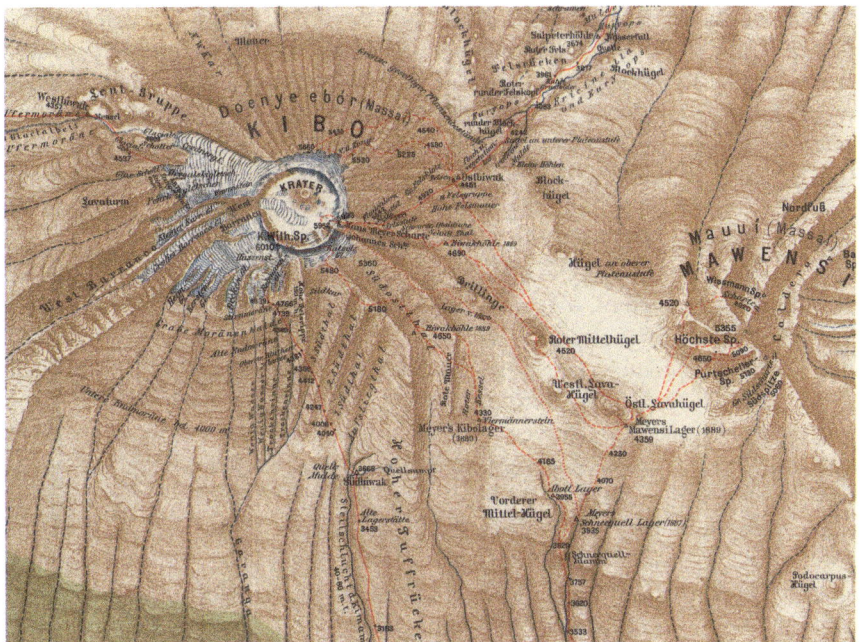

**Fig. 2** Paul Krauss *Spezialkarte des Kilimandjaro. Nach den neuesten Aufnahmen von Prof. Dr. Hans Meyer und mit Benutzung von Messungen, Entwürfen und Skizzen von Hauptmann Johannes, Dr. Carl Lent, Oberst v. Trotha, Graf Wickenburg, Dr. A. Widenmann u. a. 1:100,000* was published in 1900 as an appendix to Meyer, Hans Der Kilimandjaro; Reisen und Studien Reimer, Berlin (Courtesy Staatsbibliothek zu Berlin, Us 1109)

Meyer describes the results of his last visit in 1898 in his book 'Kilimandjaro', which was published in 1900. Within this book is the *Spezialkarte des Kilimandjaro 1:100.000*, created by Paul Krauss (Fig. 2). The first aim of this trip was the systematic research of the glaciation of the Kibo, which is now clearly demonstrated by the designation of individual glaciers. At the west side of Kibo, Meyer now identifies several ice streams with various glacial lobes separated by ridges. He christens them with names of important geographers, geologists and glaciologists: *Credner-, Drygalski-* and *Penck-Gletscher*[13] (Meyer 1900: 174).[14] Further to the southwest, Meyer identifies two glaciers in the deeply incised erosion gully of the volcanic cone, which according to the then valid nomenclature was called a Barranco.[15] Accordingly, he names them *Kleiner* and *Großer*

---

[13]Erich von Drygalski (1865–1949); Hermann Credner (1841–1913); Albrecht Penck (1858–1945).

[14]For translations of the generic names see the list at the end.

[15]On the special map of 1890, this gully is still called Westspalte (Western Branch).

*Barranco-Gletscher* (Meyer 1900: 181).[16] At the southern slope, he interprets the glacial situation, which consists of four glaciers with six glacial lobes. From east to west he sticks to his pattern of naming in honor of individuals which he deems important: *Rebmann-, Decken-, Kersten-* and *Heim-Gletscher*[17] (Meyer 1900: 222). When comparing the 1900 map with the one of 1890, one can see south of Hans Meyer Notch a further incision in the eastern crater rim. This new incision was formed as a result of deglaciation, which occurred in the short span of that decade, so Meyer named the new notch in appreciation of Kurt Johannes[18] *Johannes-Scharte* (Meyer 1900: 354).

During this expedition Meyer, again, honours his European companion with a toponym. The noted painter of mountain sceneries, Ernst Platz[19] becomes the namesake for a prominent volcano cone north of the Schira Ridge: *Platz-Kegel* (Meyer 1900: 178). Other remarkable rocks, which stick out from the ice masses of the glacier and therefore are very helpful for positioning, bear the names of the cartographers *Hassenstein* and *Ravenstein*.[20] As 'Stein' (stone) is already an element of both family names, no generic term is added. Another remarkable landscape element, which has no additional morphological description, Meyer in commemoration names after the geologist Carl Lent, who was murdered in Rombo in 1894: *Lentgruppe* (Meyer 1900: 162).[21] Hans Meyer does not only honour geoscientists by using their family names for toponyms, but also personalities of significance in the colonial context, missionaries or politicians on the map: *Wissmann-Spitze* (Meyer 1900: 155), *Liebert-Spitze* (Meyer 1900: 308), *Krapf-Hügel, Volkens-Hügel* (Meyer 1900: 114), *Bismarck-Hügel, Moltke-Stein*.[22] From the literature it is not always clear, whether these are neologisms were created by Hans Meyer or whether he used older sources. The individual objects are prominent landmarks like hills, cones or rocks, but only some are located in the summit area or

---

[16]Jaeger (1909: 129) speaks out in favour of the term *Breschengletscher*.

[17]Johann Rebmann (1820–1876) 'Discoverer' of Kilimanjaro; Carl Claus von der Decken (1833–1865) explorer; Otto Kersten (1839–1900), geographer; Albert Heim (1847–1939) geologist, glaciologist.

[18]Hauptmann (captain) Kurt Johannes (1864–1913) climbed Kilimanjaro in 1898 to Kibo's rim. He made his own mapping surveys available to Hans Meyer.

[19]Ernst Platz (1867–1940).

[20]Bruno Hassenstein (1839–1902). When talking about Ravenstein a whole dynasty of cartographers has to be mentioned: Friedrich August (1809–1881), Ernst-Georg (1834–1913), Ludwig (1838–1915), Simon (1844–1932), Hans (1866–1936), probably referred to is Ernst-Georg. The observations of Geierraben, White-necked raven (*Corvus albicollis*), may have been decisive for this naming, which inspired the word game.

[21]Carl Lent (1867–1894).

[22]Hermann von Wissmann (1853–1905) Africa explorer, Reich Commissioner and Governor of German East Africa; Eduard von Liebert (1850–1934) Governor of German East Africa; Johan Ludwig Krapf (1810–1881) missionary and philologist; Georg Volkens (1855–1917) geobotanist; Otto von Bismarck (1815–1898) Reich Chancellor; supposedly Helmuth von Moltke (1800–1891) General Field Marshal, Chief of the Prussian General Staff.

in especially exposed places. They are to be found across the map sheet, mostly offside the much-used routes.

Hans Meyer may have had different motivations for using the names of glaciologists of merit. On the one hand, this research field had just emerged, but already produced solid theories and scientific consensus. On the other hand, it could be interpreted as an indication of where he sees his own place within this young discipline. Whereas the first Europeans, who had seen the Kilimanjaro, spoke of a snow or firm ice cover, Meyer was able to prove that indeed there is a tropical glaciation in East Africa almost exactly on the equator, which put him on the same level as the other glaciologists.[23]

For the sake of completeness, it is important to include a word on *Baumann-Hügel* (*Baumann Hill*) and the *Kersten-Hügel* (*Kersten Hill*), two more exploration pioneers, who are honoured this way. Since 1890, the volcanic cones southeast of the Kilimanjaro massif can be found on the maps in various forms, since 1910 they are set in parentheses under the endonyms *Boro* and *Vilima Viwili*.[24] These two examples show an early turn away from the use of exonym neologisms on the official colonial maps and a tendency towards existing endonyms. This leads to the question: how durable are the names found in the Kilimanjaro peak region?

## 3 International Use of German Names

Of particular interest is the question, to which degree these inscriptions found international recognition. An initial answer might be found in the official German map series *Karte von Deutsch-Ostafrika 1:300.000* (Fig. 3a). After just 16 years, the 35 map sheets of this series were completed in 1911 (Passarge 1912; Brunner 1989, 2004; Demhardt 2000: 172 ff.). At the time of its production, this map series was deemed of such high quality that even before the transfer of the territory to the League of Nations the British General Staff and the British Survey began reprinting the maps (Geographical Section General Staff 1915 ff.). With appreciation, a note on the map sheets points out: 'NOTE: Copied from a German map on the same scale' (Fig. 3b). A comparison of these two editions provides a good perspective on the sustainability of German toponyms. The Kilimanjaro region is on sheet B 5. At first sight, the map image of the British map series is identical to the German map series, as could be expected. When taking a closer look, however, one discovers that there are considerably less toponyms, also less endonyms. In the British

---

[23]'Jedenfalls erscheinen mir diese Ausblicke weitreichend genug, um die Entdeckung der einstigen großen Kibovergletscherung für das wichtigste Ergebnis meiner diesjährigen Expedition zu halten' (Meyer 1900: 227). 'In any case these views seem to be far-reaching enough, in order to convince me that the discovery of the former large glaciation of Kibo is the most important event of my expedition in this year' (translated by the author).

[24]For example, Heymons (1891 First naming), Höhnel (1892) and Hassenstein (1893), the endonym names e.g. Sprigade and Moisel (1910: 18).

**(a)**

**(b)**

Fig. 3  a *Karte von Deutsch Ostafrika 1:300,000*. Sheet B 5 Kilimandscharo was produced in 1911 by Paul Sprigade and Max Moisel (Courtesy Staatsbibliothek zu Berlin, Kart. C 16739-B 5). **b** The Geographic Section General Staff reprinted the German official maps from 1915–1917: German East Africa 1:300,000. Sheet Kilimanjaro B 5 (Courtesy Staatsbibliothek zu Berlin, Kart. C 16740/50-B 5)

edition, there is no equivalent to the name of the summit, the Kaiser-Wilhelm-Spitze, it remains unnamed and is simply indicated by a point and the indication of its altitude.[25] In any case, the summit region is represented nearly without toponyms, only the *Credner Glacier* is recorded. This may be due to the small scale, since in the German edition there are also only few indications in abridged spelling: *K.Wilh.-Sp.*, *Ratzel-Gl.*, *Credner-Gl.* and *H.-Meyer-Sch^te.*[26]

However, toponyms related to individuals like *Wissmann Peak*, *Purtscheller Peak*, *Liebert Peak*, *Krapf Hill* and even the *Bismarck Hill*, can be found on the map. The British paid tribute to some representatives of the German colonial empire. Scientific merits or significance in colonial politics may be the reasons, why personal names are retained as a kind of acceptance. The generic terms of the designation are translated, as other toponyms of the map sheet. It must be noted that there could not have been any form of final edit on the names, since there are designations like *Platz-Kegel* or *Lentgruppe*. Finally, there are more examples, also regarding toponyms, which do not include personal names, such as *Rote Mauer*, *Bastions Bach*, *Schiranadel*, *Europäer-Rücken* or *Galuma-Höhle,* yet with umlauts, where even the generic terms have not been translated.

Toponyms from the German colonial period were also used in the map series 1:50,000, which after World War II came about as a result of surveying flights carried out by the Royal Air Force and others on behalf of the Directorate of Overseas Surveys.[27] The section of interest here was published in 1963 as 'Sheet 56/2 Kilimanjaro'. On a map of this scale all names listed already have generic terms translated into English. It shows that Phase (g) 'Replacement of German names by English names after WW I', as postulated by Ormeling (2003: 50), can clearly be determined, but with a certain delay. The personal toponyms in the Kilimanjaro region, created by Meyer, were not only still in use, with the obvious exception of the summit, which lost its dedication to the German Emperor, even the way to inscribe names of scientists or other personalities of merit is still being continued. This can be seen on large-scale maps which were published after 1900.

In 1906–1907, the geographer Fritz Jaeger and his cousin Eduard Oehler undertook a prolonged expedition to Mount Kilimanjaro, publishing the cartographic observations in two maps (Jaeger 1909). On the 'Kartenskizze des westlichen Kibo' (c. 1:40,000) one can identify the following new toponyms related to individuals: *Lentgrat* as eastern continuation of the *Lentgruppe*, *Kleiner Penck-Gletscher* as separate glacier tongue, *Uhlig-Gletscher* as separate glacier between *Penck-Gletscher* and *Kl.-Barranco-Gletscher*, *Oehler-Grat* and next to it the *Oehler-Tal* or *Hans Meyer Grat* (Fig. 4). The latter is a lava stream, which from

---

[25]Officially the name exists until the Independence of Tansania in 1961. In this connection a new name is given as late as 1963, since then the toponym is, strange in a linguistic sense, as it is a mixture of an endonym and an English generic term: *Uhuru-Peak* (*Freedom Peak*).

[26]Kaiser-Wilhelm-Spitze (Kaiser Wilhelm's Peak), Ratzel-Gletscher (Ratzel Glacier), Credner-Gletscher (Credner Glacier), Hans-Meyer-Scharte (Hans Meyer Notch).

[27]D.O.S. 422 East Africa 1:50.000 (Tanganyika); Series Y 742. Flyings occured in the 1950s and 1960s, the map sheets were published since 1963.

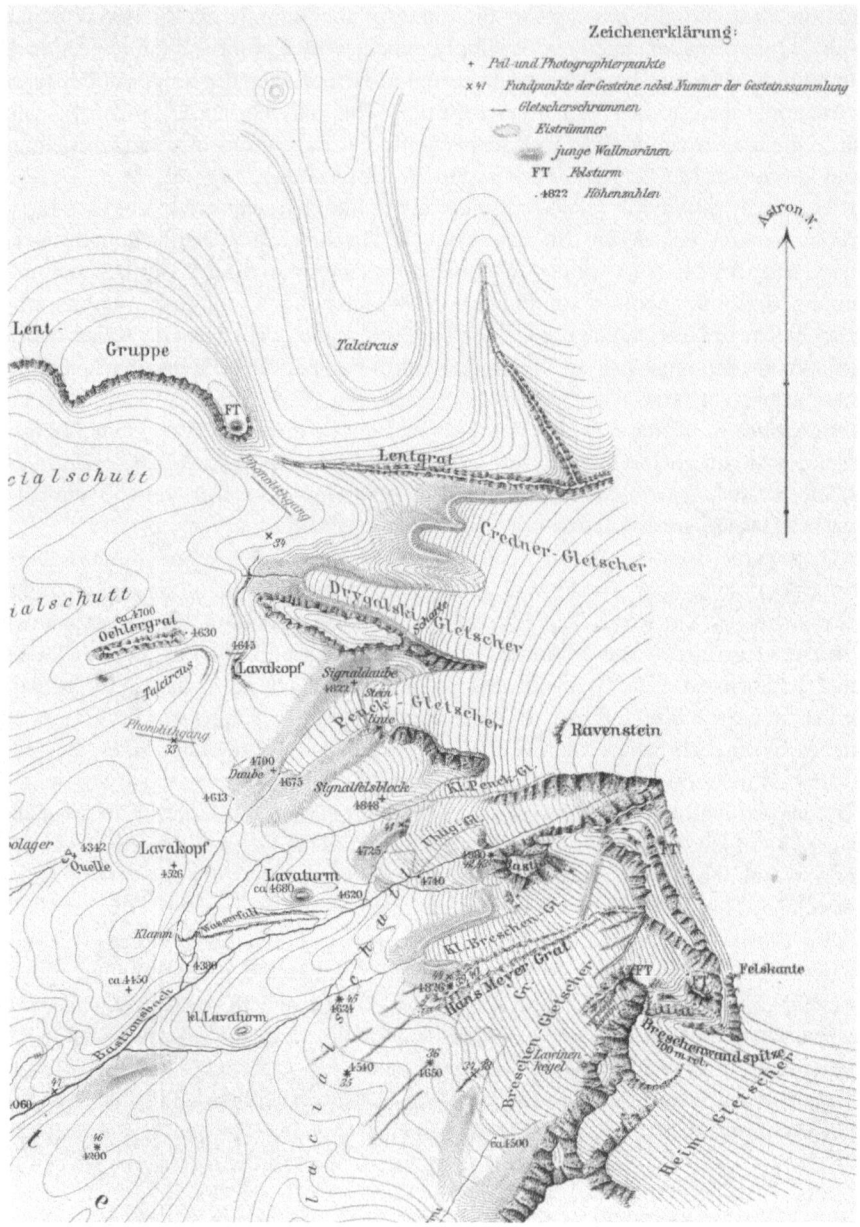

**Fig. 4** Fritz Jaeger published in 1909 the sketch map *Kartenskizze des Westlichen Kibo. Nach eigenen Kompaß-Aufnahmen und Photographien von Eduard Oehler konstruiert von Fritz Jaeger, gezeichnet von W. Rux* as map 2 in the scale of about 1:40,000 based on his cartographic observations (Courtesy Staatsbibliothek zu Berlin, Kart. C 16960/90)

afar had correctly been identified by Meyer in 1898. This led Jaeger to name this ridge after his predecessor and fellow geographer (Jaeger 1909: 128 ff.). With the toponyms *Oehler-Grat* and *Oehler-Tal* the honouring of travel companions is continued, too.

In 1912, Oehler took part in another ascent to Mount Kilimanjaro, this time with the geographer Fritz Klute. Because of World War I, the results of that expedition were only published in 1920 and 1921 respectively (Fig. 5).[28] On the relevant map *Karte der Hochregion des Kilimandscharo-Gebirges* (1:50,000), the glacier which previously was *Penck-Gletscher* now is named *Großer Penck-Gletscher*, in order to clearly distinguish it from the *Kleiner Penck-Gletscher*. The new toponyms used by Jaeger and Klute are a direct result of improved and thus more precise surveying abilities of glacial phenomena, but also because of the changing glacier landscape caused by the already discovered melting activities. By the retreat of the ice masses new landscape elements become visible elements, which before were prominently set apart from the ice mantle, but now were no longer easily recognizable in the rugged environment. Thus, soon the *Ravenstein* and the *Hassenstein* disappeared from the maps. In the area of Mawensi, their map now shows the toponyms *Neumann-Turm* and *Neumann-Tal*, which can possibly be attributed to the ornithologist Oscar Neumann (1867–1946). The summit, until then simply called *Höchste Spitze* [=highest peak], now is given the name *Hans-Meyer-Spitze*.

Research on glacial history and ice mass budget at Kilimanjaro still uses the toponyms introduced by Meyer, Jaeger and Klute (Cullen et al. 2013, S. 423), but as a result of extensive investigations, new survey methods and the morphodynamic developments, other glaciers could be defined. One of these, *Furtwängler-Gletscher*, which is located in the interior of Kibo, is named after the mountaineer Walter Furtwängler, who in 1912 was the first to ski down from the Kilimanjaro summit. Meyer had still considered this glacier as being connected to the Barranco-Gletscher.[29] Finally, in 1954 the government of Tanganyika named the inner cinder cone of the Kibo *Reusch-Crater*, because they intended to commemorate the 25th ascent of the summit by the missionary, ethnologist and mountain climber Richard Reusch.[30]

In summarizing, it can be stated that Hans Meyer, when naming individual and prominent landscape elements, which did not exclusively serve orientation purposes such as way marking or camp[31] for an expedition, applied personal preferences. Thus, he named the glaciers at Kibo, which he examined more closely, after glaciologists or discoverers of merit. He also honoured his travel companions, other

---

[28]As map supplement in Klute (1920) and also in Klute (1921).

[29]The Arrow Glacier is a remnant of the Kleine Barranco-Glacier, whereas Diamond- and Balletto-Glacier are above Heim-Glacier near Breschenwand (Cullen et al 2013: 424 ff.).

[30]On his ascent in 1926 he discovered the frozen cadaver of a leopard. In 1936 Ernest Hemingway used this motif in his short story *The Snows of Kilimanjaro* (Hemingway 2004).

[31]For caves, which served as his expedition camps and where he discovered traces of a temporary use by indigenous hunters, he used endonym names of his carriers or named them after local chiefs.

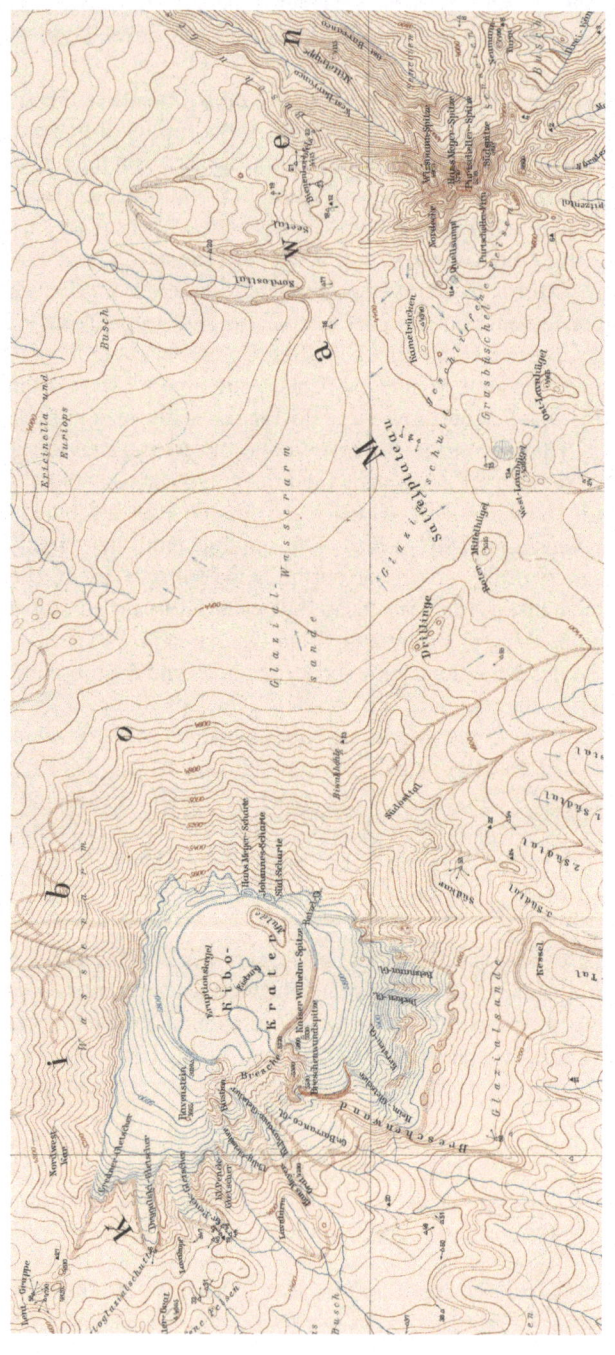

**Fig. 5** Fritz Klute and Eduard Oehler created their *Karte der Hochregion des Kilimandscharo-Gebirges nach stereophotogrammetrischen Aufnahmen, flüchtigen Triangulationen u. Krokis* [...]. *Konstruiert u. gezeichnet unter teilweiser Benutzung des vorhandenen Materials von Fritz Schröder unter Leitung u. Mitarbeit von Fritz Klute. 1:50,000* in 1912. The publication of this map, however, was only possible after WWI (Courtesy Staatsbibliothek zu Berlin, Kart. GfE J 5-497)

scientists or colonial administrators by making them namesakes of mountain peaks, although mostly in somehow remote territories. In part, this procedure was later continued by scientists of other nations, especially when naming newly emerged glaciers.

Finally, one should also look at modern hiking maps to see whether personified toponyms continue to exist until today. Depending on the scale, one can indeed find a selection of these names with English translations of the generic terms.[32] In the area of Schira Ridge, these maps show the toponym *Klute Peak*, a name which does not appear on any of the dicussed German maps. Furthermore, on some of these maps the Bismarck Towers, or in a wrong spelling the Bismark Towers, are marked at the southeastern rim of the crater, without there previously having existed an equivalent on German maps. They document, however, already described problems of errors in translation because of insufficient knowledge and/or understanding. As a result of the ignorance regarding the naming practice and the etymological derivation, the cone, for example, which is named after the mountain painter Ernst Platz is translated literally as: *Cone Place*. The source of this misunderstanding is official, the toponym comes from 'Sheet 56/1 West Hai', published in 1963. It is part of the already-mentioned map series *East Africa 1:50.000 (Tanganyika) series Y 742* from the Direcorate of Overseas Surveys.

| German Generic | Names English Translation |
|---|---|
| Bach | Creek |
| Bresche/Barranco | Breach |
| Gletscher | Glacier |
| Grat/Rücken/Kamm | Ridge |
| Gruppe | Group |
| Hügel | Hill |
| Höhle | Cave |
| Kegel | Cone |
| Kette | Range |
| Krater | Crater |
| Lager | Camp |
| Mauer | Wall |
| Nadel | Needle |
| Scharte | Notch |
| Spitze | Peak |
| Stein | Stone |
| Tal | Valley |
| Turm | Tower |

---

[32]See: Loch (2007), Greulich (2008), Wirth (2011) and Szyczak (2012).

# References

Baumann O (1894) Originalkarte des nördlichen Deutsch-Ostafrika für das Deutsche-Antisklaverei-Komitee nach eigenen Aufnahmen u. Ortsbestimmungen constuiert von Dr. Oscar Baumann. Mit Benutzung der Englischen Grenztriangulierung, der Originalaufnahmen von Dr. G.A. Fischer (1885/6), Kapitän Spring u. Leutnant Werther, sowie unter Berücksichtigung aller vorhandenen Materialien bearbeitet und gezeichnet von Dr. B. Hassenstein. 1:600,000. Petermanns Geographische Mitteilungen/ Ergänzungsheft No. 111, Tafel 1. https://archive.thulb.uni-jena.de/hisbest/rsc/viewer/HisBest_derivate_00015931/Mittheilungen_Perthes_Ergbl_129602507_1894_111_0059.tif. Accessed 27 November 2018

Brunner K (1989) Erstbesteigung und erste Routen des Kilimandscharo. Kartograph Nachr 39:216–222

Brunner K (2004) Frühe Karten des Kilimandscharo: Ein Beitrag zur Expeditionskartographie. Cartograph Helv 30:3–9. https://www.e-periodica.ch/digbib/view?pid=chl-001:2004:29::61#61. Accessed 27 November 2018

Crom W (2003) Was ist Kolonialkartographie? Mitteilungen Freundeskreis für Cartographica 16(17):31–33

Cullen NJ, Sirguey P, Mölg T, Kaser G, Winkler M, Fitzsimons SJ (2013) A century of ice retreat on Kilimanjaro: the mapping reloaded. The Cryosphere 7: 419–431. https://www.the-cryosphere.net/7/419/2013/tc-7-419-2013.pdf. Accessed 27 November 2018

Directorate of Overseas Surveys (ed) (1964) 56/1 West Hai; 56/2 Kilimanjaro. In: East Africa 1:50,000 (Tanganyika), series Y 742. London

Demhardt IJ (2000) Die Entschleierung Afrikas: Deutsche Kartenbeiträge von August Petermann bis zum kolonialkartographischen Institut. Klett/Perthes, Gotha

Demhardt IJ (2006) Kolonialkartographie. In: König V (ed) Vermessen: Kartographie der Tropen. Ethnologisches Museum, Berlin, pp 60–65

Eckert M (1924) Die Bedeutung der deutschen Kolonialkartographie. Verhandlungen des Deutschen Kolonialkongresses, pp 436–454

Finsterwalder R, Hueber E (1943) Vermessungswesen und Kartographie in Afrika. de Gruyter, Berlin

Geographic Section General Staff (ed) (1915–1917) Kilimanjaro B 5. In: German East Africa 1:300,000. Ordnance Survey, Southampton

Greulich S (2008) Kilimanjaro, Kibo: climbing and trekking map 1:80,000. Climbing-Map Company, Bern

Hafeneder R (2008) Deutsche Kolonialkartographie 1884–1919. Amt für Geoinformationswesen der Bundeswehr, Euskirchen

Hamann C, Honold A (2011) Kilimandscharo: Die Geschichte eines afrikanischen Berges. Wagenbach, Berlin

Hassenstein B (1864) Das Gebiet der Schneeberge Kilima-Ndscharo und Kenia in Ostafrika. Nach den Reiseberichten der Missionäre Krapf & Rebmann, 1844–53, und auf Grund der Forschungen K van der Decken's, 1861 u. 1862, von Burton, Speke (1857), Owen (1824) u. A. 1:1,500,000. Petermanns Geographische Mitteilungen [10]: Tafel 16. https://zs.thulb.uni-jena.de/receive/jportal_jparticle_00512878. Accessed 27 November 2018

Hassenstein B (1868) Das Gebiet der beiden Reisen des Baron C. C. von der Decken zum Schneeberg Kilima-Ndscharo in den Jahren 1861 and 1862. 1:1,000,000. Kraatz, Berlin

Hassenstein B (1893) Spezialkarte des Kilima-Ndscharo- und des Meru-Gebietes. Nach Breitenbestimmungen, Routenaufnahmen, Winkelmessungen & von Dr. Hans Meyer, Lieut. v. Höhnel, Dr. Oscar Baumann u. A. 1:350,000. Petermanns Geographische Mitteilungen 39: Tafel 7. https://zs.thulb.uni-jena.de/rsc/viewer/jportal_derivate_00256036/ThULB_129489824_1893_Perthes_0112.tif?logicalDiv=jportal_jparticle_00512213. Accessed 27 November 2018

Heymons M (1891) Routen-Skizze der v. Wissmann'schen Kilimandscharo-Expedition 1891. Nach dem Itinerar des Lieut. Heymons unter Zugrundelegung der Aufnahmen von Lieut. v. Höhnel und Dr. H. Meyer. 1:500,000. Mitteilungen aus den deutschen Schutzgebieten Bd. IV, Tafel IX

Hemingway E (2004) The snows of Kilimanjaro: six stories. Reclam, Stuttgart

Höhnel L (1892) Karte des Forschungs-Gebietes der Graf Samuel Teleki'schen Expedition in Ostafrika 1887–88. Mit Berücksichtigung neuerer Aufnahmen vornehmlich nach den eigenen Messungen. 1:1,000,000. In: Höhnel L Zum Rudolph-See und Stephanie-See: die Forschungsreise des Grafen Samuel Teleki in Ost-Aequatorial-Africa 1887–1888, Hölder, Wien, Tafel I

Jaeger F (1909) Forschungen in den Hochregionen des Kilimandscharo. Mitteilungen aus den deutschen Schutzgebieten 22:113–197

Kiepert H (1863) Karte des Schneegebirges Kilima-Ndjaro aufgenommen von Baron C. von der Decken auf seiner ersten Reise. 1:500,000. Zeitschrift für allgemeine Erdkunde NF XV, Tafel V. http://www.digizeitschriften.de/download/PPN391365622_1863_0015/PPN391365622_ 1863_0015___log65.pdf. Accessed 27 November 2018

Kiepert R (1895) Aequatorial-Ost-Afrika 1:3,000,000. In: Deutscher Kolonial-Atlas: für den Gebrauch in den Schutzgebieten. Reimer, Berlin

Klute F (1920) Ergebnisse der Forschungen am Kilimandscharo 1912. Reimer, Berlin

Klute F (1921) Die stereogrammetrische Aufnahme der Hochregion des Kilimandscharo. Zeitschrift der Gesellschaft für Erdkunde, pp 144–151

Krapf JL (1858) Reisen in Ost-Afrika ausgeführt in den Jahren 1837–55 zur Beförderung der ostafrikanischen Erd- und Missionskunde. Teil 1: Des Verfassers Erlebnisse, Missionsthätigkeit und Reisen in Nord- und Süd-Ost-Afrika (Abessinien und die Aequator-Gegenden). Teil 2: Meine größere Reisen in Ostafrika. Stroh, Stuttgart

Krapf JL (1860) Travels, researches, and missionary labours, during an eighteen years' residence in eastern Africa together with journeys to Jagga, Usambara, Ukambara, Shoa, Abessinia, and Khartum and a coasting voyage from Mombaz to Cape Delgade. Trübner, London. https:// books.google.de/books?id=z78NAAAAQAAJ&hl=de&source=gbs_navlinks_s. Accessed 28 September 2018

Loch H (2007) Kilimanjaro Trekking-Karte. 1:50,000. Rotter, München

Meyer H (1887) Der Kilimandscharo. Eine provisorische Skizze von Dr. Hans Meyer. Petermanns Geographische Mitteilungen Tafel 19. https://zs.thulb.uni-jena.de/rsc/viewer/jportal_derivate_ 00256031/ThULB_129489824_1887_Perthes_0400.tif. Accessed 27 November 2018

Meyer H (1891) Across East African Glaciers: an account of the first ascent of Kilimanjaro (trans: Calder EHS). Longmans, London. https://archive.org/details/acrosseastafrica00meye/page/n0. Accessed 28 September 2018. German edition: Meyer H (1890) Ostafrikanische Gletscherfahrten. Duncker & Humblot, Leipzig

Meyer H (1900) Der Kilimandjaro: Reisen und Studien. Reimer, Berlin

Obst E (1921) Die deutsche Kolonialkartographie. In: Praesent H (ed) Beiträge zur deutschen Kartographie. Akademische Verlagsgesellschaft, Leipzig, pp 98–118

Ormeling F (2003) Place name change models and European expansion. In: Liebenberg E (ed) Proceedings of the Symposium on the History of Cartography in Africa. Copy Master, Pretoria, pp 49–50

Passarge S (1912) Die Vollendung der grossen Karte von Deutsch-Ostafrika im Maßstab 1:300,000. Deutsche Kolonialzeitung 29:2–3

Petermann A (1859) Originalkarte von Burton's u. Speke's Entdeckungen in Inner Afrika 1857 u. 1858. Nebst Angaben aller übrigen im Bereich der Karte von Europäischen Reisenden zurückgelegten Routen. Mit Benutzung einer Originalzeichnung von Capt. JH Speke. 1:7,000,000. Petermann's Geographische Mitteilungen, Taf. 15. https://zs.thulb.uni-jena.de/ rsc/viewer/jportal_derivate_00260681/ThULB_129489816_1859_Perthes_0421.tif. Accessed 27 November 2018

Pillewizer W (1941) Der Anteil der Geographie an der kartographischen Erschließung Deutsch - Ostafrikas. Jahrbuch der Kartographie: 45–175

Rebmann J (1850) Imperfect Sketch of a Map from 1 ½° North, to 10 ½° South Latitude, and from 29 to 44 degrees East Longitude, by the Missionaries of the Church Missionary Society in Eastern Africa. Rabbai Mpia

Rebmann J (1997) Tagebuch des Missionars vom 14. Februar 1848–16. Februar 1849. Veröffentlichungen des Archivs der Stadt Gerlingen, Gerlingen. http://www.johannes-rebmann-stiftung.de/cms/wp-content/uploads/2016/11/tagebuch.pdf. Accessed 27 November 2018

Simo D (2002) Anschauungen eines Berges: Der Kilimandjaro und seine Bedeutungen. In: Arlt H (ed) Realität und Virtualität der Berge. Röhrig, St. Ingbert, pp 55–62

Sprigade P, Moisel M (1910) Kilimandscharo. In: Kolonialabteilung des Auswärtigen Amtes (ed) Großer Deutscher Kolonialatlas 1:1,000,000 (1901–1914). Reimer, Berlin

Sprigade P, Moisel M (1911) B 5 Kilimandscharo. In: Karte von Deutsch Ostafrika 1:300,000. Reimer, Berlin

Sprigade P, Moisel M (1914) Die Aufnahmemethoden in den deutschen Schutzgebieten und die deutsche Kolonial-Kartographie. Zeitschrift der Gesellschaft für Erdkunde zu Berlin: 527–545 http://www.digizeitschriften.de/dms/img/?PID=PPN391365657_1914%7CLOG_0181. Accessed 27 November 2018

Sriguey P, Cullen NJ (2014) A century of photogrammetry on Kilimanjaro. In: FIG Congress Engaging the Challenges—Enhancing the Relevance. Kuala Lumpur 16–21 June 2014. http://www.fig.net/resources/proceedings/fig_proceedings/fig2014/papers/ts08b/TS08B_sirguey_cullen_6959.pdf. Accessed 27 November 2018

Szyczak M (2012) Africa—the highest peaks: Kilimanjaro, Mount Kenya, Rwenzori. Different scales. terraQuest, Warsaw

Uhlig C (1909) Die ostafrikanische Bruchstufe und die angrenzenden Gebiete zwischen den Seen Magad und Lawa ja Mweri sowie dem Westfuss des Meru. Mitteilungen aus den Deutschen Schutzgebieten, Erg.-Heft 2. Teil I: Die Karte. Hirt & Sohn, Leipzig

Voigt I (2012) Die 'Schneckenkarte': Mission, Kartographie und transkulturelle Wissensaushandlung in Ostafrika um 1850. Cartograph Helv 45:27–38

Wirth M (2011) Kilimanjaro National Park. 1:100,000. Harms-ic-Verlag, Kandel

**Wolfgang Crom** is a geographer and librarian, and the head of the Map Department of the Staatsbibliothek zu Berlin. He studied geography, pedology, botany and ethnology at Rheinische Friedrich-Wilhelms-University in Bonn, then undertook postgraduate studies in librarianship with focus on map curatorship in Tübingen and Köln. He served as subject librarian for geography and maps at Württembergische Landesbibliothek in Stuttgart. Since 2000, he has been head of the Map Department at the Staatsbibliothek zu Berlin. He is Speaker of the German map curators group and a board member of several cartographic commissions.

# The French Map of Beirut (1936)

**Jack Keilo**

**Abstract** Toponymic inscriptions are an 'authorized version' of history written on space. This article aims to explore the toponyms on a French map of Beirut published in 1936 to show how France, as a sovereign power, transformed her 'Lebanese policy' into place names and thus created a different reality, in rupture with the past. This reality still endures today on the map. The new polity was created under the *mission protectrice* of France. The 'mission' is read on the map through the names of Gouraud, Foch, Pétain, and other generals of World War I, and by key features of the French Republic ('The Marseillaise', 'the French', 'Paris', and so on). With Lebanon being a 'refuge for minorities', the 1936 map of Beirut has thoroughfares named after saints, ulemas, and religious figures of Christians and of Muslims ('rue patriarche Hoyek', 'rue Ibn Arabi', and 'rue Abou Bakr'). In 1918, political martyrdom was introduced to political discourse, but also to the map; thus the main square of the city is renamed 'Place des Martyrs', with numerous streets named after intellectuals hanged by the Ottomans and considered martyrs of the new Republic. These three 'toponymic systems' are in discontinuity with the toponymic past of Beirut. These toponymic dynamics still shape the map of Beirut; no constitution change or 'toponymic cleansing' happened after Independence in 1943. There are more 'martyrs' and religious figures added to the map and mandate army generals are still commemorated. Mandate-made maps continue to shape Beiruti place names today.

J. Keilo (✉)
Université Paris-Sorbonne, Paris, France
e-mail: keilojack@hotmail.com

© Springer Nature Switzerland AG 2020                                      247
A. J. Kent et al. (eds.), *Mapping Empires: Colonial Cartographies of Land and Sea*,
Lecture Notes in Geoinformation and Cartography,
https://doi.org/10.1007/978-3-030-23447-8_14

# 1   Introduction

In October 1918, French troops disembarked in Beirut, a possession of the Ottoman Caliphate at the time, and replaced the British Imperial troops that had occupied the city for a week (Davie 2001: 71–72). Two years later, On 31 August, General Henri Gouraud[1] created the new political entity of Greater Lebanon by his Decree N° 318 (Haut-Commissariat 1920: 132–134). The newly-formed polity united the coastal cities of Beirut, Tripoli, Sidon, and Tyre with Mount Lebanon.[2] The next day Gouraud proclaimed the State of Greater Lebanon from his residency in Beirut, the capital and the seat of the French High Commission. Subsequently, the League of Nations granted the mandate over Lebanon and Syria to France in 1922.[3]

France, as the new sovereign power over Syria and Lebanon, mapped Beirut and codified its place names. This chapter will focus on the toponyms[4] of the 1:10,000 map of Beirut that was published in November 1936, by the Topography Bureau of the French Army of the Levant.[5] Using a qualitative approach, it will explore some aspects of the toponymic systems on the map. The chapter aims to show how France transformed its 'Lebanese policy' into place names on the map and how colonial mapping does not only shape territory, but also sets the dynamics of toponymic systems for future post-independence maps.[6]

---

[1]High Commissioner of France and Commander-in-Chief of its Army in the Levant.

[2]For more information see (Ammoun 1997; Salibi 2002; Kassir 2010; Najjar 2014).

[3]Lebanon was not a *colonie* of France, but under mandate. Thus in this article 'colonization' is defined in its broader sense: the *mise en tutelle d'un territoire sous-développé et sous-peuplé par les ressortissants d'une métrople* according to the French Centre national de ressources textuelles et lexicales. The Class A Mandate on Syria and Lebanon was covered by the Covenant of the League of Nations of 1919 (SDN 2011).

[4]This article follows the distinction between the prototype and the image, as put forward by Theodore the Studite and Paul Evdokimov: the image is always dissimilar to its protoype 'in essence', and similar to it'in hypostasis' (Evdokimov 1997: 52). Toponyms are considered as 'images' of their prototypes; they reproduce the hypostasis, not the essence. For example, 'Louis IX of France has a *rue Saint Louis*' means that King Louis IX of France has a certain 'image' of himself that is inscribed or marked on the map of Beirut.

[5]The map is at the archives of the Hebrew University of Jerusalem and a scan is available from: http://historic-cities.huji.ac.il/lebanon/beirut/maps/tfl_1936_beirut.html Courtesy of The National Library of Israel, Eran Laor Cartographic Collection, Shapell Family Digitization Project and The Hebrew University of Jerusalem, Department of Geography—Historic Cities Research Project.

[6]This study is partially based on toponymic systems analyses made in the context of a doctoral thesis defended at the Sorbonne in 2018: *The centre and the name, readings in Beirut's toponymy*. The abstract is available from: http://www.theses.fr/2018SORUL067.

# 2   Toponymic Inscriptions as Ideology Written on Space

In this chapter, toponyms[7] are considered as revelators of the set of values of the power controlling the map. In a capital city, itself an 'organized remembrance' (Arendt and Canovan 1998: 198), toponyms are revelators of the state's 'ruling socio-political order and its particular "theory of the world"' and a 'narration without villains' of the official version of national narrative (Azaryahu 1996). They are the insertion of ideology in the banality of daily life as a factor of concretization of hegemonic structures of power and authority (ibid.).

In capital cities, toponyms are an integral part of 'the performative, preservative functions' of a nation (Daum and Mauch 2005: 18–19):

> *Performative functions*: the ability to stage events that put the political mission of a state and the idea of national identity on display [...]. A distinct cultural task of the capitals also lies in their *preservative functions*. Capitals serve as nation-states' repositories of memory. There are prominent, though rarely the only, *lieux de mémoire* within nation-states. Capitals thus serve as 'hinges'. They mediate between the nation-state's past, present, and envisaged future.

Toponymic systems[8] can be even more revealing of political ideology of the power introducing them to the map. For the French, Beirut was not only the capital city of the state of Greater Lebanon, but also as the seat of the French High Commissioner and Commander-in-Chief for the entire Levant, and 'showcase' of France (Saliba 2004).

# 3   The French Shift of Beiruti Place Names

Before 1918, most Beiruti toponyms stemmed from traditional functional place names. During the first half of the nineteenth century, local architecture and planning were mostly functional (Hallak 1987; Davie 2001: 17–20, 2005) and thus was toponymy: *Sahat el Khubz* (Bread Square), *Sahat el Saraya* (Serail Square), *Sahat Bayt Trad* (The Trad Family's Square), *Souk el Hayyakin* (Weavers' Souk), *Mahallet el Gharbiyyeh* (West Side), *Tariq Trablous* (Tripoli Road), and so on. The *Maydan el Madfaa'* (Cannons' Maydan) was the only 'commemorative' name, used since the 1770s (Khalaf 2006).

With the expansion of the city, commemorative place names began to appear on the map. For example the Majidiye quarter, with its eponymous mosque, was named after Abdulmejid I, Ottoman caliph from 1839 to 1861. Similarly, the

---

[7]Toponyms, for this study are every name attached to a place, e.g. a street name, a river or water-surface name, a quarter name, a city name, or a green-space name.

[8]In this paper toponymic systems are defined as interacting toponyms present in more or less the same space, and usually written, re-written, controlled and run by political power.

Hamidiye Square honoured Abdülhamid II, Ottoman caliph from 1876 to 1909.[9] After the 1908 coup d'État staged by the Committee of Union and Progress, Hamidiye was renamed Union Square. Still, most Beiruti place names were traditional, as shown in the Baedeker map of 1912.[10]

By the end of October 1918, the Ottoman Empire surrendered and signed the Treaty of Mudros with the Allies, establishing an occupied territory in the Levant. After declaring the state of Greater Lebanon in September 1920, French Mandatory authorities proceeded to write a constitution and codify the fundamental principles of the new political entity. In 1926 a special committee appointed by France wrote the Lebanese Constitution (Michel Chiha Foundation 2017). This codification of fundamental text between 1920 and 1926 was accompanied by a parallel 'codification' of the map. An 'official' 1:10,000 map of the city was published in June 1920 by the Topography Bureau of the French Army of the Levant,[11] three months before the proclamation of Greater Lebanon. In this map, the new political order was reflected by the replacement of old toponyms by French military, political, and geographical names (Cheikho 1920).

## 4   Examining the French Map of Beirut (1936)

By 1936, Lebanon had become a republic with a constitution and a president. However, French Mandatory authorities continued to control the map. During the 1920s French Mandatory authorities continued the reconstruction of the city centre that had begun under the Ottomans in 1915 (Ghorayeb 2014: 21–36). The French built the Place de l'Étoile and considered it the 'oriental' heart of the city (Davie 2003, 2005). An updated 1:10,000 map of 1936 was published by the Topography Bureau of the French Army of the Levant, and later used to elaborate other plans of Beirut during World War II.[12] French sovereignty, French *mission protectrice* of the 'mountain-refuge' that Lebanon is supposed to be, and the ideology of the new nation are all written as toponyms on the 1936 map (Fig. 1).

---

[9]The Löytved map of 1876 shows some commemorative trends but not these two toponyms. This map, drawn by Julius Löytved (Vice-Consul of Denmark), is a south-up map and one of its original copies is located within the archives of the Bibliothèque nationale de France and available from: https://gallica.bnf.fr/ark:/12148/btv1b8494564f.

[10]The Baedeker map of Beirut was drawn by Karl Baedeker in a 'Handbook for Travellers, 5th Edition' in 1912 and is available from: http://legacy.lib.utexas.edu/maps/lebanon.html

[11]The map of Beirut 1920 is in the archives of the Bibliothèque nationale de France and is available from: https://gallica.bnf.fr/ark:/12148/btv1b53066704k/

[12]Some historical Beiruti maps are available at the Archives militaires de la Défense in Vincennes and at the archives of the IGN in Saint-Mandé. The author visited the two archives during his doctoral research on Beiruti toponyms.

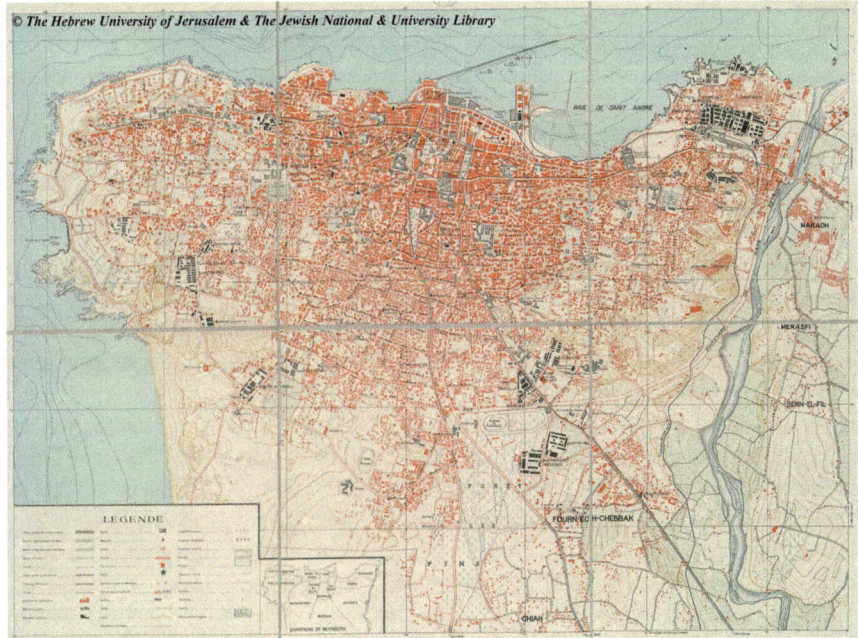

**Fig. 1** Map of Beirut, 1936 (Courtesy of The Hebrew University of Jerusalem)

## 4.1   The Sovereign Power on the Map

French sovereignty is expressed on the map by a toponymic system. A central street bears the name of *rue Clemenceau* after Georges Clemenceau, French Prime Minister of World War I and of the Versailles Peace Conference. Mandatory Authorities replaced the old *rue de la Prusse* by *rue Georges Picot*, honouring the diplomat, high commissioner and one of the signatories of the Sykes-Picot Agreement.[13] Near the Port,[14] the *rue de la Marseillaise* transforms the national hymn of the French Republic and one of its symbols into a street name. Not far, a *rue des Français* turns the French themselves into a cartographic feature. Other streets bear the name of French cities: the longest Beiruti street, running along its western maritime façade (by then still non-urbanized) is called *rue de Paris*. Arterial *rues d'Alger*, *de Bordeaux*, and *de Lyon* are named after the eponymous French cities.[15] The longest battle of World War I is commemorated by a *rue*

---

[13]The Sykes-Picot Agreement was a secret 1916 agreement between France and the United Kingdom to define and to share the spheres of influence in the Levant.

[14]By that time the Port of Beirut was the most important of the Eastern Mediterranean façade and the main entry point to Beirut (For more information see Fawaz (1984) and Issawi (1977).

[15]During the whole French presence in Lebanon (1918–1945) Algeria was, officially, a French *département*, a part of France.

*Verdun*. The ally of France in the battle has the *rue de Londres*. After World War I, France officially went to the Levant to 'protect minorities'. Its protective mission is also codified on the map.

## 4.2   The Military Mission Protectrice of France

In 1919, Georges Clemenceau promised a national home for Maronites in his correspondence with Patriarch Elias Hoyek (Rondot 1954; Tadié 2016: 46). At the same time, the Patriarch, heading a delegation of Lebanese notables to the Versailles Peace Conference, requested an independent Lebanese 'Phoenician' state with French protection (Ippolito 2011; Kaufman 2014: 85–86). Thus, the French mission in the Levant was described not as the traditional *mission civilisatrice*, but as a *mission protectrice*, that is, a protective mission of minorities (Commission de publication des documents diplomatiques français 2004: XLVIII), based on the 'Catholic connection' between France and Lebanon (Laurens 1991). The mission was used as a reason to justify French military interventions in Syria (Pinta 1995: 94–95; Heyberger 2018). The military visibility of France (Davie 2001: 73) was written on the map since 1920 and on the first map of Beirut, the names of living military leaders adorned major thoroughfares of the city. On the 1936 map, this military toponymic presence was kept, even if Lebanon was its own republic by then.

The *rue de l'Armée* runs by the Grand Serail (The High Commission's head-quarters.) Another street by the Port, *rue des Libérateurs*, commemorates the 'liberation' of the Levant from Ottoman rule. The High Commissioner in the Levant who established the Mandate, General Henri Gouraud, has a *rue Gouraud*,[16] one of the most famous streets of Beirut that connects the old city to its eastern quarters.[17] To the south east of the city centre, the governor of Lebanon and of the isle of Arwad, Albert Trabaud, is commemorated by a *rue Trabaud*.

Other generals, not necessarily affiliated to the French Army of the Levant, are toponymized on the 1936 map. World War I French officers, Ferdinand Foch and Philippe Pétain, had a *rue du maréchal Foch*, *quartier Foch*, and *rue du maréchal Pétain*[18] since 1920. Another General and High Commissioner of the Levant (1923–1924), Maxime Weygand, is honoured by a *rue Weygand*, the official address of Beirut Municipality, and a *quartier Weygand* at the city's eastern entry.

---

[16]In other words, General Henri Gouraud commemorated himself by putting himself on the map of Beirut.

[17]Another Mandate officer was commemorated on the 1920 map; Colonel Émile Niéger, who played an important role in 'pacifying' the southern parts of Lebanon (Khoury 2004), has a *rue Colonel Niéger* connecting the Grand Serail to the south-western quarters of the city. The street is renamed *rue [Maurice] Barrès* in the 1936 map.

[18]Later and in 1945, *rue Pétain* became *rue France* and is still on the map of Beirut today.

Since 1920, a British World War I hero, General Edmund Allenby,[19] has a very central *rue Allenby* in the heart of Beirut. Allenby was the first of the Allies to enter the Levant and then the city (Magazine Circulation Co. 1918), and such was his fame after World War I that two contemporaneous eminent Arab-speaking poets, Elia Abu Madi and Ahmed Shawqi, praised him in their writings.[20] Today, Beirut is the only Arab city that keeps the souvenir of Allenby's march on its present map.

The French Third Republic was known for its leanings towards the *laïcité* and its conflict with the Catholic Church following the 1905 law on the separation of Churches and State. Yet Mandatory Authorities introduced the names of two warrior patron saints of France to the map of Beirut. On the eastern hills of the city a *rue Saint Louis* was named after the Crusader king of France. Its naming provoked some objections from locals who did not want 'foreign saints' (Davie 2001: 73). In 1920, the French Republic reestablished its diplomatic relations with the Holy See and a French special ambassador participated in the canonization of Joan of Arc in Rome by Pope Benedict XV. Later, Pope Pius XI declared her a secondary patron saint of France, assuring that the French nation is the *Fille aînée de l'Église*.[21] A *rue Jeanne d'Arc* appears on the map of 1936 and still exists in the Hamra quarter.

## 4.3   Lebanon as the Refuge for Minorities and Confessions

Lebanon was considered as a 'mountain-refuge' (Lammens 1921) and a country of communities and minorities (Rondot *op cit*: 85–87). Beiruti toponymy was meant to reflect the mosaic of faiths in Lebanon with its officially-recognized 17 religious groups. The 1936 map highlights this diversity. Besides Joan of Arc and Louis IX, two other French saints are toponymized: St Charles and St Vincent (de Paul). Six locally-venerated saints enjoy a visible presence on the map and their names adorn important thoroughfares: Mar[22] Nicholas of Myra, Mar Demetrius [Mitr] of Thessalonica, prophet Mar Elias of Tishba, Mar Michael the Archangel,[23] Mar Antonios [Anthony] of Egypt and of Padua, and Mar Maron the Hermit. A saint of Eastern Churches and an important figure in the history of Beirut *via* the Roman Law School (Badre 2016) and in legitimizing Beirut's role as capital (Emereau 1915: 422–424), the Roman emperor Justinian has a *rue Justinien*. Curiously, the 1920 map names it *rue de la République*, but before 1936 the French Mandatory

---

[19]Field Marshal and 1st Viscount Allenby of Megiddo in 1919, see Hughes (2011).

[20]See Abu Madi (1996: 241, 595–597) and Shawqi (2005).

[21]For more information see Leo XIII (1884), the entry 'Benoît XV' in Ambrogi and Le Tourneau (2017), and Drago and Tawil (2017).

[22]*Mar* is a Syriac word meaning 'saint' or 'venerated', found in many Beiruti hagiotoponyms.

[23]Saint Michael the Archangel gives his names to a street and to the Camp St Michel of Armenian refugees.

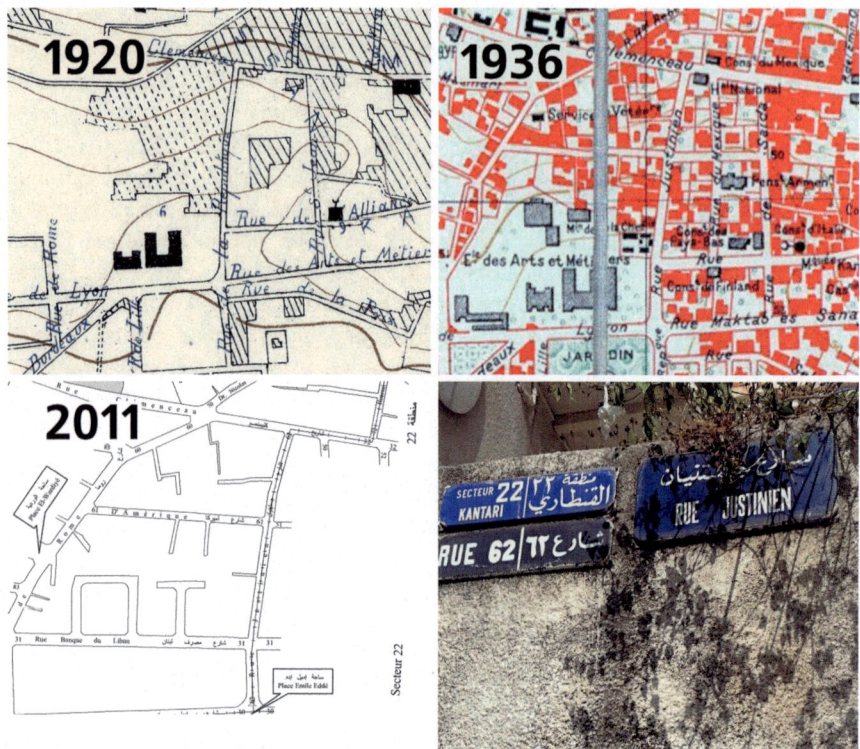

**Fig. 2** From République to Justinian on the map of Beirut (Courtesy of the BnF, The Hebrew University, and the Municipality of Beirut; image by the author)

authorities found the figure of the Roman emperor so important that they renamed the biggest part of rue de la République after him (Fig. 2).

Different Christian institutions lend their name to the entire street where they are located. The *rue de l'archevêché orthodoxe, du patriarcat syrien* [syriaque], *rue du Saint-Sauveur* [Mokhallisiyé], *rue de l'Hôtel-Dieu, rue de la Faculté Saint-Joseph, rue de l'Hôpital orthodoxe, rue des Saints-Coeurs, rue des Capucins*, and *rue de Sagesse* are some examples. Finally, two localities named *quartier arménien* appear on the eastern side of the city to mark the settlement of refugees from the 1915 Genocide (Kévorkian et al. 2007: 119–122). Patriarch Elias Hoyek himself is commemorated by an eminent *rue patriarche Hoyek*, connecting the courthouse to the sea façade. Other Christian Greek Orthodox archbishops of Beirut are commemorated by the *rue du Monseigneur Messarra* and *rue du Monseigneur Ghofrail*.

In Muslim-majority quarters of Beirut, to the south and south-west of the city centre, toponyms emanating from Islamic traditions appear on the map. Streets

**Fig. 3** A detail of the map of Beirut 1936 with the Place des Martyrs, rues Gouraud, Foch, Allenby, Patriarche Hoyek, amongst others (Courtesy of The Hebrew University)

carrying the names of *Sidi*[24] *Hassan* [Imam Hassan bin Ali], *Saïde Fatmé* [Fatima, daughter of Mohamed, founder of Islam], *Saïde Khadija* [first wife of Mohamed], *Saïde Aisha* [wife of Mohamed], *Ibn Arabi* [a soufi holy man], *Umayyads* and *Abbasids* [first and second dynasties of caliphs], *Abu Bakr*, *Omar*, and *Othman* (the three first caliphs) were added to the Beiruti toponyms by the Mandatory Authorities.[25]

## 4.4 Political Martyrs of the New Nation

The state of Greater Lebanon was formed by merging the city of Beirut with Mount Lebanon (Davie 1996: 71–105). Historical figures of the mountain were added to the map of Beirut. Fakhreddine the Maanid, an Ottoman figure of the newly-introduced national myth (Abu Fakhr and Salibi 2012) has a beautiful street in the city centre. Another important figure of the Mountain history, Emir Bashir II, has a

---

[24]In Arabic *sidi* is 'my lord' and *saïde* is 'my lady', both words are used as veneration titles for members of the House of Mohamed.

[25]In a city where notable Muslim families claim descent from the Prophet or from first caliphs (Hallak 1987, 2010) the choice of toponyms among members of Mohamed's household and caliphs can make sense.

central street. But the toponymic system of national martyrs, set by the French, is the most eminent on the Beiruti map (Fig. 3).

In August 1915 and May 1916, in Beirut and Damascus, Ottoman authorities executed about forty Syro-Lebanese intellectuals and political activists (of different Christian and Muslim denominations), accused of high treason and of collaborating with the Allies (Ajay 1974; Kassir 2010: 284; Zachs 2012). Upon the arrival of the French, names of some of the executed, considered martyrs, appeared on the map, with a *rue des Martyrs* to the south of the Place des Canons, Beirut's central square (Khalaf 2006).

Since 1930, a statue 'to celebrate understanding between Christians and Muslims' (Kassir *op cit*) was erected in the central square of Beirut, *place des Martyrs*, as it is named on the 1936 map and on present-day maps. Numerous streets are named after the martyrs: *rues Abdel Wahab Englizi, Saïd Akl, Chafic Mouayed* [Azem], *Abdel-Karim El-Khalil, Petro Paoli*, and *Abdel-Ghani Arayssi*, amongst others. The 1936 map includes up to 15 toponyms of the *shouhada*[26]; half of them Damascenes.

## 5 France Yesterday, Lebanon Today: A Continuity of Toponymic Dynamics

After the Lebanese Independence of 1943, the name of General Georges Catroux (one of the last high commissioners) was added to the map. Later, General de Gaulle was also added to the map, to adorn the western maritime façade of the city and form the'corniche de Beyrouth' along with the *rue de Paris*. The most recent toponym related to French sovereignty is the *rue Jacques Chirac*, added to the map by the end of the 2000s.

In 1951 the Municipal Council of Beirut decided to rename the 'foreign military names' [*sic*] after local celebrities (Lisan Al Hal 1951). However, the renaming was never accepted by the Ministry of the Interior and the 'foreign military names' remain on the map of Beirut. Some writers have recently voiced their opposition to the 'ever-present colonisation in the streets of Beirut' (Al-Kasim 2006; Mohsen 2008). An increasing number of religious figures are commemorated on the map—Christians and Muslims of every denomination and official religion of Lebanon. Different map features carry the names of local clergymen, e.g., Patriarch Paul Meouchy, Archbishop Ignace Moubarac, Antoine Salhani SJ, Mufti Hassan Khalid, Cheikh Mohammed Toufik Khalid, and Cheikh Mustapha Naja. After Independence, more martyrs of 1915 and 1916 were commemorated, and were later augmented by other national martyrs, considered to be heroes of the whole Lebanese nation. The last to be added are Rafic Hariri, Gibran Tueni, Samir Kassir, and some others who were assassinated between 2005 and 2008.

---

[26]The Arabic word for martyrs.

Toponymic systems introduced by the French Mandate are still operational on the map. These include martyrs, religious figures and even French generals' names. The city did not undergo a 'toponymic purge' and some toponyms were gradually changed but without any rupture with the toponymic system that was established by the French Mandatory authorities. The Lebanese Constitution (the founding text of Lebanon), was never abrogated or changed, but only modified upon Independence (Koch 2005). The reality of this fundamental text is indeed reflected by the continuity of the text on maps of the Lebanese capital, where colonial-era place names were not re-written, but left in place.

Is the Beiruti example to be generalized? Damascus, the neighbouring sister city of Beirut and capital of Syria, has undergone numerous toponymic purges since Syria gained independence in 1946 (Keilo 2015) and a cleansing of all French mandatory names. Yet the toponymic system of national martyrs, introduced at the same time in Beirut, is still expanding and new martyrs are being added to the map. In the Damascene case, in spite of toponymic cleansing, some toponymic dynamics were retained. In the Beiruti case, the colonial mandatory power has not only radically changed toponyms on the map but also created toponymic systems and established a commemoration dynamic that still governs the map of Beirut today: The Beiruti and Damascene cases remind us of the invented tradition of Hobsbawm and Ranger (Hobsbawm and Ranger 1992, 1) or even of the invention of memory inscribed in space (Said 2000). Lebanon itself was created by a Decree of Gouraud, in the name of the French Republic. Does this explain the importance of the 'foreign military names'? If the rupture is expressed through the toponyms imposed during the Mandate, the *raison d'être* of the Lebanese Republic, can we consider this to be a viable explanation of the presence of these toponyms and their toponymic systems on the map today? The Beiruti case is worthy of attention by comparing it to others in order to understand how colonial mapping outlives colonialism and its dynamics and systems still draw the map and dictate its outline.

# References

Abu Madi E (1996) Diwan Elia Abu Madi [Sha'er al Mahjar al akbar]. Dar al Aouda, Beyrouth

Abu Fakhr S, Salibi K (2012) The Heretic sage: dialogue with Kamal Salibi [Al Hartuqi al hakim, hiwar ma'a Kamal Salibi]. Al Mu'assasa al Arabiya lil Dirasat wal Nashr, Beirut

Ajay N (1974) Political intrigue and suppression in Lebanon during world war I. Int J Middle East Stud 5:140–160. https://doi.org/10.1017/s0020743800027793

Al-Kasim F (2006) French passion for Lebanon [Al Guira al Faransiyya a'la Lubnan]. In: Aljazeera. Available via Aljazeera.net. http://www.aljazeera.net/programs/opposite-direction/2006/3/19/%D8%A7%D9%84%D8%BA%D9%8A%D8%B1%D8%A9-%D8%A7%D9%84%D9%81%D8%B1%D9%86%D8%B3%D9%8A%D8%A9-%D8%B9%D9%84%D9%89-%D9%84%D8%A8%D9%86%D8%A7%D9%86. Accessed 23 September 2017

Ambrogi P, Le Tourneau D (2017) Dictionnaire encyclopédique de Jeanne d'Arc. Desclée de Brouwer, Paris

Ammoun D (1997) Histoire du Liban contemporain. Fayard

Arendt H, Canovan M (1998) The human condition. University of Chicago Press, Chicago

Azaryahu M (1996) The power of commemorative street names. Environ Plann D Soc Space 14:311–330. https://doi.org/10.1068/d140311

Badre L (2016) The Greek Orthodox Cathedral of Saint George in Beirut, Lebanon: the archaeological excavations and crypt museum. J East Mediterr Archaeol Herit Stud 4:72–97

Cheikho SJL (1920) New names for Beirut street, a critical point of view [Shaware' Beirut wa asma'ouha al jadida, nazra intiqadiya]. Al Mashrek 1025–1031

Commission de publication des documents diplomatiques français (2004) Documents diplomatiques Francais 1921. Peter Lang, Brussels

Daum A, Mauch C (eds) (2005) Berlin–Washington, 1800–2000: Capital cities, cultural representation, and national identities. The German Historical Institute, Washington, DC; Cambridge University Press

Davie M (1996) Beyrouth et ses faubourgs (1840–1940). Centre d'études et de recherches sur le Moyen-Orient contemporain (CERMOC), Beirut

Davie M (2001) Beyrouth 1825–1975, un siècle et demi d'urbanisme. Ordre des Ingénieurs et Architectes de Beyrouth, Beyrouth

Davie M (2003) Beirut and the Etoile area: an exclusively French project? In: Volait M, Nasr J (eds) Imported vs exported urbanism. Wiley, pp 206–229

Davie MF (2005) Communautés, quartiers et métiers à Beyrouth en 1923. Tempora Annales d'histoire et d'archéologie, pp 315–354

Drago G, Emmanuel T (eds) (2017) France & Saint-Siège: accords diplomatiques en vigueur. Les éditions du Cerf, Paris

Emereau A (1915) Bulletin de Droit, L'enseignement du Droit à Beyrouth. Echos d'Orient 17 (108):422–431

Evdokimov P (1997) L'amour fou de Dieu. Éditions du Seuil, Paris

Fawaz L (1984) The city and the mountain: Beirut's political radius in the nineteenth century as revealed in the crisis of 1860. Int J Middle East Stud 16:489–495. https://doi.org/10.1017/s002074380002852x

Ghorayeb M (2014) Beyrouth sous mandat français, Construction d'une ville moderne. Karthala, Paris

Hallak H (1987) Social, economic and political history of Ottoman Beirut and its Wilayet during the 19th century [Al Tarikh al ijtima'i wal iqtisadi wal siyasi fi Bayrouth wal Wilayat al Othmania fi al Qarn 19]. Al Dar Al Jamiya, Beirut

Hallak H (2010) Encyclopedia of Beiruti families, vol I [Maousou'at al A'ailat al Beyrouthiyya, al juz' al awwal]. Dar Al Nahda Al Arabiya, Beirut

Haut-Commissariat (1920) Recueil des actes administratifs du Haut-Commissariat de la République française en Syrie et au Liban, Année 1919–1920, vol 1. Imprimerie Jeanne d'Arc, Beirut

Heyberger B (2018) La France et la protection des chrétiens maronites. Généald'une représentation. Relat Int 173:13–30. https://doi.org/10.3917/ri.173.0013

Hobsbawm E, Ranger T (1992) The invention of tradition. Cambridge University Press, Cambridge

Hughes M (2011) Allenby, Edmund Henry Hynman, first viscount Allenby of Megiddo (1861–1936). Oxford Dictionary of National Biography

Ippolito C (2011) Naissance d'une nation: La Revue Phénicienne au Liban en 1919. In: Tadié B, Mansanti C (eds) Revues modernistes, revues engagées: (1900–1939). Presses universitaires de Rennes, Rennes, pp 39–49

Issawi C (1977) British trade and the rise of Beirut, 1830–1860. Int J Middle East Stud 8:91–101. https://doi.org/10.1017/s0020743800026775

Kassir S (2010) Beirut. University of California Press, Berkeley

Kaufman A (2014) Reviving phoenicia: the search for identity in Lebanon. Tauris, London

Keilo J (2015) La Syrie et la guerre des noms des lieux. In: Stadnicki R (ed) Villes arabes, cités rebelles Éditions du Cygne, Paris, pp 34–41

Kévorkian R, Nordiguian L, Tachjian V (2007) Les Arméniens, 1917–1939, La quête d'un refuge. RMN & Presses de l'université Saint-Joseph, Paris

Khalaf S (2006) Heart of Beirut, reclaiming the Bourj. Saqi Books, London

Khoury G (2004) Sélim Takla 1895–1945. Une contribution à l'indépendance du Liban. Karthala, Paris

Koch C (2005) La constitution libanaise de 1926 à Taëf, entre démocratie de concurrence et démocratie consensuelle. Égypte/Monde arabe Troisième Série 2:159–190. https://doi.org/10. 4000/ema.1739

Lammens H (1921) La Syrie: précis historique (two volumes). Imprimerie catholique, Beirut

Laurens H (1991) Le Liban et l'Occident. Récit d'un parcours. Vingtième Siècle Revue d'Histoire 32:25–32. https://doi.org/10.2307/3769995

Leo XIII (1884) Nobilissima Gallorum Gens. In: Libreria Editrice Vaticana. Available via the Holy See http://w2.vatican.va/content/leo-xiii/en/encyclicals/documents/hf_l-xiii_enc_08021884_ nobilissima-gallorum-gens.html. Accessed 23 December 2018

Lisan Al Hal (1951) New names for streets of Beirut [Asma'a jadida li shaware'e Beyrouth]. Lisan Al Hal 17482, Beirut

Magazine Circulation Co. (1918) Liberty's victorious conflict: a photographic history of the World War. Magazine Circulation Co., Chicago

Michel Chiha Foundation (2017) Organic Law Commission. http://www.michelchiha.org/ political-career/organic-law-commision/3/1/. Accessed 14 January 2017

Mohsen M (2008) Colonisation is always present in the land register [Al Isti'mar ma zala fi al zakira al a'aqariyya]. In: Al Akhbar. Available via Al Akhbar. http://www.al-akhbar.com/node/ 104169. Accessed 23 September 2017

Najjar A (2014) Dictionnaire amoureux du Liban. Plon, Paris

Pinta P (1995) Le Liban. Karthala, Paris

Rondot P (1954) Les structures socio-politiques de la nation libanaise. Revue française de science politique 4(1):80–104

Said E (2000) Invention, memory, and place. Crit Inq 26(2):175–192. https://doi.org/10.1086/ 448963

Saliba R (2004) Beirut city center recovery: the Foch-Allenby and Etoille conservation area. Steidl, Göttingen

Salibi K (2002) A bird on an Oak tree [Taer 'ala sindiana]. Al Shoroq Publishers, Amman

SDN (2011) Mandat pour la Syrie et le Liban 1922. Available via Digithèque MJP. http://mjp. univ-perp.fr/constit/sy1922.htm. Accessed 10 August 2018

Shawqi A (2005) Who works a lot will be promised to rest [Ou'iddat al rahatou al koubra li man ta'iba] 1922. Available via Adab. http://www.adab.com/modules.php?name=Sh3er&doWhat= shqas&qid=70134. Accessed 14 February 2017

Tadié B (2016) Revues modernistes, revues engagées: (1900–1939). Presses universitaires de Rennes

Zachs F (2012) Transformations of a memory of Tyranny in Syria: from Jamal Pasha to'Id al-Shuhada', 1914–2000. Middle East Stud 48:73–88. https://doi.org/10.1080/00263206.2012. 644459

**Jack Keilo** is a doctoral student and is a trained engineer and cartographer. His research focuses on the toponymy of capitals and seats of government, particularly the relationship between the state and toponyms.

# Mapmakers

# Military or Missionary Map? The First Topographic Map of Northern New Spain (1725–1729)

Mirela Altić

**Abstract** The first topographic map of northern New Spain appeared as part of the military inspection of the borderlands carried out by Brigadier General Pedro de Rivera y Villalón (1724–1728). Compiled by the military engineer Francisco Álvarez Barreiro between 1725 and 1729, this remarkable manuscript map, comprising five sheets and one overall map is known as the earliest official military map of the northern Spanish borderlands. However, apart from the northern edge of the Spanish Empire in New Mexico and Texas, the map also covers the vast area of Sonora, Sinaloa, Nayarit, Nueva Vizcaya, Extremadura and Nuevo León, reaching all the way to central Mexico. Although based on an original field survey and compiled with the clear military purpose of reinforcing the borderlands, the map shows a strong resemblance to Jesuit maps of the same region. In its style of presentation of the relief and symbolization used for the settlements, Álvarez Barreiro's map looks like a rather typical missionary map. How did that come about, and did the Jesuits contribute to its content? Based on original research of the sources of military and Jesuit provenance, the paper analyses the role the Jesuits played in the appearance of this map, as well as how this map affected the subsequent Jesuit mapping of the region. Moreover, using this example, I discuss how Jesuit mapping influenced the early military cartography (and vice versa) in general.

M. Altić (✉)
Institute of Social Sciences, Zagreb, Croatia
e-mail: mirela.altic@gmail.com

© Springer Nature Switzerland AG 2020
A. J. Kent et al. (eds.), *Mapping Empires: Colonial Cartographies of Land and Sea*,
Lecture Notes in Geoinformation and Cartography,
https://doi.org/10.1007/978-3-030-23447-8_15

# 1   Introduction

Brigadier General Pedro de Rivera y Villalón (c. 1664–1744) and his inspection of the Spanish borderlands, which he conducted from 1724 to 1728, are well known to historians. Rivera's military report, known as the *Proyecto*, and especially its third part, which makes the initial version of the military regulations, later to be known as the *Reglamento de 1729*, were the subject of many scholarly papers.[1] Rivera himself significantly contributed to this fact. While serving as governor of Guatemala, he prepared the first printed edition of his diary that was published in 1736. The next edition of Rivera's diary was published in Mexico in 1945, under the editorship of Guillermo Porras. This version was soon followed by Vito Alessio Robles's edition of the *Diario*, which, for the first time, besides the diary, included both the *Proyecto* and the *Reglamento*. Although Rivera mentioned Álvarez Barreiro and his six maps in his diary, the author did not pay much attention to him, leaving the mapmaker in the shadow of Rivera's imperial merits. Rivera kept silence on the existence of Barreiro's geographical description as well. Consequently, in contrast to Rivera's work, which became well known relatively early, the importance of the maps that emerged as part of the inspection went unnoticed.

The first to draw attention to Barreiro's maps and geographical description was Wroth (1951). He examined all the manuscripts related to Rivera's inspection that are kept in the Archivo General de Indias in Seville (AGI). Namely, the earlier works on Rivera (Porras 1945 and Robles 1946) were based on the copies that are kept in the Archivo General de la Nación and the Archivo Histórico Militar, both in Mexico. The manuscript copy found in AGI was the original one that was sent to the Spanish King Philip V and to the Council of the Indies on 2 March 1730.[2] That is the only copy that, apart from the *Proyecto* and the *Reglamento*, contains Francisco Álvarez Barreiro's geographic report sent to the Viceroy on 10 February 1730. Moreover, five of six sheets of Barreiro's map were identified in AGI that were part of his report. In 1988, the first English edition of the *Proyecto* and the *Reglamento*, prepared by Thomas H. Naylor and Charles W. Polzer, appeared, containing for the first time Francisco Álvarez Barreiro's geographical description and the reproductions of five sheets of his map. Even then, Barreiro's sheets showing a large part of today's Mexico and parts of New Mexico (USA) did not attract much attention of map historians. Only when the missing sixth sheet, actually an overall map of 1728, was brought to scholarly attention in 1992,

---

[1]Rivera's *Proyecto* is composed of three parts. The first part portrays the status of the presidios as Rivera found them, the second briefly recounts Rivera's immediate reforms at each presidio, and the third incorporates his recommendations for military reform. In its essence, the third part of the *Proyecto* was the precursor of the *Reglamento de 1729*, the military regulations for New Spain proclaimed by the Marqués de Casafuerte, Viceroy of New Spain (Naylor and Polzer 1988: 18).

[2]AGI, Audiencia de Guadalajara, 144.

the interest in Barreiro's work suddenly increased.[3] However, once again, the full value of Barreiro's work remained in the shadow. Most attention was given to his survey of Texas and to the overall map that includes the southern parts of the United States, leaving the rest of Barreiro's sheets neglected.

## 2 The Historical Background of Barreiro's Cartographic Campaign

The northern borderlands of New Spain had always been a restless region that was difficult to control. The main tools of Spanish power there were missions and presidios. The missions controlled the labour supply for royal mines and protected the native population. On the other hand, the presidios offered protection to both missions and Spanish colonial ventures, especially silver and gold mines (Naylor and Polzer 1988: 2). By the late seventeenth century, the question was raised whether the existing presidios were being operated properly, and whether they could provide appropriate protection to Spanish settlers and their lucrative mining business. Until the 1720s, the case against presidial mismanagement had grown to enormous proportions. The presidios became too costly, inefficient, and exploited by their own stuff. Military officers stationed in the presidios were often more concerned with their mine revenues than with the military protection of the area in which they were stationed. Well known is the case from 1712, when the captain of the Sonoran presidio of Fronteras left his post and went to live near the new royal mine of Nacozari in order to concentrate on his mining venture (Polzer and Sheridan 1997: 279–283). The negligence of the military staff began to jeopardize the effectiveness of Spanish control over the whole region, urging the military reorganization of the borderlands. Last but not least, the French threat to the Spanish possessions became more real after New Orleans was founded in 1718 (not accidentally, that same year, the Spaniards occupied San Antonio).[4]

The arrival of Juan de Acugña, the Marqués de Casafuerte, the newly appointed Viceroy and Captain General of New Spain, in 1722 proved to be a turning point. As a soldier and administrator born in Lima, Casafuerte understood the delicate situation in the borderlands. Already in 1723, he appealed to the king, Philip V, asking for permission to conduct a major inspection. The king responded

---

[3]Strangely enough, Barreiro's 1728 overall map has been owned by the Hispanic Society of America since 1907. It was re-discovered only in 1992, when it was exhibited at the exhibition 'Maps, Charts, Globes: Five Centuries of Exploration', from 26 February to 8 May 1992. Even stranger, a later copy, which was based upon Barreiro's 1728 map kept in the British Library (MS. 17,650.b), and made by Don Luis de Surville in 1770, has been known at least since 1912 when it was listed in the catalogue of the Lowery Collection (Phillips 1912: 538).

[4]The presidios of New Spain were not erected in some strategic line of defence from an outside enemy. Prior to the end of the seventeenth century, the only recognized threat came from rebellious native groups who needed to be kept under control.

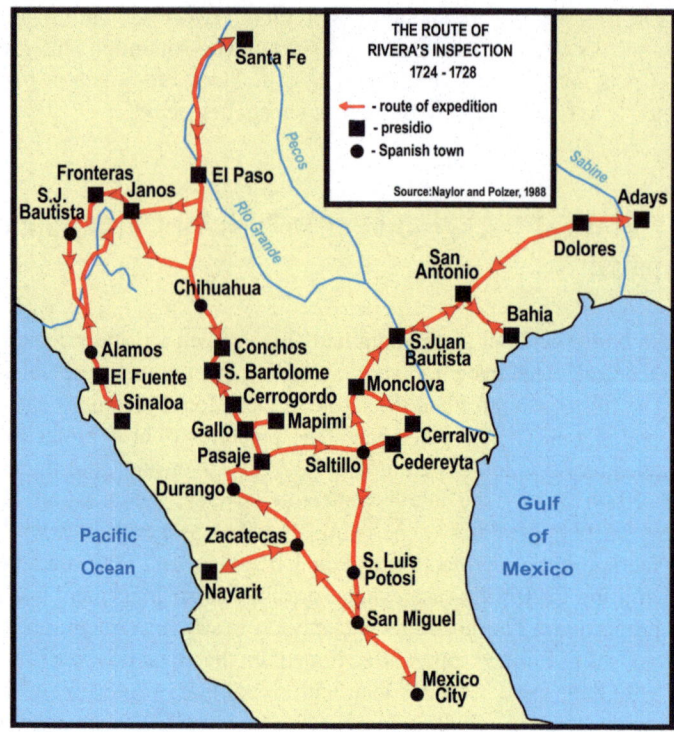

**Fig. 1** The route of the military inspection of 1724–1728, led by Pedro de Rivera y Villalón and his cartographer Francisco Álvarez Barreiro (drawn by the author)

affirmatively on 19 February 1724, naming Pedro de Rivera y Villalón as a brigadier general to carry out the inspection.[5] Casafuerte provided Rivera with detailed instructions on what needed to be done, arranged in twenty-six articles. According to the instructions, a cartographic campaign was an integral part of the military inspection. Article twenty-four charged Rivera with describing and mapping the whole region. For that purpose, Rivera was accompanied by Francisco Álvarez Barreiro, an experienced engineer and cartographer. Fully supplied and accompanied by a military escort, one royal scribe, and two clerical assistants, on 21 November 1724, Rivera and Barreiro were ready to leave Mexico City and start their four-year long trip through the borderlands that would eventually cover nearly 8000 miles. The theatre for the first military mapping campaign of New Spain was set (Fig. 1).

---

[5]AGI, México 690 (Casafuerte to the king on 25 May 1723, and the king's cédula of 19 February 1724).

# 3  Early Mapping of New Spain's Borderlands

Although the mapping of New Spain began even before Spanish expansion into this land started, the mapmaking development of this region was rather slow. Thanks to the joint efforts of navigators, cartographers, and cosmographers, during the first half of the sixteenth century, distinctive contours were given to America. However, within this vast landmass almost everything remained to be done. As time went by and the colonial powers started to take roots in New Spain, a need for more detail maps appeared. That especially refers to the northern borderlands whose mapping was an important geostrategic issue. In the northern borderlands, such as Pimería Alta, Sonora, Sinaloa, Tarahumara, and Nueva Vizcaya, the Jesuits were by far the most active cartographers in the area. The cartographers like an anonymous Jesuit (the 1662 map of Sonora and Sinaloa), Ivan Rattkay (the 1683 map of Tarahumara), and Adam Gilg (the 1692 map of Sonora) produced the first regional maps of the northern borderlands that were based on field observations. Their mapping activities in the interior of the land, as well as the publication of their accounts, enabled the appearance of syntheses on the whole region. One of the first such syntheses was the 1691 map of New Spain by Carlos de Sigüenza y Góngora, a Mexican professor of mathematics and Royal Cosmographer of the Realm. The appearance of Sigüenza y Góngora's map coincides with the beginning of a stronger development of Jesuit mapping that, with the arrival of Eusebio Francisco Kino, would become the most distinctive, but also the most productive branch of cartography in the territory of New Spain. Thus, until the early eighteenth century, the mapping of New Spain's borderlands almost exclusively depended on the missionaries. Apart from missionary needs, in the lack of other maps, Jesuit cartography served all other governmental purposes, including those of military affairs.

The precondition for the engagement of military authorities in the mapping of New Spain appeared only in 1711, when the Spanish Royal Corps of Engineers (*El Real Cuerpo de Ingenieros Militares*) was founded as one of the first manifestations of Bourbon reform. The Corps was established by Jorge Próspero Verboom, a Flemish-born military engineer in the service of the king of Spain. Educated at the Spanish Military Engineering Academy of Brussels (founded in 1675 and closed in 1706), Verboom was a prolific cartographer who understood well the military implications of mapmaking skills. The institutional framework of the Corps enabled the formal education of the first professional military cartographers. The requirements for entry into the Corps were stringent—all engineers had to have previous military training and status. Once accepted, they entered a rigid system of rank and class overseen by the engineer general. In 1718, Verboom ordered a set of rules which defined the engineers' duties that included mapping and making plans for ports and towns; drawings of fortifications, barracks, and other royal installations; and designs for roads, bridges and hydraulic projects (Fireman 1977: 33–34). At the time when the Corps were created, other military schools were established in Barcelona, Pamplona and Cádiz with three-year programs that included much instruction in cartography as well as mathematics. After the formal foundation of the Corps, the number of engineers grew steadily, until in 1728 there were about 128 engineers at work on a variety of projects (Buisseret 2005: 53). Some of them worked in New Spain. In the period until 1720, only six military engineers were listed in New Spain, while in the period from 1721 to 1763, no less than thirty military engineers were active in the province (Moncada Maya 2011: 4).

## 4  Francisco Álvarez Barreiro: The First Professional Military Mapmaker of New Spain

One of the trained engineers who ended up in New Spain was Francisco Álvarez Barreiro.[6] Born in 1701, in his youth Barreiro served in Naples and in Spain. We do not know what level of military education he had received before his arrival in the New World, but since he was referred to as a military engineer from his earliest days in America, we can conclude that he had already been military trained. Moreover, he signed his maps as well as his final report of 1730 as Lieutenant Colonel of Infantry and Chief Engineer of the Province of Texas, which implies that he was meanwhile recruited into the Spanish Royal Corps of Engineers. No less significantly, Barreiro arrived in New Spain in the company of the viceroy Marqués de Valero in 1716, which suggests that he was assigned with some very important task (Capel 1983: 31). Soon after his arrival in the New World, Barreiro was appointed military engineer for the 1718–1719 expedition of Governor Martín de Alarcón. The expedition was charged with the task of founding religious, military, and civilian settlements on the San Antonio River as well as with resupplying the missions in East Texas. On that occasion Barreiro apparently assisted in the construction of the famous chapel for the San Antonio de Valero Mission (later known as The Alamo) (Wagner 1967: 82). It is not known whether Barreiro produced any cartographic works in that period of his life.

In 1720, due to a general order, his career in New Spain was shortly interrupted. In obedience to the general order, which required all Spaniards in Mexico whose wives were in Spain to return home, Barreiro was obliged to return. However, his services were needed in Mexico once again, and he was back in 1724, this time to join the military inspection led by Pedro de Rivera y Villalón. For the next four years he accompanied Rivera's unit, surveying the terrain and compiling five relatively large-scale map sheets and one overall map of New Spain, designated by numbers from 1 to 6. Formally, he finished his campaign on 4 June 1728, when he re-joined Rivera's inspection caravan at San Luis Potosí. However, as his fifth sheet is dated 1729, he obviously continued his mapping even after the inspection was formally over.

---

[6]In his service record he was noted as Francisco de Barreyro y Alvarez, but since he was signing his works as Francisco Álvarez Barreiro, this spelling of his name became standardized. AGI, Relación de Servicios, 23 April 1722, Indiferente 141.

Soon after Barreiro had finalized six sheets for the purpose of Rivera's military inspection, he received yet another assignment from Casafuerte. This time he headed to the Pacific coast to map the town of Acapulco with its harbour and fort. Based on that survey, Barreiro compiled a semi-bird's-eye view plan of Acapulco, which also included the soundings of Acapulco Bay.[7] Its designation, *plano numero 7*, could suggest that this plan was actually a continuation of his previous mapping campaign of New Spain, which he now extended to the south Mexican coast. The assumption is additionally confirmed by the fact that the Acapulco plan was part of the same package of documents that Casafuerte sent to the King on 2 March 1730, together with Rivera's report and other Barreiro's map sheets of New Spain.[8]

After 1730, there are no records yet found that testify to Barreiro's work. Nevertheless, he was most probably the first professional military cartographer who worked in New Spain. Although several military engineers (mainly educated at the Academy of Mathematics, founded in Madrid in 1584) worked in New Spain before him, Barreiro's predecessors were almost exclusively concerned with the construction and maintenance of fortifications. Barreiro was the first educated military engineer to carry out the extensive mapping of almost the entire province, thus opening the way to other engineers such as Miquel Constançó (1739–1814), Nicolás de Lafora (c. 1730–d. unknown), and José de Urrutia (c. 1678–1741), who would be acknowledged as the most distinguished military cartographers of New Spain.

# 5 Barreiro's Survey and Maps

Although Rivera and Barreiro travelled together, when reaching their destination, they separated. Rivera would usually stay at a presidio and conduct a military inspection, while Barreiro travelled around, making a survey and drafting his maps. Pursuant to Article 24 of Casafuerte's instruction, Barreiro's primary duties were to conduct the survey, determine the latitude and longitude of each place, compile plans and maps of the provinces, and make a calculation of the provincial surface area and its boundaries. He also acted as observer of the lands and peoples he

---

[7] *Plano Topographico y Hydrográphico de el Puerto de Acapulco, de su poblado y Real Fuerza de San Diego situado en la costa de el Mar del Sur de la Nueva España en los 16 grados y 40 minutos de Latitud Boreal y en los 268 y 47 minutos de Longitud al respecto de el Meridiano de la Isla de Thenerife* [...] *Francisco Álvarez Barreiro* (scale: c. 1:983). Manuscript in colour; 44 × 49 cm. AGI, MP-MEXICO,125.

[8] It is clear from Casafuerte's letter that the plan of Acapulco appeared as part of defence strategy. Casafuerte explained to the king, 'Plan 7 is that of the Castle of Acapulco, in which it is known that although it has some defects of regularity, its fortification is very sufficient for defence against any threat or attack which may be offered; it never appears that any great one is to be feared, because the enemies would have to approach via the Southern Sea, and it would be a very remote and rare matter that they should bring sufficient forces to make this fortress be surrendered' (cf. translation of the letter in Rivera 1995: 54).

visited as well as census taker. For that purpose, Barreiro had been enjoined to determine the number and identify the nation of inhabitants, as well as appraise their disposition within the province. He was also to record the product which each territory produced, the varieties of timber, as well as to describe the climate, flora and fauna, and the quality of the soil. In the twenty-six folio pages of Barreiro's report, these data are presented in a fixed and regular order with the aim of providing a better understanding of his maps.[9] The report is divided into six sections that correspond to each sheet of the map.

Unfortunately, Bareirro's report provides no information on the techniques of the survey he applied, the instruments he used or on the way he compiled his maps. As in the case of most of the exploratory records of that time, the methods of field observation and mapping were not explicitly described because these procedures were considered to be commonly known. The usual navigational instruments consisted of a compass, which was equipped with a gnomon so that it could be used as a sundial, an astrolabe, sometimes a quadrant, a telescope, and a navigational handbook that included tables with latitude, longitude, and magnetic declination. Some information regarding the field measurements can be found in Rivera's diary in which he describes their movement, day by day. Each of Rivera's daily entries contains data on their heading and the distance they had travelled given in customary Spanish leagues (the equivalent of 2.6 miles). All distances had to be estimated, mostly based upon travel time. When they reached some larger settlement, a measurement of the longitude and latitude would be taken as well by 'observing the sun' (cf. entry of 16 August 1727), which confirms that they used appropriate astronomical instruments.

Based on the survey, six maps appeared—five detailed sheets and one overall map—which all together accompanied Barreiro's report. The sheets are numbered from 1 to 6 following their chronological appearance (with the exception of sheet 5, which was finalized in 1729, i.e. after sheet 6). Although not compiled at the same scale, Barreiro's sheets 1–5 were designed to line up (or overlap), so they provide continuous map coverage of the whole area. All detailed sheets as well as the overall map refer to Tenerife (Pico de Teide) as their prime meridian. They are accompanied by a graticule of latitude and longitude (except sheet 1) and a scale, and oriented by a compass rose. Though part of the same series, each sheet has its own title, so it can also be used separately. Such an approach testifies to the author's extraordinary sense of utilitarianism. Furthermore, each sheet has its own explanation key that is adjusted to the geographic content of the sheet in question (yet, the same type of objects on different sheets is marked by the same symbol). The boundaries of each province are delineated with special attention. According to Barreiro, the demarcation of the provinces was carried out under Rivera's direction.

---

[9]*Descripciones de las provincias internas de esta Nueva España, que sirven para la más clara inteligencia de los planos o mapas que las acompañan.* AGI, Audiencia de Guadalajara 144, folios 309–318.

For that purpose, the viceroy provided them with almost three thousand pages of documentation concerning the frontier and provincial boundaries to prepare Rivera and Barreiro for what they would have to face in the field.

## 5.1  Mapping the New Kingdom of Toledo (Nayarit)—Sheet 1

The expedition members started their journey in Mexico City on 21 November 1724. They travelled fairly lightly. They spent Christmas in Zacatecas and did not leave until 22 January. The first subject of their inspection was Nuevo Toledo de Nayarit. According to Barreiro's report, the survey of Nayarit lasted from 10 February to 2 March 1725.[10] Based on the survey and the information collected, Barreiro drafted his first sheet which he designated as Plano numero 1 (Fig. 2).[11] As it was the first province to be inspected and mapped, the pattern had yet to be established. This is confirmed by Barreiro's map of Nayarit, which is somewhat different from his subsequent maps. It is compiled at a far larger scale than the others, c. 1:500,000, and is the only sheet oriented with west at the top of the map. It is not accompanied by a graticule of latitude and longitude.

That was one of the youngest provinces that was incorporated into Spanish administration by force in 1722 and since then was administered by the Jesuits. Barreiro noted the native settlements that were actually Jesuit missions (Jesús María, Santa Gertrudis, Santa Teresa, San Francisco de Paula, San Pedro de Ixcatlán, San Juan Bautista, Nuestra Señora de Dolores, and Santa Rita de Peyotán) and the two presidios (La Santísima Trinidad and San Ignacio de Guaynamota). The whole presentation was quite original as no previous maps existed for the region that would document its newly formed human geography.[12] The presentation of mountain features of the region situated in the Sierra Madre Occidental is especially impressive, while the detailed hydrography of the Río San Pedro and the Río Jesús María with their tributaries is presented for the first time. Barreiro pointed out in his description that the province was administered by the Jesuits and there was no

---

[10]The duration of Barreiro's surveys does not coincide with the duration of Rivera's inspections. For example, Rivera's inspections of Nayarit lasted for two months, while Barreiro's survey was completed in one month. Barreiro probably used the remaining time to finalize his map and his geographic description of the province.

[11]*Plano Corográphico de el Nuevo Reyno de Toledo, Provincia de S[a]n Joseph de el Nayarit cuya Capital, que es la Mesa de el Tonat o Sol, se halla situada en los 22 grados y 23 minutos de Latitud Boreal y en los 262 de Longitud, tomado el primer Meridiano en la Isla de S[an]ta Cruz de Tenerife [...] Francisco Álvarez Barreiro. Escala de ocho leguas españolas* [1:500,000]. Mesa del Tonati: 1725. Manuscript in colour; 49 × 65.5 cm. AGI, MP-MEXICO, 120.

[12]The first known Jesuit map that shows Nayarit appeared in 1745 as part of the Relation by Salvador Ignacio Bustamante.

**Fig. 2** *Plano Corográphico de el Nuevo Reyno de Toledo*, drafted by Francisco Álvarez Barreiro in 1725 at the scale of 1:500,000 (Courtesy of AGI, Seville)

Spanish population in it. He identified the two native groups living in the province, the Coras and the Tecoalmes, both counting 3683 souls. He concluded his description of the province on 2 April 1725.

## 5.2  Kingdom and Province of Nueva Vizcaya—Sheet 2

After finalizing his survey of Nayarit, Barreiro headed to the north towards Nueva Vizcaya. To map this vast province, Barreiro made his journey in four stages, exploring the region. He made the first trip from Durango (8 August–29 September 1725), and then travelled from Pasaje (13–15 October 1725), from Gallo to Parral (11 November 1725–18 January 1726), and from Conchos (11 March–22 April 1726). This route followed the El Camino Real, traversing through the eastern side of the province, and enabled Barreiro to conduct a survey of the Central Mexican Plateau, elongated between the eastern and western Sierra Madre ranges, where most of the Jesuit and Franciscan missions were situated. Besides Nueva Vizcaya, this sheet also covers the province of Culicán.

The map that covers Nueva Vizcaya is the largest sheet in Barreiro's series.[13] While working on the province of Vizcaya, Barreiro realized that the scale he applied to relatively small surfaced Nayarit needed to be adjusted. Sheet 2 is therefore compiled at a scale of 1:2,200,000. Due to the more diverse human geography of the region, the cartographic key also needed to be expanded. Besides the native settlements (Jesuit missions with the Tarahumaras, the Tepehuanes and the Chinipas, as well as Franciscan missions with the Conchos) and the presidios (Gallo, Mapimí, Pasaje, Cerrogordo, Conchos, and Janos), this sheet also includes numerous Spanish settlements (cities, villas). Among those mentioned above, his description also highlighted the importance of Durango, the villa of Saltillo, Parras, Parral, the villa of San Felipe Real de Chihuahua, Cusihuiriachic, San Bonaventura, and Casas Grandes, all correctly marked on the map. Also depicted are most of Nueva Vizcaya's lucrative mines like Urique, Batopilas, Chihuahua, Cushuiriachic, Parral, Guanacevi, Indé, Loreto, and so on. Due to its large surface, a more realistic presentation of the relief was especially difficult. Barreiro clearly described in the description attached to this sheet that the province was divided from the northwest to the southeast by the Sierra Madre into two distinct regions, yet, this is not visible on his map. However, he made a huge step forward in mapping the hydrographic network. For the first time, the exact connection between the Río Conchos and the Río Grande is presented accurately, as well as the Río Nazas with its source in the Sierra Madre Occidental. Barreiro noted 31 native groups living in the province with a total of 51,910 souls. The description of the province was finalized in Santa Fe (New Mexico) on 23 August 1726.

## 5.3 Kingdom and Province of New Mexico—Sheet 3

As the northern extension of sheet 2, sheet 3 covers New Mexico with the Río Grande del Norte which flows through its central part (Fig. 3).[14] Although covering a considerably smaller surface area, the sheet that shows New Mexico is compiled at the same scale (1:2,200,000) and according to the same explanation key as sheet 2.

---

[13]*Plano Corográphico é Hidrográphico de las Provincias de la Nueva Vizcaya y Culiacán de el número de las de Nueva España, situadas entre los 22 y 32 grados de Latitud Boreal y entre los 256 y 271 de Longitud a el respecto de el Meridiano de la Ysla de Tenerife* [...] *Francisco Álvarez Barreiro. Escala de 50 leguas españolas de 17 1/2 en grado* [1:2,200,000]. Santa Fe, 1726. Manuscript in colour; 59 × 79 cm. AGI, MP-MEXICO, 121.

[14]*Plano Corográphico del Reyno y Provincia de el Nuevo México, una de las de Nueva España, situada entre los 31 y 38 grados de Latitud Boreal y los 258 y 264 de Longitud a el respecto de el Meridiano de la Ysla de Thenerife* [...] *Francisco Álvarez Barreiro. Escala de 35 leguas españolas de 17 1/2 en grado* [1:2,200,000]. San Felipe y Santiago; 1727. Manuscript in colour; 43 × 52.5 cm. AGI, MP-MEXICO, 122.

**Fig. 3** Third sheet of Barreiro's survey of New Spain showing the province of New Mexico at the scale of 1:2,200,000 and drafted in 1727 (Courtesy of AGI, Seville)

The province consisted of two presidios (El Paso in the south and Santa Fe in the north), four Spanish towns (San Lorenzo, Albuquerque, Santa Cruz, and Bernalillo) and a large number of Franciscan missions, e.g. Socorro, San Agustín de la Isleta, Senecú, Acoma, San José de Laguna, San Felipe, Santa Ana, Santo Domingo, San José de los Jémez, San Buenaventura de Cochiti, Zia, San Ildefonso, Santa Clara, Zuni, Pecos, San Lorenzo de Picurís, Galisteo, Pojoaque, Tesuque, Nambé, San Juan, and San Gerónimo de Taos. Beyond the provincial borders, there are several notes on the Apaches, the Moquis, the Navajos, and on other native settlements that threatened the Spanish territory. This region was not unknown to his predecessors, so Barreiro could use some older maps to confirm his findings. According to Barreiro's report, there were 24 native settlements in this province that were inhabited by 14 native groups, with a total of 9647 souls. The description of the province was finalized in the presidio of Sinaloa on 12 January 1727.

**Fig. 4** Sheet covering Sonora, Ostímuri and Sinaloa, drafted by Francisco Álvarez Barreiro in 1727 at the scale of 1:2,200,000 (Courtesy of AGI, Seville)

## 5.4 Provinces of Sonora, Ostímuri and Sinaloa—Sheet 4

Sheet 4 is the western extension of sheet 2. It covers the provinces of San Juan de Sonora, San Ildefonso de Ostímuri and San Felipe y Santiago de Sinaloa (Fig. 4).[15] Barreiro made two trips for the purpose of exploring the provinces, one from Janos to Alamos (23 October to 21 December 1726) and one from Sinaloa to Gallo (11 January–20 June 1727). The rugged trip from the coastal villages to the mountain plateaus demanded a long rest, which the whole team took in Chihuahua for the entire spring of 1727. Due to the several prolonged stopovers that were

---

[15] *Plano Corographico y Hydrográphico de las tres Provincias de Sonora, Ostimuri y Sinaloa, de las internas de la Nueva España. Situadas entre los 25 y 32 grados de Latitud Boreal y entre los 251 y 259 de Longitud a el respecto de el meridiano tomado en la Ysla de Santa Cruz de Tenerife* […] *Francisco Álvarez Barreiro. Escala de 40 leguas españolas de 17 1/2 en grado* [1:2,200,000]. San Pedro del Gallo; 1727. Manuscript in colour; 43 × 48 cm. AGI, MP-MEXICO, 123.

needed to conduct a proper survey of the terrain, it proved to be one of the most demanding sections of the expedition. To map the region, Barreiro had to follow the courses of the Magdalena, the Sonora, the Yaqui, the Mayo, the Fuerte, and the Sonora Rivers along which most of the Jesuit missions and Spanish settlements were situated. All the main missions and Spanish towns are presented, yet, many of the smaller missions (and visitas) are omitted. Again, the mining sites are marked: La Soledad, San Juan, Nacozari, Motepori, Baroyeca, and Río Chico. The presentation of the relief is deficient, the hydrographic network is shown in some detail, but in accordance with older maps of the region. According to Barreiro's description, the native settlements of these three provinces were inhabited by 11 native nations with a total of 21,764 souls. The description was finalized in the presidio of Gallo of Nueva Vizcaya on 20 June 1727.

## 5.5 Provinces of the New Kingdom of Extremadura or Coahuila and Nuevo León—Sheet 5

Sheet 5 covers two provinces, the Kingdom of Extremadura or Coahuila and Nuevo León, which are situated to the east of Nueva Vizcaya (Fig. 5).[16] Barreiro based this map mainly upon the reconnaissance taken on the way from the presidio San Juan Bautista del Río Grande to Monterrey, from 30 December 1727 to 14 February 1728. The geographical content of this sheet is much scarcer than of the previous ones. The heart of the province of Extremadura is situated between the Río Saladilo with its two tributaries, the Río Sabinas and the Río Nadadores, and the Río Grande (Río Bravo), all presented correctly. The capital of the province is Monclova, which is designated as a town, presidio, and a mission. Besides the one in Monclova, three other Franciscan missions are noted in the southern part of the province: Nadadores, Candela, and Santiago. The northern boundary of the province is defined by the course of the Río Grande, which makes the gateway to Spanish Texas. Near the river, the presidio and the mission San Juan Bautista, as well as the missions San Francisco Solano and San Bernardo are correctly marked. Barreiro noted only 815 persons living in the Coahuila's missions.

For the province of Nuevo León, more comprehensive details are presented. Two presidios are shown, Bocca de Leones and Cerralvo, and several Spanish towns: Santiago de las Sabinas, Las Salinas, Cadereyta, Monterrey, Guajuco, San Mateo del Pilón (present-day Montemorelos), Linares, and El Pablillo. Mining activities are identified in Bocca de Leones and Sabinas. The hydrography includes the Río San Juan with its tributaries, the Ramos, the Pilón and the Santa Catarina Rivers. The number of natives in Nueva León was only 650, and they lived in ten

---

[16]*Plano Corográphico de los dos Reynos, el Nuevo de Extremadura o Coaguila y el Nuevo de León, Provincia de el número de las de la Nueva España, situadas entre los 23 y 31 grados de Latitud Boreal y entre los 269 y 274 de Longitud a el respecto de el Meridiano de la Isla de Santa Cruz de Tenerife […] Francisco Álvarez Barreiro. Escala de 35 leguas* [1:2,000,000]. [S.l], 1729. Manuscript in colour; 44 × 53.5 cm. AGI, MP-MEXICO, 124.

**Fig. 5** Fifth sheet of Barreiro's survey of New Spain presents Extremadura or Coahuila and Nuevo León at a scale of 1:2,000,000. The map was based on the survey of late 1727 and early 1728, but finalized in 1729 (Courtesy of AGI, Seville)

Franciscan missions (Lampazos, Guadalupe, Jesús de Río Blanco, Concepción, San Nicolás de Gualeguas, Tlaxcala, Natividad, San Antonio de los Llanos, Labradores, and San Cristóbal). The description for this region was written in the city of San Luis Potosí on 2 June 1728, but the map itself was finalized in 1729.

## 5.6   New Kingdom of Filipinas and the Province of Texas—Sheet 6

As an addendum to his detailed sheets, Barreiro produced an overall map that covers the whole region between 21° and 41° north latitude and 250°–285° longitude east of Tenerife (Fig. 6).[17] The map, which is designated as sheet 6, is based

---

[17]*Plano Corográphico é Hydrographico de las Provincias de el Nuevo México, Sonora, Ostimuri, Sinaloa, Culiacán, Nueva Vizcaya, Nayarit, Nuevo Reyno de León, Nueva Extremadura, ò Coaguila, y la del Nuevo Reyno de Philipinas, Provincia de los Tejas/por Don Francisco Alvarez*

on observations that Barreiro conducted while travelling with Rivera from San Antonio to Adays, as well as on the additional journey from San Antonio to the presidio Nuestra Señora de Loreto de la Bahía that he started on 12 November 1727. From there, accompanied by twenty soldiers, Barreiro continued his coastal reconnaissance all the way to the Neches River. Between the presidio de los Adays and the presidio San Antonio, many members of his escort became ill; several died and were buried along the roadside. Barreiro finished his survey on 23 December 1727, and then returned to the presidio San Juan Bautista from whence he proceeded to survey Coahuila and Nuevo León.

Barreiro noted for Texas that it extended between 26° and 34° north and 272° and 286° east, having no settlements, only three presidios (Nuestra Señora del Pilar de Zaragoza de los Adays, San Antonio de Béxar, and Nuestra Señora de Loreto de la Bahía), and a few pueblos of native inhabitants, situated near the San Antonio and the Adays presidios. Its territory was intersected with numerous rivers: the Nueces, the Frio, the Hondo, the Medina, the Guadalupe, the San Marcos, the Inocentes,[18] the Colorado, the Trinity, the Neches, the Sabine, and the Río de San Andrés de los Caudacho (Red River), where the French territory started. His presentation of the coastal configuration of the region is quite original. While he repeats a few errors, such as the misconception that the Guadalupe River flows into Matagorda Bay or the Nueces River into the Río Grande, the accuracy of the rest of the presentation was astonishing and was thus praised as 'the most comprehensive map of the upper Texas coast yet achieved' (Weddle 1991: 239).

Apart from Texas, the map also includes vast territories around the Gila River and the regions to the north. In that regard, Barreiro's map could be understood as an echo of an old doubt about the geographical features of Baja California. While the mouth of the Colorado River and its confluence with the Gila River are represented relatively correctly, the so-called Channel of California is erroneously extended beyond 40° north, probably to harmonize the myth of California as an island with the data on the northern course of the Colorado River.

---

*Barreiro* [scale c. 1:6,000,000]. [S.l], 1728. Manuscript in colour; 44 × 25.8 cm. The Hispanic Society of America Collection. For reproduction, cf. Rivera 1995: following p. 126. The 1770 copy of the map, drawn by Luis de Surville, has the same title and an additional subtitle: '*reducido y delineado por Don Luis de Surville, en 4 de Julio del año de 1770*'. Manuscript in colour; 46 × 27 cm. Surville was a Spanish military engineer who worked as assistant archivist at the Spanish Archive of the Indies. The map is a true copy of the 1728 original with the exception of the cartouche. Surville replaced the original cartouche containing the Spanish coat of arms with a completely inadequate one that was obviously taken from a map that referred to the Ottoman Empire.

[18]The history and naming of the river is somewhat unclear. Some researchers consider the Río de los Inocentes to be an early Spanish name for the San Marcos River, given to it by Alarcón on his return trip from East Texas in December 1718. Some others believe that, in 1689, members of Alonso De León's expedition gave the name San Marcos to the first considerable river east of the Guadalupe, which scholars now believe to have been either the Colorado River or the Navidad River.

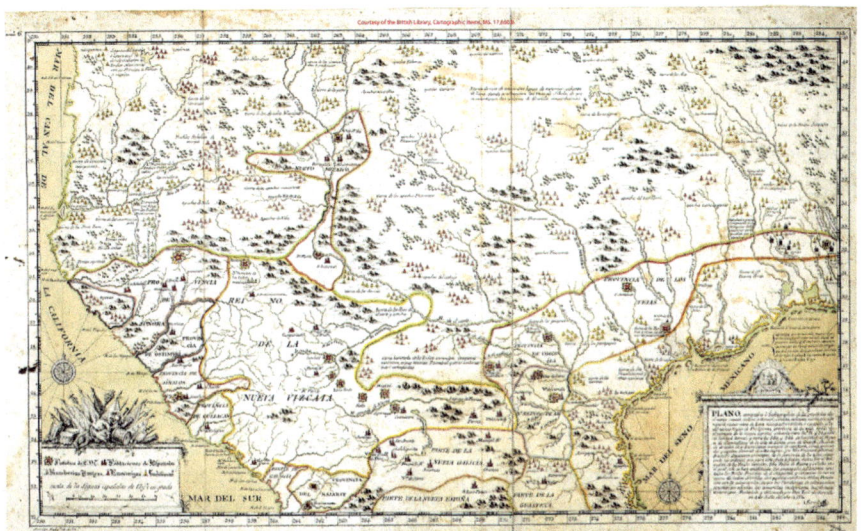

**Fig. 6** Barreiro's overall map (sheet 6) of 1728 that contains all the northern borderlands of New Spain, including New Mexico and Texas, as well as the territories in the north that were not under Spanish rule. This is a copy of Barreiro's original, drafted by Luis de Surville in 1770 (Courtesy of the British Library)

This map, although drawn in the same style as other Barreiro's sheets, was different in its purpose. This concerns not only its scale, which is about 1:6,000,000. While Barreiro's detailed sheets aimed to present the territorial organization of the northern borderlands with their physical and human geographies, showing how to reorganize and protect the borderlands from the inside, the overall map has another purpose. A different discourse of the overall map is well reflected through the revised explanation key—mines and missions are now excluded (with the exception of those in eastern Texas); only presidios and some Spanish towns are represented. At the same time, the native nations outside immediate Spanish control are categorized into three groups: friends (*rancherías amigos*), enemies (*enemigos*), and indifferent (*indiferentes*), thus revealing a much deeper concern for the outside threats than for those that came from the inside. By extending the presentation towards the north, into the territories that were not under Spanish control, this map should be understood as a geostrategic document for further territorial expansion. The strong statement of the map is further reinforced by the lavish cartouche accompanied by a large coat of arms of King Philip V of Spain. Without any doubt, sheet 6 was compiled as an overview for the royal authorities who needed a brief but clear insight into the Spanish dominions in America and their possibilities for further territorial development.

# 6   Barreiro's Sources and Role Models

No doubt, Barreiro's maps are based on his direct field observations. He spent almost four years exploring the terrain, observing the land and its peoples, speaking with local officials and collecting the geographic data. Yet, in many aspects, his sheets do not look like typical military topographic maps of the era (cf. the Spanish topographic map of Bahía de la Paz of 1739).[19] A strong echo of some other sources he used (or was influenced by) are clearly visible on his maps.

The presentation of relief is one of the most complex but also most unique elements of Barreiro's sheets. Presented using the pictorial technique of semi-bird's-eye view glyphs, which vary in their density and shape, it reflects Barreiro's attempt to present the configuration of the terrain without conducting any exact measurements of relief. His depiction of the relief is based upon the 'à la vue' method by which cartographers depicted what they saw, but also what they felt. The dramatic landscape is highlighted by the intense colours and sharp shapes of mountain massifs, which are illuminated from above, emphasizing the impression that the landscapes made on the mapmaker. While the use of bird's-eye view molehills for representing orography was quite conventional on numerous maps of that era, Barreiro's impressionistic style of mountain glyphs shows more resemblance to sixteenth-century Spanish *pinturas*, which were characterized by a similar pictorial presentation influenced by indigenous cartography.[20] That kind of cartographic syncretism was quite common for Jesuit maps as well. To some extent, the same applies to the presentation of vegetation and native huts, which Barreiro also depicted in a pictorial style.

Strangely enough, one of the most important elements of military topographic maps, a road network, is omitted from all Barreiro's sheets. Although he was familiar with the El Camino Real and other road communications he used in his travels, there is no sign of them on his maps. Despite their secrecy, Spanish military maps of the era regularly included roads and trails (cf. the aforementioned 1739 map of Bahía de la Paz, or the 1758 map of Santander).[21] Nevertheless, neither of the Jesuit (nor the Franciscan) maps of the early-eighteenth century portrayed roads.

This insight into the human geography of Barreiro's maps gives us even more clues about his possible sources. He categorized the settlements into three categories, Spanish towns (*poblaciones de españoles*), forts (*presidios*), and native

---

[19]*Seno de la Bahia de la Paz en la California en su Costa Oriental con puerto de difícil entrada de vientos del Poniente la delinio con perfiles y profundidades* by Ferdinand Konščak and Rafael Villar del Val (1739). Ink on paper; 27.5 × 27.5 cm. Biblioteca Nacional de España, Madrid, MR/42/620.

[20]For more information on the early presentation of orography, cf. Manuel Morato-Moreno (2017).

[21]*Mapa General ychnographico de la nueva colonia de Santander by* López de la Cámara Alta (Mexico City, 1758). Manuscript in colour; 125 × 230 cm. British Library, Add MS 17657.

settlements (*pueblos de indios*).[22] He also marked the mining sites (marking R[1]). Regarding the symbolization of settlements, Barreiro's maps show a strong resemblance to Jesuit maps of that era. Missions are designated by a circle with a cross—in contrast to Spanish towns marked by a church symbol—which fits the conventions of Jesuit cartography exactly. Barreiro's explanation key is actually very similar to the one used by Eusebio Kino on his 1710 map *Nuevo Reyno de la Nueva Navara*.[23] Subsequently, Spanish military cartography would discard such symbolization, especially for missions, and mark them by more neutral symbols without a cross (cf. the explanation key of Miquel Constançó's 1779 map of New Spain).[24]

A further insight into written Jesuit documents confirms that Barreiro was not just well informed about previous Jesuit mapping of the region. The Jesuits played an active part in his mapping endeavour, acting as his informants. The Jesuit Relations of New Spain testify to the fact that Rivera was in direct contact with Jesuit and Franciscan fathers in the field. In his letter of 2 February 1727, Rivera praises the efforts of Jesuit missionaries who maintain colonial power in this remote part of the Spanish dominion and asks the viceroy to support them more generously in the future (Alegre 1960: IV: 331–333). Moreover, he acknowledges them for their contribution to the success of his military inspection. Namely, besides the presidios, Rivera visited numerous missions, and was therefore in intense communication with Jesuit and Franciscan fathers who ran those missions, as well as with their superiors. That was exactly how Rivera and Barreiro managed to collect so much detailed information not only on the native populations, their number and their customs, but also on the geography of the parts of the provinces they did not visit. The influence of the geographical knowledge provided by the Jesuits was reflected not only through the resemblance of Barreiro's maps to some of the Jesuit maps, primarily to Kino's 1710 map. Also, the style of Barreiro's report (geographical description), as well as the part of Rivera's diary, which refers to the physical and human geography of the region, show a strong Jesuit influence. Their strict structure, somewhat repetitive, the definition of the extension of the provinces by their latitude and longitude, extensive information on the natives (sometimes more detailed than the one on military personnel), as well as formal statistics on the missions resemble a Jesuit account more than a military report.

---

[22]Yet, he made no difference between Jesuit or Franciscan missions as well as between settlements of baptized and unbaptized natives.

[23]*Nuevo Reyno de la Nueva Navara con sus confinates obros Reynos* by Eusebio Kino (Paris: d'Anville, 1724). Copperplate; 33.5 × 46 cm. Bibliothèque nationale de France, Cartes et plans, GE DD-2987 (8881).

[24]*Carta ó Mapa Geográfico de una gran parte del Reino de N. E.* [Nueva España], *comprendido entre los 19 y 42 grados de latitud Septentrional y entre 249 y 289 grados de longitud del Meridiano de Tenerife, formado de orden del Exc[elentísi]mo S[eño]r B[eilí]o Fr[ey] D[o]n Ant [oni]o Maria Bucarely y Vrsúa p[ar]a indicar la division del Virreinato de México y de las Provincias internas erigidas en Comandancia General en virtud de Reales Órdenes el año 177[9]* […] *Construyólo el Ingeniero D[o]n Mig[ue]l Constansó.* AGI, MP-MEXICO, 346.

# 7 Concluding Remarks

Barreiro's map series that was based on his direct field observation and survey represents the first cartographic representation of the northern Spanish borderlands compiled for strictly military purposes. It is an original work that significantly improved the geographical knowledge of the whole region. He managed to correct many misunderstandings and inaccuracies present on older maps, mostly those of Jesuit origin. That particularly refers to the mathematical base of the map, which is completely original. The accuracy of longitude and latitude calculations far exceeds the measurements of his cartographic predecessors. He was obviously equipped with instruments and surveying knowledge unavailable to the Jesuit fathers mapping the region before him.

Yet, despite his good equipment and expertise, he was strongly dependent on missionaries, who were his guides and informants. Their geographical and carto-graphical knowledge is woven into Barreiro's maps along with the knowledge he acquired from his own observations. His maps therefore reflect a strong influence of Jesuit cartography, not only in their visual style but also in their geographical content that was provided to a great extent by missionaries in direct contact with members of the military survey. As a result, though compiled for clearly military purposes, Barreiro's maps were a mixture of military and missionary information, thus falling into the category of hybrid maps in terms of their content and iconography. In that regard, his maps should be considered a transition phase from missionary to military mapping. Barreiro's maps are not the only example of such a phenomenon; José de Escandón's 1747 map of the Sierra Gorda (Nuevo Santander) contains the same blend of military and missionary (Franciscan) geographic data presented in the style of missionary cartography (Fig. 7).[25]

The cooperation achieved between the Jesuits and the military authorities during the military and administrative reorganization of New Spain left a deep mark on the development of its cartography. From the mid-eighteenth century, the exchange of knowledge between the missionaries and the military authorities, already common in seventeenth-century New France and Brazil, would become more frequent in the lands of the Spanish Crown. A 1758 military map is particularly indicative in this respect.[26] It originated from the cooperation between the missionaries and the military authorities during the military survey of the new colony of Nuevo Santander led by Agustín López de la Cámara Alta, a Lieutenant Colonel of royal infantry and engineers. The military map that appeared as a result of the latter

---

[25]Mapa de la Sierra Gorda y Costa del Seno Mexicano desde la Cuidad de Queretaro/José de Escandón. [S.l, c. 1747]. Manuscript; 77 × 59 cm. Library of Congress Geography and Map Division Washington, DC, G4410 1747.E8. In 1746, Escandón was commissioned to inspect the country between Tampico and the San Antonio River, later known as Nuevo Santander. In January 1747 he sent seven divisions into the area, and in October he presented a colonization plan. His map appeared as part of this colonization endeavour.

[26]Mapa General Ychnographico de la Nueva Colonia de Santander…/López de la Cámara Alta. Mexico City: 1758. Manuscript in colour; 125 × 290 cm. British Library, Add MS 17657.

**Fig. 7** José de Escandón's map of Nuevo Santander of 1747, characterized by a mixture of missionary and military styles (Courtesy of the Library of Congress)

survey was drafted by the Franciscan missionary Francisco José de Haró and had all the features of a topographic map. Nevertheless, such a high level of cooperation is still the exception, as is confirmed by the fact that the above-mentioned map, undoubtedly drafted by José de Haró, was formally signed by Cámara Alta. The influence between missionary and early Spanish military cartography was mutual; missionaries influenced the content and visual style of military maps, so over time, more and more military information found its way into missionary maps. José de Haró's map of Santander, which he compiled in 1770 for the purpose of administering the Franciscan province, differs in many respects from the conventions of a typical missionary map.[27] Not only are settlements marked by military symbolization (no circles with crosses), but the representation of relief and detailed marine topography also clearly reflect the influence of military cartography (Reinhartz 2011: 95).

Due to Spanish unwillingness to publish geographical information on their oversees dominions, Barreiro's maps were never published. The only exception is his general map of 1728, which was printed in Madrid in 1803 in a redaction by Juan Lopéz.[28] However, the success of Barreiro's maps pointed to the importance of a stronger military involvement in the mapping of the borderlands. Military

---

[27]*Este Mapa comprende todas las billas y lugares de españoles haci como Missiones de indios y presidios existentes en la Provincia Nuevo Santander*/Francisco José de Haró. [Mexico, c. 1770]. University of Texas Arlington Libraries, 86–255, 50/1, X/2.

[28]*Mapa Geográfico de las Provincias al N. de Nueva España. Por D. Juan Lopez, Geógrafo de S. M. Madrid, año de 1803*/Francisco Alvarez Barreiro. Madrid: Juan Lopez, 1803. Copperplate: hand coloured. Barry Lawrence Ruderman Antique Maps, ID 33864ct.

mapmakers and their maps, such as José de Haró's map of 1758, Nicolás de Lafora's and José de Urrútia's map of 1769[29] or Miquel Constançó's map of 1779, would soon dominate and start producing topographic maps that would meet all the requirements of military cartography.

# References

Alegre FJ (1956–1960) In: Burrus EJ, Zbillaga F (eds) Historia de la Provincia de la Compagñia de Jesús de Nueva España, 4 vols. Institutum Historicum S.J., Rome
Buisseret D (2005) Spanish military engineers in the new world before 1750. In: Reinhartz D, Saxon GD (eds) Mapping and empire: soldier-engineers on the Southwestern Frontier. University of Texas Press, Austin, pp 44–56
Capel H, García Lanceta L, Omar J, Olivé F, Quesada S, Rodríguez A, Sánchez J-E, Tello R (1983) Los Ingenieros militares en España, siglo XVIII: repertorio biográfico e inventario de su labor científica y espacia. Edicions Universitat Barcelona, Barcelona
Fireman JR (1977) The Spanish royal corps of engineers in the Western Borderlands: instrument of Bourbon reform, 1764 to 1815. A.H. Clark Company, Glendale
Moncada Maya JO (2011) La Cartografía Española en America durante el siglo XVIII: La actuación de los Ingenieros Militares. In: Anais do I Simpósio Brasileiro de Cartografia Histórica, pp 1–15
Morato-Moreno M (2017) Orígenes de la representación topográfica del terreno en algunos mapas hispanoamericanos del siglo XVI. Bol Asoc Geógr Espa 73:175–199
Naylor TH, Polzer CW (eds) (1988) Pedro de Rivera and the military regulations for Northern New Spain, 1724–1729. University of Arizona Press, Tucson
Phillips PL (1912) The Lowery Collection. A descriptive list of maps of the Spanish possessions within the present limits of the United States, 1502–1820. Government Printing Office, Washington
Polzer CW, Sheridan TE (1997) The Presidio and Militia on the Northern Frontier of New Spain: a documentary history. Volume two, part one: the Californias and Sinaloa-Sonora 1700–1765. University of Arizona Press, Tucson
Reinhartz D (2011) Mapping New Spain Borderlands. In: Dym J, Offen K (eds) Mapping the Latin America: a cartographic reader. University of Chicago Press, Chicago, London, pp 93–97
Rivera P (1736) Diario. Y derrotero de lo caminado, visto, y obcervado en el discurso de la visita general de precidios, situados en las provincias ynternas de Nueva España: que de orden de Su Magestad executô d. Pedro de Rivera, brigadier de los reales exercitos. Haviendo transitado por los Reinos del Nuevo de Toledo, el de la Nueva Galicia, el de la Nueva Vizcaya, el de la Nueva Mexico, el de la Nueva Estremadura, el de las Nuevas Philipinas, el del Nuevo de Leon. Las provincias, de Sonora, Ostimuri, Sinaloa, y Guasteca. Sebastian de Arebalo, Guathemala
Rivera P (1945) Diario y derrotero de lo caminado visto, y obcervado en el discurso de la visita general de precidios, situados en las provincias ynternas de Nueva España, que de orden de Su Magestad executó d. Pedro de Rivera, brigadier de los reales exercitos haviendo transitado por los reinos del Nuevo de Toledo, el de la Nueva Galicia el de la Nueva Viscaya, el de la Nueva Mexico, el de la Nueva Estremadura, el de las Nuevas Philipinas, el del Nuevo de León, las provincias, de Sonora, Ostimuri, Sinaloa, y Guasteca, 1724–1728. Introduction and notes by Guillermo Porras Muñoz. Porrua Hermanos, México

---

[29]*Mapa, que comprende la Frontera, de los Dominios del Rey, en la America Septentrional/* Nicolás de Lafora, José de Urrútia. Madrid: 1769. Colour manuscript on 4 sheets; 63 × 160 cm. Library of Congress Geography and Map Division Washington, DC, G4410 1769.U7 TIL.

Rivera P (1946) Diario y derrotero de lo caminado: visto y observado en la visita que hizo a los presidios de la Neuva España Septentrional el brigadier Pedro de Rivera. Con una introduction y notas por Vito Alessio Robles. Taller Autográfico, México

Rivera P (1995) Imaginary Kingdom Texas as Seen by the Rivera and Rubi Military Expeditions, 1727 and 1767. Edited and with an introduction by Jack Jackson, annotations by William C. Foster. Texas State Historical Association, Austin

Wagner HR (1967) The Spanish Southwest, 1542–1794. Arno Press, New York

Weddle RS (1991) The French Thorn: Rival explorers in the Spanish Sea, 1682–1762. Texas A&M University Press, College Station

Wroth LC (1951) Frontier Presidios of New Spain. Books, maps, and a selection of manuscripts relating to the Rivera expedition of 1724–1728. Paper of the Bibliographical Society of America, vol 45, pp 191–218

**Mirela Altić** is a Chief Research Fellow at the Institute of Social Sciences in Zagreb, Croatia. In the Department of History, University of Zagreb, Dr Altic holds the rank of Full Professor and lectures in the history of cartography and in historical geography. Besides her specialization in South Eastern and Central European map history, over the last few years she has published extensively on the Jesuit cartography of the Americas and conducts research in European and American archives and libraries. She is the author of eighteen books, numerous scholarly papers and a contributor to The History of Cartography Project. She was awarded the David Woodward Memorial Fellowship at the University of Wisconsin, Madison for 2013–14. In 2016, she was invited by the American Geographical Society Library to give an annual talk on Maps and America: The Arthur Holzheimer Lecture Series with the title 'Encounters in the New World: Jesuit Cartography of the Americas' and in 2017, she was an invited speaker at the David Rumsey Map Centre at Stanford University, with a talk on 'Jesuit Cartography of Americas: a comparative case study of Baja California, Tarahumara and the Amazon'. She is currently the Vice-Chair of the ICA Commission on the History of Cartography and Vice President and President Elect of the Society for the History of Discoveries.

# 'Dead on Arrival': The Unused Cartographic Legacy of Carl Friedrich Reimer

Jeroen Bos

**Abstract** After the Fourth Anglo-Dutch War (1780–1784), which ended disastrously for the Dutch East India Company (VOC), the need to reform was strongly felt. The Board of Directors (*Heren XVII*) asked for state support. This resulted in the formation of an independent Military Commission, with the mandate of reporting on the (military) state of affairs in the East, surveying the settlements and making plans for their improvement. The Prussian-born Carl Friedrich Reimer was employed as the main surveyor and military engineer. He had already been in the VOC's service for two decades before he was given this important task and became a confidant of Governor-General Arnold Willem Alting. The Governor-General was very skeptical towards the activities of the Military Commission, which operated fully outside the Company's established chain of command. By maneuvering Reimer into the Commission, Alting had eyes and ears in its affairs. Next to observing, surveying, drawing plans and writing recommendations, Reimer would also inform Alting about the journeys. Every major Dutch settlement from South Africa to the Moluccas was visited by the Military Commission, forming a unique view on the (military) state of affairs of the Dutch presence in Asia in around 1790. Together with the various recommendations that were accompanied by the excellent military maps by Reimer, the Dutch could make a fresh start in their imperial ambitions. However, when the Commission Fleet returned to the Republic in 1793 and all the reports and maps were transferred, the political constellation no longer had an eye for the overseas troubles of the VOC. The young and revolutionary French Republic just declared war. As such, the cartographic legacy of CF Reimer was 'dead on arrival'.

J. Bos (✉)
Leiden University Libraries, Leiden, The Netherlands
e-mail: j.bos@library.leidenuniv.nl

© Springer Nature Switzerland AG 2020

A. J. Kent et al. (eds.), *Mapping Empires: Colonial Cartographies of Land and Sea*,
Lecture Notes in Geoinformation and Cartography,
https://doi.org/10.1007/978-3-030-23447-8_16

# 1   Introduction

'*The Stadtholder, whose marriage was solemnized on 4 October in Berlin, was ceremoniously welcomed in the city of The Hague. In Amsterdam there were festivities as well. Some cannons were fired and bells were ringing. There was joyous music and plays*'.[1]

The fourth of October 1767 was a special day in the history of the Dutch Republic. In Berlin the young Stadtholder Willem V (1748–1806) married the Prussian princess Frederica Sophia Wilhemina (1751–1820). As was common, this was an arranged marriage, tying bonds between the House of Orange and the House of Hohenzollern. Throughout the country festivities were organized to celebrate this moment. The city of Amsterdam was no exception with illuminated buildings, fireworks, music and plays.

Amsterdam, though no longer the bustling metropole of the seventeenth century, still attracted many fortune seekers from Scandinavia and the German lands. It is very well conceivable that the protagonist of this contribution, a Prussian man named Carl Friedrich Reimer, was wandering the streets of Amsterdam while the festivities to celebrate the royal marriage took place. In contrast to his more fortunate compatriot, Carl Friedrich belonged to the anonymous German crowd, looking for job opportunities in Holland. When all other options failed there was always employment to be found at the long-distance trading companies: the Dutch West Indian Company or the Dutch East India Company (abbreviated VOC, after *Vereenigde Oost-Indische Compagnie*). At the main office, the austere *Oostindisch Huis* in the Oude Hoogstraat (Fig. 1), he either enlisted himself, or was collectively enlisted by a broker, in the rank of common soldier for a wage of nine guilders per month. His ship, *Vlietlust*, sailed from the island of Texel on 20 December 1767 bound for Ceylon.

Coming from the town of Königsberg (present-day Kaliningrad), Reimer followed in the footsteps of many anonymous Germans before him, seeking a better life and maybe a little adventure in the East.[2] Unlike so many of them though, Reimer would not go unnoticed in history. Amsterdam was the last grand European city he saw in his life. He fully embraced his career, spanning almost thirty years, within the VOC. From 'mere' soldier he climbed the hierarchical ladder and eventually distinguished himself as a military engineer, enjoying the necessary patronage from superiors along the line. He eventually died in Batavia (present-day Jakarta) in 1796 as Director of Fortifications and Inspector of Waterworks in the Dutch East Indies.

This chapter follows CF Reimer in the service of the VOC, especially during three decisive years in his career. As attaché to the Military Commission to the East (1789–1793), Reimer surveyed many of the fortifications of the VOC and drew plans and maps accordingly. He is by far the most productive in situ military

---

[1]Jacob Bicker-Raye in his manuscript chronology of Amsterdam, 9 November 1767.

[2]For the life and work of CF Reimer, see van Gerven (2002).

**Fig. 1** Main office of the VOC's Amsterdam branch at the Oude Hoogstraat by R Vinkeles, 1768 (Courtesy Stadsarchief Amsterdam)

mapmaker the VOC ever employed. Still, his cartographic legacy is largely unknown. The reasons why his production was never fully exploited, and as such 'dead on arrival', will be revealed.

## 2   Formative Years

As soon as the Dutch East India ship *Vlietlust* arrived at Ceylon in July 1768, Reimer was part of the island's military. Within a year he was employed as *derde chirurgijn* (third surgeon) and his wages were raised to sixteen guilders. We know virtually nothing of the life and education of the young Prussian prior to his enlistment with the Company, but this seems to suggest that he had at least a rudimentary knowledge of *materia medica*, maybe even having followed classes in medicine or botany. Five years later, in 1774, he was promoted once more and now held the rank of *onderchirurgijn*, or junior surgeon, earning twenty four guilders per month (De Silva and Beumer 1988: 460–461).

**Fig. 2** Formal visit by the envoys of the King of Candy to the Ceylon Governor in Colombo by CF Reimer, 1772 (Courtesy Rijksmuseum, Amsterdam, inv. nr RP-T-1904-18)

In this early period of his career in Dutch colonial service, Reimer also found pleasure in drawing. A very famous watercolour by his hand is now in the possession of the Rijksmuseum in Amsterdam, called *de Afbeelding der plegtige audientie verleent door Zijne Weledele Grootachtbare de Heer Gouverneur en Directeur van 't Eyland Ceylon, aan 't jaarlijkse Gesandschap van den Koning van Candia, in den jaare 1772*. It translates as 'Depiction of the formal visit, granted by the Honorable Governor and Director of the Island of Ceylon, granted to the Envoy of the King of Candy, in the year 1772' (Fig. 2).[3] Research has confirmed the lifelike portrayal of the people in this watercolour. Although sometimes VOC staff developed drawing skills in the East, the overall quality suggests education in Europe (Zandvliet 2002a: 129–131).

A set of two watercolours are attributed to Reimer that depicts the South Indian town of Chidambaram from roughly the same period.[4] The town is known for the enormous Hindu temple complex. From his later life it is known that Reimer had a personal interest in ancient Hindu architecture, so we can with near-certainty attribute this set—also in the Rijksmuseum—to Reimer. Whether the Prussian painted them in situ or copied an unknown original is still subject of investigation (Zandvliet 2002a: 239–241). In the years 1773–1774, a detachment of the Ceylon army was sent to the Indian mainland to support the Dutch settlement of

---

[3]Rijksmuseum, Amsterdam, inv. nr RP-T-1904-18.

[4]Rijksmuseum, Amsterdam, inv. nr RP-T-1904-19 and RP-T-1904-20.

Negapatnam at the Coromandel Coast. In the 1770s and 1780s, the Mysore Kingdom under Hyder Ali and, later, Tipu Sultan, was a constant threat for the European presence in this part of India. It is known that Reimer stayed at the Coast somewhere in the 1770s, probably as surgeon to the Ceylon military detachment at Negapatnam.

An anonymous map of the city and direct hinterland of Negapatnam lies in the collection of the Scheepvaartmuseum in Amsterdam (Fig. 3) (Gommans et al. 2010: 344–345).[5] This needs further investigation, but Reimer was possibly involved in making this map. However, the cartographic style, lettering, and also the scale bar—in *Schreeden* rather than the more commonly used *Rijnlandse Roeden*—differ from the later known maps by Reimer. Yet, it could be a collective work for which Reimer did the surveying and perhaps he even drew the sketches. It would explain his promotion in November 1777 to *eerste landmeter* or head surveyor in the rank of *vaandrig ingenieur*. The Ceylon government was always on the lookout for skilled surveyors since the Dutch East India Company was not only a merchant, but also a ruler on the island. The Dutch occupied large parts of the coastline and some inland regions and accurate maps of the regions were required for tax purposes. It seems that Reimer was employed to help survey the lands, although no maps by his hand are known from this period.

## 3   Promotion in Wartime

In 1782, Reimer was promoted as *fabriek* at the city of Colombo. A *fabriek* was an architect and main supervisor over the artisans working at the Company buildings, such as the warehouses, offices, ship wharves, hospitals, fortifications as well as public and religious housing. This marks the middle stage of Reimer's career and it is necessary to consider the bigger picture and discuss some geo-political matters in order to understand his further career more meaningfully.

In the year Reimer was promoted to *fabriek*, the Dutch fought a war, namely the Fourth Anglo-Dutch War (1780–1784). It painfully reaffirmed the weak position of the Dutch Republic and overseas the Dutch were humiliated. Several settlements in India were taken without a blow. Only the support of well-paid French allies prevented the takeover of Ceylon and of the Cape Colony. It was clear to friend and foe: the Dutch position in the East was frail. The Company Board of Directors, called the *Heeren XVII*, or Gentlemen Seventeen, had to swallow their pride and ask the state for financial and military support. The *Staten-Generaal* and Stadtholder decided to send a naval fleet to Asia. Commodore Jacob Pieter van Braam (1737–1803) came to the aide of the VOC and restored its prestige, especially among Asian opponents (Enthoven 2002; Bruijn 2003; Knaap et al. 2015: 156–159). When the commodore returned to the Republic, he sounded the alarm regarding the

---

[5]Scheepvaartmuseum, Amsterdam, inv. nr. SNSM a0145 (211) [0005].

**Fig. 3** City and hinterland of Negapatnam, artist(s) yet unknown, ca. 1773–1774 (Scheepvaartmuseum, Amsterdam, inv. nr. SNSM a0145 (211)-0005)

deplorable state of defence in Asia. Van Braam found a willing ear in Joan Cornelis van der Hoop (1742–1825), secretary of the Amsterdam branch of the Dutch Navy (*Admiraliteit*). Slowly, but surely, a plan was devised to inspect the Dutch overseas settlements by an independent Military Commission. This Commission would be installed by the Stadtholder, who was believed to be the only person who could break the deadlock and forcefully introduce the much-needed (military) reforms.

## 4   (Vain) Attempts to Reform

The Board of Directors was not entirely oblivious to the deplorable state of the military defence. After the war, in 1785, the brothers Van de Graaff were appointed as VOC governors. Cornelis Jacob van de Graaff (1734–1812) would head the Cape Colony. His younger sibling Willem Jacob van de Graaff (1736–1804) would rule the Ceylon government (Tates 2018). Both brothers enjoyed a military education in the Republic. Cornelis would begin as engineer in the Dutch Army to eventually become General Inspector of Fortifications in Holland, before his appointment as governor. Willem followed a less military path. As early as 1755 he was sent to Ceylon as merchant for the VOC. Jacob Cornelis was explicitly instructed to modernize the fortifications at the Cape. Also, a military school would be established where the technical military staff would be trained. After graduation, the new recruits were deployed all over Asia. Willem had comparable ambitions. As soon as he was installed in post he worked on projects to improve the main settlements in Ceylon.

These ambitious brothers would find a powerful opponent in Governor-General Willem Arnold Alting (1724–1800), presiding in Batavia with his Council of the Dutch East Indies, collectively known as the *Hoge Regering* or High Government (Fig. 4). Alting took control in 1780, right at the start of the Fourth Dutch-Anglo War. He would stay Governor-General until the demise of the Dutch East India Company in 1795. Historiography has judged harshly on the role of Alting in the unstoppable decline of the VOC. He is called a conservative manager, a model of the *Ancien Régime* ruler who would or could not yield to calls for reform (van Putten 2002: 189).

These factors may indeed have played a role in the passive stance of Batavia against proposed (military) reforms. However, a more selfish motive was fear. Alting, and his closest allies in the Council, probably feared that the ambitious Van de Graaff brothers might succeed and gain support in the Dutch Republic. It would be a matter of time before they pursued the highest post in VOC hierarchy: that of Alting himself. To counter this imaginary attack, Alting and his Council predictably blocked and discredited the projects coming from the Cape Colony and Ceylon. The resulting stalemate could only be broken by an external force.

The need for a Military Commission was high, but its creation faced many difficulties. To start with, the political system in the Dutch Republic was in chaos. A near civil war broke out between the so-called *patriotten* and *orangisten* (van Nimwegen 2017). In short, the *patriotten* opposed the influence of the Stadtholder

**Fig. 4** Portrait of
Governor-General Willem
Arnold Alting by JFA
Tischbein (Courtesy
Rijksmuseum, Amsterdam,
inv. nr. SK-A-3785)

and wanted a regime change or, at the very least, major reforms in government. The *orangisten* were loyal to the House of Orange and required no change at all, pushing instead for a more powerful Stadtholder. The discussion was mainly fought over anonymous pamphlets, but in 1787, tensions reached a violent climax. Order was restored only after intervention from Prussia. Secondly, the necessity of a Military Commission was questioned by many. Joan Cornelis van der Hoop had to convince several persons, not in the least the Stadtholder. Van der Hoop tried to sweeten the message stating that a successful inspection tour by this Commission 'would bring Glory to the Office of the Stadtholder', stealing the thunder from the *patriotten*. Thirdly, it would prove quite an ordeal to find able men willing to tour through Asia for several years. Among his own circles Van der Hoop recruited the naval officers Jan Olpher Vaillant (1751–1800) and Christiaan Anthony VerHuell (1760–1832) to head the Commission (Dörr 1988; Landheer 2006). But no artillery and infantry officers could be confirmed. Eventually, Johan Frederik Levinus Graevestein was persuaded to become an infantry officer to supervise the subordinate army officers, all of whom were Prussians.

## 5   The Military Commission to the East (1789–1793)

After a long delay, the Commission's fleet of two ships, the *Zephir* and the *Havick*, finally set sail for Asia in February 1789. Vaillant, VerHuell and Graevestein would only return to Holland in June 1793 (Appendix). Reimer's name circulated in the Board Room of the Amsterdam branch of the Dutch East India Company and it was thanks to commodore Van Braam that he was considered for a position as main surveyor and mapmaker of the Military Commission.

How did this come to be? Van Braam and Reimer had met each other in Ceylon in August 1785 (Odegard 2017). The commodore was ending his Asian tour of duty and was on his way back to the Dutch Republic. Reimer had just transferred to Batavia, where he was soon incorporated in the circles of Willem Arnold Alting. At Ceylon, Reimer made plans for the construction of modified fortifications on the island of Ceylon. Reimer showed these plans to Van Braam, who was deeply impressed by the quality of his mapmaking. When the projects made by Reimer were presented to Willem van de Graaff, however, they were rejected. Reimer was a victim of the power struggle between the Van de Graaff brothers and Alting. Moreover, Reimer was a *fabriek*, an artisan. The Van de Graaff brothers represented the new class: educated military engineers and Reimer lacked formal training. Besides, his plans were based on Prussian principles of fortification. All were insuperable reasons to be rejected, according to Van de Graaff.

The Ceylon governor sought expertise from the French engineers at Pondicherry. He invited Chevalier De la Lustriere and his assistant De la Goupillière to inspect the Dutch settlement on Ceylon during the years 1786–1787 (Fig. 5). Reimer criticized the French plans for being too costly and too rigid. The French system required the complete reconstruction of new fortifications without considering the existing infrastructure or the local terrain. This meant that the Dutch settlements were defenseless during the years of their demolition and reconstruction. As such, according to Reimer, the French plans were impracticable fantasies. Furthermore, Reimer thought it was unwise that foreign engineers, working for a competing Company, were given full access. In the eyes of Reimer, it was thanks to Van de Graaff that the French were now completely aware of the Dutch defenses and the weaknesses at Ceylon. Alting was glad of this sharp critique by Reimer, which meant that progress of Van de Graaff's projects could be blocked. *Heeren XVII* agreed and a stalemate was the result. Reimer could get his revenge thanks to the mediation of Van Braam; because the commodore advocated that Reimer would assist the Military Commission, he could now present his projects directly to Vaillant, VerHuell and Graevestein.

For Alting, who was very skeptical towards the Military Commission, it was very convenient that Reimer would be attached to it. Alting ordered Reimer to cooperate with the commissioners, but at the same time to report to the High Government. Reimer decided to keep a journal in which he wrote the most memorable events that took place during the inspection tour. An extract was given to Alting to inform the Governor-General about the tour. This unique manuscript is now kept at the National Archives of The Netherlands (Fig. 6).[6] Reimer travelled to Ceylon where he was to await the ships which came from the Cape Colony. Here he faced Willem van de Graaff and his team of engineers, which Reimer criticized so

---

[6]NL-HaNa, 1.10.03, inv. nr. 87.

**Fig. 5** Copy of the project plan for the defenses of Trincomale; original by De La Lustière, copy made for analysis and criticism by CF Reimer (Courtesy Nationaal Archief, inv. nr. VEL1016A)

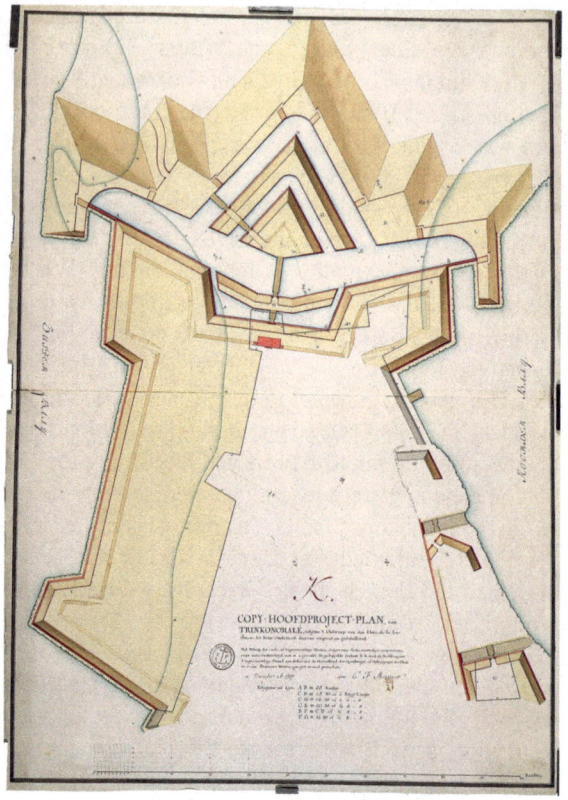

harshly. In his journal he wrote that it took a while before the awkward situation subsided and normal relations between the two were established.

The inspection tour may be divided into three parts: the first being the inspection of the Cape Colony, the second the inspection of the so-called *Westerkwartieren* or all provinces west of Batavia, and the third the inspection of the possessions in what contemporaries called the *Grote Oost*, or Great East (the settlements in what can now be roughly called Indonesia and Malaysia).

Reimer fulfilled a dual role. For the Military Commission he measured, surveyed and made maps of the several settlements. At the same time, he kept the High Government informed of the main events during the tour. Reimer now had the ideal opportunity and was very aware that if his projects and plans were judged favourably by the naval and military command in the Dutch Republic, he could really leave his mark. With this in mind he set himself on a mission to produce as many maps as possible. His name is mentioned on several project plans, although it must be stated that Reimer had help from the Prussian engineers and other officers on board ship. In his journal, which Alting would read, Reimer never mentions their names.

**Fig. 6** Extract of the diary
kept by CF Reimer; presented
to WA Alting and the Council
of the Dutch East Indies, 2
volumes, 1789–1792
(Courtesy Nationaal Archief,
inv. nr. 1.10.03, inv. nr. 87)

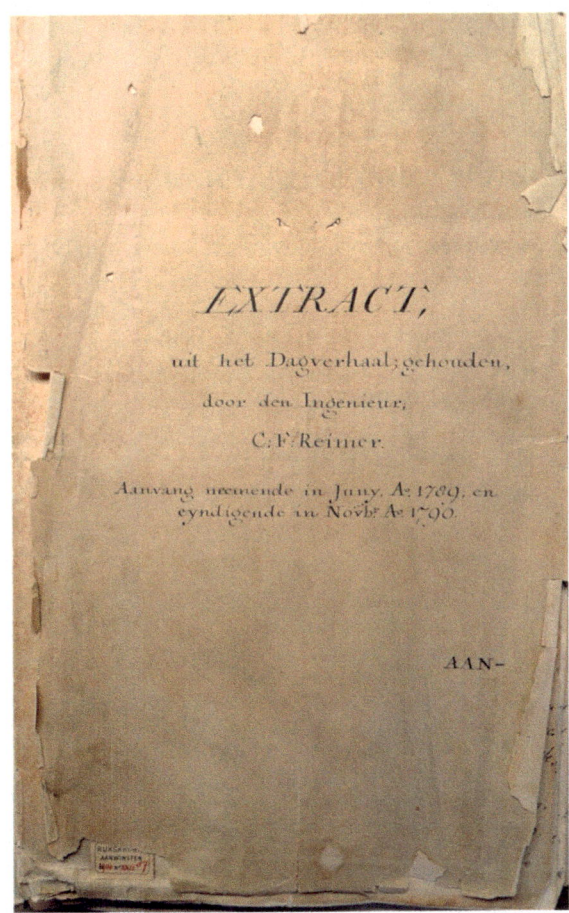

## 6   A Closer Look: Madras and Riau

Reimer and his accomplices produced around fifty maps and plans of the Dutch
settlements in Africa and Asia (see Appendix). Most of them were reproduced in
the seven volumes comprising the series *Comprehensive Atlas of the Dutch United
East India Company* (2006–2010). Although careful archival research was done for
this enormous project, a military plan by Reimer of the English defense works at
Madras (present-day Chennai) was missed.[7] This attempt at espionage by the
Military Commission deserves proper attention. Secondly, some focus will be
placed on the activities in the Riau archipelago. This region would come to have

---

[7]NL-HaNa, 1.10.03, inv. nr. 76.

much strategic importance in the early nineteenth century. The commissioners sensed this future interest and made considerable efforts to survey and report on it.

## 6.1 Espionage at Madras

Although the Commission worked to a very tight schedule, the commissioners also found time to plan stopovers with European competitors. When they inspected the Coromandel Coast, they visited the Danish at Tranquebar and the English at Madras in September 1790. It would prove to be a great opportunity to gather information on their strengths and weaknesses. Commissioners Vaillant, VerHuell and Graevestein tasked Reimer to discretely appraise the defences of Fort St George. Interestingly, Reimer initially objected because he was not sure that, as a VOC servant, he was allowed to set foot in the English port city. After the commissioners reassured him that he now felt under jurisdiction of the Military Commission, he was persuaded. A regular full survey with instruments was, of course, completely out of the question. The English hosts were fully aware of the hidden agenda of their Dutch guests and only allowed them guided walks through town and around the defences. It is for this reason that Reimer delivered a plan (and supplementary report) on which some blank areas can be seen (Fig. 7). He noted that the parts which he did not get to see

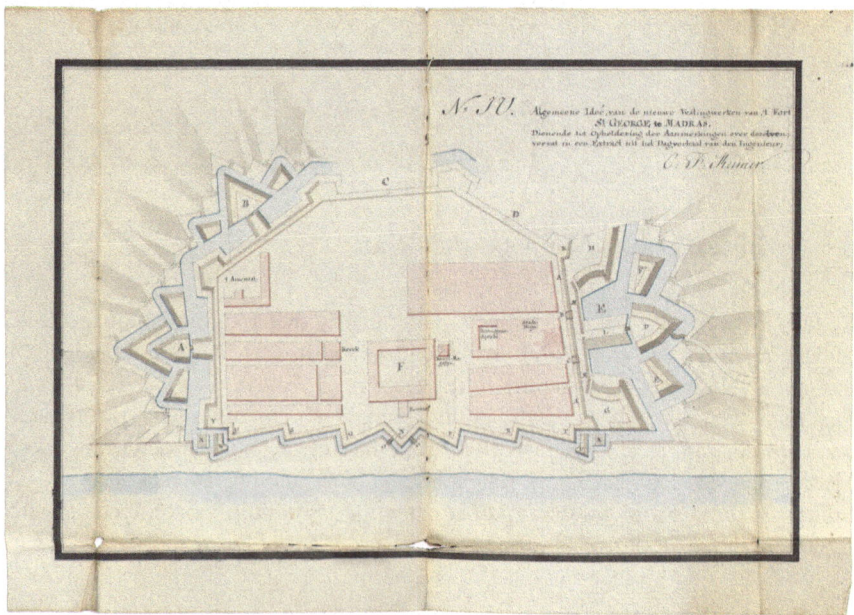

**Fig. 7** Plan of the defenses of Fort St George (Madras) by CF Reimer, 1790 (Nationaal Archief, inv. nr. 1.10.03, inv. nr. 76)

were intentionally left unfilled. Reimer considered himself a professional military engineer and he wrote that accuracy should be the leading principle in mapmaking and cartography. In the report he apologized more than once for the incompleteness of the plan he had produced. Incomplete as it may have been, the accuracy of the parts he did draw is admirable. Both his quality and pace of work were noted and appreciated. On various occasions the commissioners Vaillant, VerHuell and Graevestein wrote positively on his work to their superiors. They were not being merely polite, since the other engineers and officers tasked with drawing plans were scolded in the same letters for their lack of competence, effort or quality.

In the manuscript intended for the High Government, Reimer also wrote interesting facts about the botanical activities of Dr James Anderson, a Scottish physician in EIC service. In the Company garden at Madras he (unsuccessfully) tried to grow *opuntia*, a cactus which was the host plant for the highly sought cochineal insect. This little bug provided intense red dye, indigenous to Oaxaca in Mexico, and was exclusively imported to Europe by the Spanish, who firmly kept this profitable commodity to themselves. Under the Director of Kew, Joseph Banks (1743–1820), an imperial scheme was set up to break this Spanish monopoly (Butler Greenfield 2005). The English tried to get living cochineals and transplant the bugs to Madras, where it was thought that the climate was comparable to Oaxaca.

Anderson was ordered to grow opuntia, which the bugs liked, and to wait for the right shipment of the insect (which never arrived). Reimer was invited to the garden by Dr Anderson, who probably recognized that Reimer was a fellow medical practitioner, and perhaps also as another 'amateur'; a person in pursuit of knowledge for the sake of it, or its 'usefulness'. The Prussian was unimpressed by this experiment, stating that 'further investigative research needs to be done in order to conclude that transplanting the Spanish cochineal to another climate does not negatively affect its purity'. He was, however, shocked by the revelation that in another garden in Madras the English tried to cultivate cinnamon. According to Dr Anderson, the English managed to acquire a living tree via a certain Madame Leitts (possibly Lights), who received it from her Dutch acquaintances on Ceylon. If this were true, Reimer argued, and if the cinnamon proved to be of good quality, the Dutch had lost one of their most profitable commodities to their English competitors. A worrying observation indeed.

By drawing the plan of the military defenses of Fort St George, accompanied by a written report of 22 sheets, and the separate description of his visit and observations at the botanical garden, Reimer clearly committed espionage as he took on the role of information broker, passing classified information to his superiors.

## 6.2  Opportunities at Riau

In the Malay world, the Dutch contented themselves for over a century with Malacca, the port city they conquered from the Portuguese in 1641. The conquest turned out to be more important strategically than commercially. From Malacca, control could extend over the ruler of Johor and European settlements in the Straits of Malacca could be blocked. This situation changed after 1760 when Sultan Suleiman (1722–1760) died and the succession was fought out between pro- and anti-Bugis parties. Eventually, the Bugis of Riau and Selangor saw in Raja Haji a leader to start the rebellion against the Dutch hegemony. The whole event led to unintentional Dutch involvement in the region. Haji and his Bugis were defeated in 1784 by commodore Van Braam. The operation was not so much executed to punish the Bugis—although they were considered a great nuisance by the Dutch—as it was an attempt to repel growing English penetration into the Straits. The Dutch (justly) suspected them to open a post at Riau. Their suspicion was further raised when, in negotiations after the Fourth Anglo-Dutch War, an English offer was made to transfer the captured port city of Negapatnam at the Coromandel Coast for the right to settle at Riau (Tarling 1962; Lewis 1995).

In response, the VOC immediately began to erect a provisional fortification at Riau in 1785. When the Military Commission inspected the place in December 1791 and January 1792, they took time to survey the harbour, the settlement and its vicinity. The commissioners wanted to assess the reasons for English interest and to understand why the Dutch should give more attention to establishing a more solid presence in the region. They felt future opportunities would be gained from this location, and sought several motives why a permanent and firm fortification was admissible. The commissioners tasked their naval officers to draw a map of the harbour and sailing routes to Riau (Fig. 8), while Reimer produced a topographical overview (Fig. 9) as well as a military plan with a projected extension of the defences (Fig. 10).

The fact that this hitherto insignificant location in the Dutch Empire attracted so much attention, not only in terms of cartographic productivity but also in the correspondence between the commissioners, Batavia and the home government, meant that much importance was given to controlling the Straits of Malacca. This had political consequences with developments in the nineteenth century such as the rise of Singapore. Reimer contributed to the debate, not only with his maps, but also by adding a description of the cultivation and extraction of gambir.[8] He wrote several sheets about this overlooked commodity, which was locally used as a medicine, a food additive and as a dye. It would be interesting, according to Reimer, to investigate its potential profitability.

---

[8]NL-HaNa, 1.10.03, inv. nr. 87, fol. 99-104.

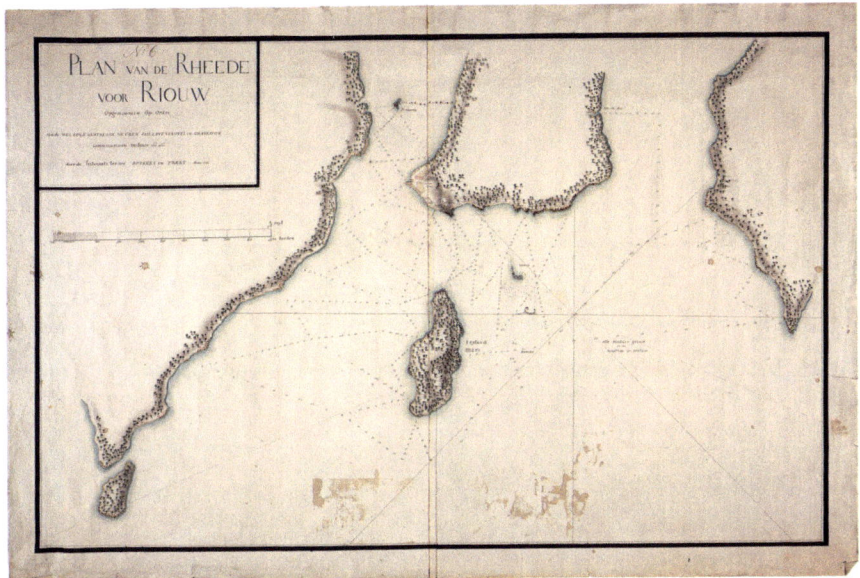

**Fig. 8** Maritime map of the roadstead of Riau and its sailing routes by AA Buyskes and AC Twent, 1791 (Courtesy Nationaal Archief, inv. nr. VEL0370)

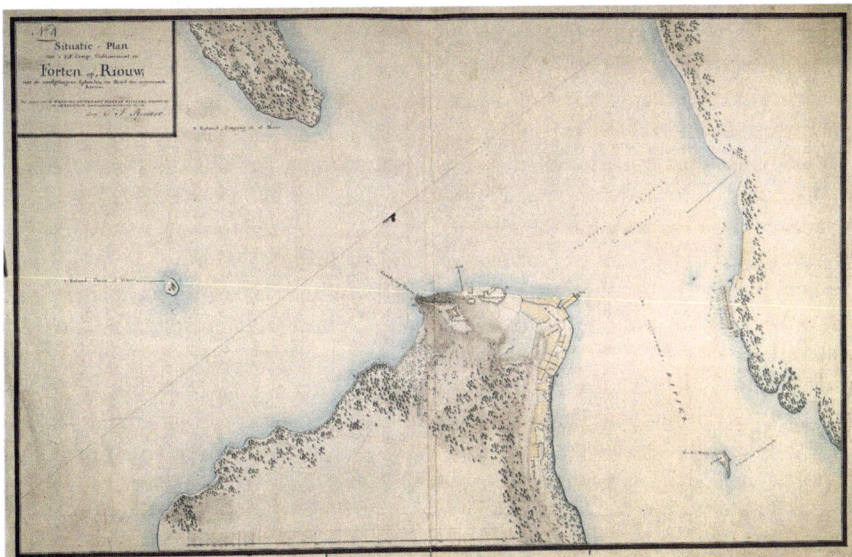

**Fig. 9** Topographical map of the town and hinterland of Riau by CF Reimer, 1791 (Courtesy Nationaal Archief, inv. nr. VEL1151)

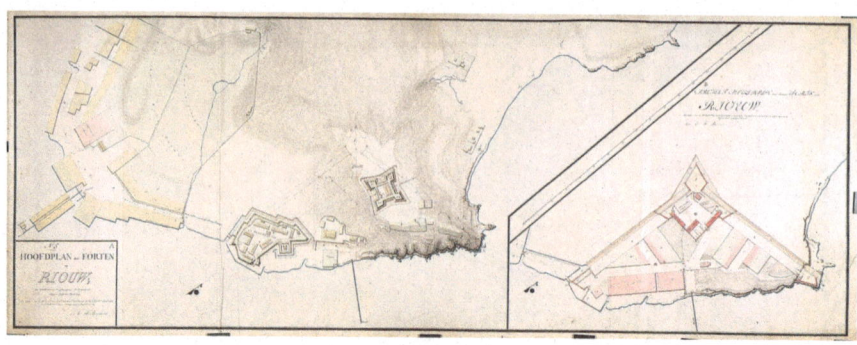

**Fig. 10** Military plan of the actual situation and projected new fortification at Riau by CF Reimer, 1791 (Courtesy Nationaal Archief, inv. nr. VEL1152)

# 7  Epilogue

Reimer's huge cartographic production was, however, of no avail. For geopolitical reasons, the deplorable state of the Dutch East India Company—which had a firm agenda in the late 1780s—could not have cared less in the year the Commission returned. The ships called at Texel in June 1793, while in February the young French Republic declared war with the Dutch Stadtholder and the King of England. All the correspondence, analyses, plans, maps and reports produced by or for the Commission were kindly received and stored away.

Alting must have been pleased with this *status quo* outcome. The activities of the Van de Graaff brothers were also neutralized. We can doubt if Alting was ever committed to fully modernizing the military defence of the Dutch East India Company. Reimer was promoted to a political position in Galle, Ceylon, which he kindly declined for lack of commercial and political skills. He wrote to Alting that he needed time to work out the sketches from the inspection tour. A year later he was promoted as Director of Fortifications and Inspector of Waterworks, overseeing the activities in the whole territory of the VOC. With a lack of spectacular results, this was probably only a 'paper promotion'. Reimer also spent time to write a voluminous report on the health situation in Batavia and unpublished notes on ancient Hindu architecture. He died in January 1796.

Reimer's cartographic legacy is enormous. He truly mapped an empire, even if only an empire in serious decline. After the French successfully invaded the Dutch Republic in early 1795, the overseas settlements were taken by the French or English, only to be restored in 1816. By this time, the maps of Carl Friedrich Reimer and the reports of the Military Commission were considered outdated. They were stored in the archives (Meilink-Roelofsz et al. 1992; Balk 2007: 141–142) and were never to be examined again.

## Dedication

In loving memory of Toke Bos-Beerens. My late wife accompanied me to the Mapping Empires symposium in Oxford, September 2018. It turned out to be our final foreign journey. She died of pancreatic cancer in October 2018. Vaarwel, mijn liefste. Je was mijn grootste fan. En ik de jouwe.

## Appendix

In this appendix the Military Commission is followed chronologically. For each location, the presently known plans, maps and charts in the collections of Nationaal Archief are listed by inventory number, followed by the volume/page number of its reproduction in the facsimile series *Comprehensive Atlas of the Dutch United East India Company* (7 vols, 2006–2010), if applicable.

| Date | Location | Maker(s) | Type | Inv. nr(s) | Reproduced |
|------|----------|----------|------|------------|------------|
| 1789 (May–Oct); 1792–1793 (Aug–Feb) | Cape of Good Hope | AC Twent | Charts | VEL 196 (Bay of Content); VEL 197 (Bay of Algoa) | V, 342; V, 343 |
| 1789 (Dec); 1790 (Feb–Apr) | Colombo | P Elias, AA Schenk | Chart | VEL 240 | IV, 100 |
| 1789 (Dec); 1790 (Feb–Apr) | Colombo | P Elias (copy after Schenk) | Plan | VEL 980 | IV, 100 |
| 1789 (Dec); 1790 (Feb–Apr) | Colombo | GE Schenk (copy after Reimer); AA Buyskes (copy after Reimer) | Plans | VEL 978; VEL 979 (identical copy of VEL 978) | IV, 99; n/a |
| 1789 (Dec); 1790 (Feb–Apr) | Colombo | B Matthijsz; GE Schenk; P Elias; CF Reitz (copies after Reimer) | Designs | VEL 974; VEL 975; VEL 976; VEL 977 A/B/C | n/a; IV, 101; n/a; IV, 102–103 |
| 1789 (Dec); 1790 (Feb–Apr) | Kotta (vicinity Colombo) | P Elias | Plan | VEL 981 | IV, 122 |
| 1790 (Jan–Feb) | Cochin | AA Buyskes | Plan | VEL 907 | VI, 260 |
| 1790 (Jan–Feb) | Cochin | GE Schenk (copy after Von Krause); D van Lier (copy after Von Krause) | Plans | VEL 905; VEL 906 | VI, 255; VI, 256 |

(continued)

(continued)

| Date | Location | Maker(s) | Type | Inv. nr(s) | Reproduced |
|---|---|---|---|---|---|
| 1790 (Jan–Feb) | Cochin | P Elias (copy after Reimer); A Heidenreich (copy after Reimer); Van Lijnden (copy after Reimer) | Designs | VEL 908; VEL 909; VEL 910 | VI, 261– 263; VI, 264–266; VI, 261 |
| 1790 (Apr–Jun) | Galle | AA Buyskes; P Elias | Chart | VEL 248; VEL 249 (identical copy of VEL 248) | IV, 198– 199; n/a |
| 1790 (Apr–Jun) | Galle | CF Reimer, GE Schenk | Plan | VEL 1071 | IV, 200 |
| 1790 (Apr–Jun) | Galle | P Elias (copy after Reimer); D Matthijsz (copy after Reimer); D Matthijsz (copy after Reimer); P Elias (copy after Reimer) | Designs | VEL 1065; VEL 1066; VEL 1067; VEL 1068 | IV, 201; n/a; IV, 202; IV 202–203 |
| 1790 (Apr–Jun) | Designs to fortify 'Oenewatte' (=Unawatuna) | D Matthijsz (copy after Reimer); P Elias (copy after Reimer) | Designs | VEL 1069; VEL 1070 (identical copy of VEL 1069) | IV, 204; n/a |
| 1790 (Jun–Aug) | Trincomalee | PJ Tency (copy after Reimer) | Plan | VEL 1025 | IV, 352 |
| 1790 (Jun–Aug) | Trincomalee | GE Schenk (copy after Reimer); D Matthijsz (copy after Reimer); D Matthijsz (copy after Reimer); P Elias (copy after Reimer); PJ Tency (copy after Reimer; D Matthijsz, P Elias (copy after Reimer | Designs | VEL 1019; VEL 1020; VEL 1021; VEL 1022; VEL 1023; VEL 1024 | IV, 353; n/a; n/a; IV, 351; IV, 354–355 |

(continued)

(continued)

| Date | Location | Maker(s) | Type | Inv. nr(s) | Reproduced |
|------|----------|----------|------|------------|------------|
| 1790 (Jun–Aug) | Trincomalee (Fort Oostenburg) | DC Belcke | Plan | VEL 1044 | IV, 370–371 |
| 1790 (Jun–Aug) | Trincomalee (Fort Oostenburg) | GE Schenk (copy after Reimer); D Matthijsz (copy after Reimer); P Elias (copy after Reimer) | Designs | VEL 1040; VEL 1041 (identical copy of VEL 1040); VEL 1042 | IV, 372; n/a; IV, 372 |
| 1790 (Jun–Aug) | Design to cut the 'Kaalenberg' (=Ostenburg Ridge) | GE Schenk (copy after Reimer) | Design | VEL 1043 | IV, 373 |
| 1790 (Sep) | Madras | CF Reimer | Plan | NA 1.10.03, inv. nr. 76 | n/a |
| 1790 (Oct); 1791 (Nov–Dec) | Malacca | AC Twent | Chart | VELH 125 | III, 108 |
| 1790 (Oct); 1791 (Nov–Dec) | Malacca | CF Reimer | Plan | VEL 1112 | III, 108–109 |
| 1790 (Oct); 1791 (Nov–Dec) | Malacca | P Elias (copy after Reimer) | Design | VEL 1113 | III, 114–115 |
| 1791 (Feb) | Ambon | CFA Volbarth (copy after Reimer) | Chart | VEL 481 | III, 276–277 |
| 1791 (Feb) | Ambon | P Elias | Map | VEL 1326 | III, 275 |
| 1791 (Feb) | Ambon | CF Reimer | Map | VEL 1330 | III, 282–283 |
| 1791 (Feb) | Ambon (Fort Nieuw Victoria) | CF Reimer | Plans | VEL 1331; VEL 1336; VEL1337 | III, 284; III, 285; III, 285 |
| 1791 (Feb) | Ambon (lesser fortifications) | P Elias (copy after Reimer); CF Reimer | Designs | VEL 1335; VEL 1338 | III, 296; III, 296 |
| 1791 (Mar–Apr) | Banda Islands | CF Reimer, AA Buyskes; MJ de Man | Chart | VEL 484; VELH 247 (identical copy of VEL 484) | III, 332–333; n/a |
| | Banda Islands | CF Reimer | Map | VEL 1361 | III, 338 |
| 1791 (Mar–Apr) | Banda Islands (Fort Belgica) | CF Reimer | Plan | VEL 1362 | III, 339 |

(continued)

(continued)

| Date | Location | Maker(s) | Type | Inv. nr(s) | Reproduced |
|------|----------|----------|------|------------|------------|
| 1791 (May–Jun) | Ternate | CF Reimer | Map | VEL 478 | III, 252–253 |
| 1791 (May–Jun) | Ternate (Fort Oranje) | CF Reimer | Plan | VEL 1315 | III, 254 |
| 1791 (May–Jun) | Ternate (Fort Oranje) | CF Reimer | Designs | VEL 1316; VEL 1317 | III, 255; III, 254 |
| 1791 (May–Jun) | Ternate (new fort 'Kajoe Mejrah') | CF Reimer | Design | VEL 1318 | III, 256 |
| 1791 (Nov–Dec) | Makassar | CF Reimer, AA Buyskes; MJ de Man | Chart | VEL 460; VELH 544 (identical copy of VEL 460) | III, 174–175; n/a |
| 1791 (Nov–Dec) | Makassar | CF Reimer | Plan | VEL 1309 | III, 178 |
| 1791 (Nov–Dec) | Makassar | CF Reimer | Design | VEL 1308 | III, 179 |
| 1791 (Dec) | Riau | AA Buyskes, AC Twent | Charts | VEL 370; VELH 206 (identical copy of VEL 370) | III, 104 |
| 1791 (Dec) | Riau | CF Reimer | Map | VEL 1151 | III, 105 |
| 1791 (Dec) | Riau | CF Reimer | Design | VEL 1152 | III, 106–107 |

*Sources* Zandvliet (1987), Landheer (2006), Van Diessen (2006)–2010. *Updated* Bos (2018)

# References

Balk GL (ed) (2007) The archives of the Dutch East India Company (VOC) and the local institutions in Batavia (Jakarta). Brill, Leiden

Bos J (2018) Unknown knowledge: the travel diary of Carl Friedrich Reimer, 1789–1792. In: D'Angelo F (ed) The scientific dialogue linking America, Asia and Europe between the 12th and the 20th century. Associationi Culturale Viaggiatori, Naples, pp 82–100

Bruijn JR (2003) Facing a new world: The Dutch navy goes overseas (c. 1750–c. 1850). In: Moore B, van Nierop H (eds) Colonial empires compared: Britain and the Netherlands, 1750–1850. Ashgate, Burlington, pp 113–128

Butler Greenfield A (2005) A perfect red: empire, espionage, and the quest for the color of desire. Harper, New York

De Silva RK, Beumer WGM (1988) Illustrations and views of Dutch Ceylon 1602–1795. Serendib Publications, London

Dörr S (1988) De kundige kapitein: brieven en bescheiden betrekking hebbende op Jan Olphert Vaillant, kapitein-ter-zee (1751–1800). Walburg, Zutphen

Enthoven V (2002) Van steunpilaar tot blok aan het been: De Verenigde Oost-Indische Compagnie en de Unie. In: Knaap GJ, Teitler G (eds) De Verenigde Oost-Indische Compagnie tussen oorlog en diplomatie. KITLV, Leiden

Gommans JLL, Bos J, Kruijtzer GC (2010) Comprehensive atlas of the Dutch United East India Company, part VI: India, Persia, Arabian Peninsula. Asia Maior/Atlas Maior, Voorburg

Knaap G, den Heijer H, de Jong M (2015) Oorlogen overzee: Militair optreden door compagnie en staat, buiten Europa 1595–1814. Boom, Amsterdam

Landheer T (2006) Oranje of Napoleon?: de wisselvallige levensloop van Christiaan Antonij Ver Huell. Matrijs, Utrecht

Lewis D (1995) Jan Compagnie in the Straits of Malacca 1641–1795. Ohio University Press, Athens

Meilink-Roelofsz MAP, Raben R, Spijkerman H (eds) (1992) The archives of the Dutch East India, Company, 1602–1795. SDU, 's-Gravenhage

Odegard E (2017) Vergeefse voorstellen: de projecten tot verbetering van de fortificaties van Colombo, Galle en Trincomalee 1785–1790. In: Ampt K, Littel A, Paar E (eds) Verre forten, vreemde kusten. Nederlandse verdedigingswerken overzee. Sidestone Press, Leiden, pp 117–136

Tarling N (1962) Anglo-Dutch rivalry in the Malay World 1780–1824. University of Queensland Press, St Lucia

Tates S (2018) Patronagenetwerken in de VOC: De carrières van de broers Van de Graaff. (Unpublished, MA thesis. University of Leiden)

van Diessen JR (ed) (2006–2010) Comprehensive atlas of the Dutch United East India Company. Asia Maior/Atlas Maior, Voorburg

van Gerven MR (2002) C.F. Reimer, een werkzaam mensch: De Militaire Commissie naar Azië 1789–1793 (Unpublished, Leiden University MA thesis)

van Nimwegen O (2017) De Nederlandse burgeroorlog 1748–1815. Prometheus, Amsterdam

van Putten LP (2002) Ambitie en onvermogen: Gouverneurs-generaal van Nederlands-Indië 1610–1796. Uitgeverij ILCO, Rotterdam

Zandvliet K (1987) Overzeese militaire kartografie na het echec van 1780–1782: De Oost. In: Caert-Thresoor (J Hist Cartogr Neth) 6(1):1–10

Zandvliet K (2002a) De Nederlandse ontmoeting met Azië, 1600–1950. Rijksmuseum, Amsterdam

**Jeroen Bos** is a member of library staff at Leiden University Libraries and independent historical researcher. His specialty involves the early modern colonial history, especially of the so-called Dutch East Indies. He is co-author of volume VI in the *Comprehensive Atlas of the Dutch United East India Company* series and published on several topics concerning the history of the long-distance trading companies.

# Head-Hunters, Cannibals and Pirates: Surveying in the 1960s

Roy Wood

**Abstract** Although many maps and charts can now be constructed accurately on geocoded pixels, maps in the colonial period, especially at topographic scales, needed a framework of control on the ground. This chapter will build on personal experience in the immediate post-colonial period with the Directorate of Overseas Surveys in Sierra Leone and Sabah and surveying on active service in Sarawak during the 'Confrontation' war with Indonesia. The paper will outline the technical methods available at the time to control the air photography used to produce plots for the cartographers. These methods depended on theodolites and an increasing use of early electronic distance measurement systems. Reducing angle and distance measurements to coordinates required seven-figure log and trigonometrical tables and hand cranked calculating machines. The chapter will also describe the logistic challenges involved in surveys in remote areas which were usually far more demanding than the observation and computing. Access to mountain top trig points varied from strings of porters with head loads in Sierra Leone to winching into the tree tops from helicopters in Sarawak. Experiences including encounters with head hunters, pirates and cannibals. The maps produced from this work are still the basis for national series, but the adventures have been lost to GPS (Global Positioning Systems).

## 1 Introduction

The Oxford symposium focused on *Cartographies in the colonial era* but worthwhile cartography depends on some kind of positional framework. This chapter describes personal experiences in the field providing this framework in the immediate post-colonial era and, although I am not quite old enough to have been around in the real colonial days, little had changed by the time I was working in the 1960s. I will cover work as a military surveyor in Sarawak between 1964 and 1966 during the 'Confrontation' war between Malaysia and Indonesia and then on

R. Wood (✉)
Independent Researcher, Newbury, UK
e-mail: roy.wood58@btinternet.com

© Springer Nature Switzerland AG 2020

A. J. Kent et al. (eds.), *Mapping Empires: Colonial Cartographies of Land and Sea*,
Lecture Notes in Geoinformation and Cartography,
https://doi.org/10.1007/978-3-030-23447-8_17

secondment to the Directorate of Overseas Surveys (DOS) in Sierra Leone and
Sabah between 1967 and 1970. This involved some interesting challenges and, as
my title indicates, involved some interesting people.

In all three cases the main task was to complete the control for national 1:50,000
scale series mapping which had been started under the colonial administrations.
This required building trig points on hilltops and then determining their position
and height. These were then used as a framework to fix the stereoscopic aerial
photographs from which the detail and contours of the maps could be drawn. These
days photographs are still used in much the same way, but GPS means that traditional surveying is not usually needed. However, back in 1964, we relied on two
main tools; the trusty theodolite to measure horizontal and vertical angles and what
was at that time the rather new-fangled Tellurometer (Fig. 1). Provided you had a
charged car battery on your hilltop, this radar-like distance measurer could work
over distances of up to about 50 miles (80 km) to the accuracy needed. This
enabled us to escape the constraints of pure triangulation and, within technical
limits, to use traversing to take position forward by bearing and distance. However,
our points had to be intervisible. To fix heights we used vertical angles measured
with the theodolites or, in certain cases, accurate altimeters.

**Fig. 1** Using a Tellurometer (© Roy Wood 2020. All Rights Reserved)

## 2  Sarawak

My first experiences were as a military surveyor in Sarawak. This part of the north coast of Borneo, once ruled by the Brooke family—the White Rajahs of Borneo—became a British colony in 1946 and then part of the newly independent Malaysia in 1963. However, President Sukarno of Indonesia objected to the new country and started a policy of Confrontation with armed incursions across the common border. This grew into a war which lasted until 1966. At Malaysia's request, the United Kingdom provided support and at one point around 50,000 allied troops were deployed.

I was posted to the Survey Squadron in Singapore in 1964 and from there to command the field survey work in Sarawak. The unit had been helping the civil mapping programme in Sarawak for several years funded by the local government and good work had been done, particularly in astronomical fixes in remote areas. However, the trig network did not reach much beyond the first range of mountains in from the coast and, from there on, the existing maps were uncontoured and often showed only an indication of the rivers. Hostilities demanded reliable maps and moved the focus towards the disputed border so my task, with my troop of about a dozen survey non-commissioned officers (NCOs), was to extend the trig network south and east up to the border. After an engineering degree and a basic Royal Engineer course but, as yet, no specialist survey training, it was my first job.

In a situation, unthinkable today, no one put me under command of the local Brigade Headquarters and, as I had no access to phones or signals, I reported to my Squadron Commander 500 miles (800 km) away by letter. I continued to have a Malaysian government chequebook to employ a local workforce or to buy any goods or services I couldn't obtain from Army sources, so it really was an independent command. Soon after I arrived, I started to discover errors in what was shown on the poor existing maps which were significant for infantry and air operations. These included 100 square miles (160 km$^2$) shown on the wrong side of the border, a 5-mile (8 km) discontinuity along the border and a mountain, frequently obscured by cloud near a helicopter route, which was shown nearly 2000 ft (610 m) too low. The Brigade staff realized that urgent action was needed and that meant giving me support.

So, in that jungle covered and mostly uninhabited country the size of England, I had access to light aircraft to find intervisible mountain tops and helicopters to get my surveyors in and out (Fig. 2). Actually, the initial getting in was quite hairy as the technique we developed required the first man to be winched into the treetops from a hovering helicopter. He would stand in the winch strop with jungle knife at the ready and cut his way to the ground. At the 4000–6000 ft (1220–1830 m) height of the mountains we occupied, the summit trees were not usually more than 30 ft (9 m) tall but this was still a challenging operation. First man down would start to open a hole in the trees so that the rest of the party, usually two technician NCOs and half a dozen locally recruited Ibans, could winch in. They cleared the mountaintops and built trig points and helicopter pads on the summits for subsequent access. This required some courage from all concerned but particularly it

**Fig. 2** Trig reconnaissance by light aircraft (© Roy Wood 2020. All Rights Reserved)

needed the exceptional skills of the Fleet Air Arm pilots often operating near the limits for their aircraft around frequently clouded mountain tops. It was a good job that risk assessments had not been invented.

Figure 3 shows progress part way through the work. Later on we extended further south towards the border and then into the big empty space further north and east. We normally deployed four field parties at any time leapfrogging them on from mountain to mountain but, because of the need to minimize the use of helicopters, the observations had to be proved before each move. To do this the angles and distances were radioed back to our forward base where we did the rather lengthy computations, which had to take account of the curvature of the earth amongst other things. There were, of course, no computers so we relied on fat books of eight-figure trig and log tables and hand cranked calculating machines.

As work progressed towards the border, we moved in front of the forward infantry positions. The Indonesians were not likely to bother us on our mountain tops but Brigade Headquarters insisted that we had an escort. However, they did not have enough infantry to do the job and, for a while, it seemed that our work would have to stop. A reminder of their need for maps eventually led to acceptance of my suggestion that providing cartridges for my Ibans' shotguns would provide more effective defence in the jungle than a section of British soldiers. There were comments about my private army but Fig. 4 shows No. 1 Section on duty. The Ibans did of course have to train for this role but the wild boar, deer, and so on, which resulted from their live firing exercises provided a welcome addition to Army rations.

**Fig. 3** Trig diagram at the half way point (© Roy Wood 2020. All Rights Reserved)

**Fig. 4** Iban defence force (© Roy Wood 2020. All Rights Reserved)

Another challenge was to fix a large number of height points needed for the photogrammetric stage. We needed flattish areas we could identify on the air photographs and our solution depended on helicopters and precision altimeters.

Sandbanks on rivers were the only suitable features but it would have been too time consuming to land at each one. So, with the cooperation of Army Air Corps pilots we found that, if we hovered above 40 ft (12 m), the downdraft from the rotors did not affect altimeter readings in the cabin. We then made the simple calibrated plumb line shown in Fig. 5 which we mounted in the rear cabin. Taking readings was then a process of getting into hover above 40 ft (12 m), dropping the plumb line to touch the sand, reading the length of line and reading the altimeters. Meanwhile a surveyor in a high flying second helicopter identified the point on the air photographs and other surveyors read base altimeters on previously established reference points to allow for atmospheric changes. It took a bit of coordination but was very effective.

I spent nearly two years in Sarawak and one of the joys was encounters with the people of the interior. On a couple of occasions we were fortunate to meet small groups of the very elusive Punans who lived a nomadic existence in the really remote areas mainly on a diet of wild sago and whatever they could kill with their blowpipes. However, our main links were with the Ibans. We recruited a team of about twenty, who were paid from the Malaysian government fund. They came from an area a few miles above our forward base and we got to know them and their

**Fig. 5** Heighting plumb line on Scout helicopter (© Roy Wood 2020. All Rights Reserved)

**Fig. 6** Iban longhouse (© Roy Wood 2020. All Rights Reserved)

families well. Most of my surveyors spoke nearly fluent Iban and we spent many happy evenings in their longhouses (Fig. 6). However, from the baskets of skulls hanging in the rafters we had a clear reminder that these were the original Headhunters of Borneo. Taking a head was marked by a tattoo on a finger joint—one for each head. We saw a number of men with the tell-tale marks but were always told that those were Japanese taken during the war. But many of the men looked far too young for that. I was just rather glad that they were on our side.

## 3   Sierra Leone

My second experience was in Sierra Leone in West Africa when I was seconded from the Army to the DOS, part of the Overseas Development Administration which had field parties of two or three surveyors in many of the ex-British colonies. Tasks were set in the UK, but execution was left to the judgement of the field party leader who could use his local bank account to engage whatever local staff, vehicles, boats or stores he needed. Once again, the only contact with headquarters was by letter. This meant that I could continue the rather independent approach I had developed in Sarawak but, this time, independence was official policy. Tours were normally one year in a country with a work ethic allowing three or four days each month in some town to send off the observations, reports and accounts, to restock, maintain vehicles, and so on, but otherwise the surveys were driven on seven days a week. Two months leave between tours in the field helped keep the balance. Married surveyors were encouraged to take their wives but conditions were spartan. My wife, nine-month-old son and I lived under canvas and moved as and where the work took us.

We were there in 1968, seven years after independence. The country, which is a bit over half the area of England, had started well but tribal disputes and a refusal to accept an election result led to a military government headed by a brigadier who had been a year ahead of me at Sandhurst. I was then a captain and initially thought perhaps I was in the wrong army. However, a couple of months after we arrived, he was displaced in a mutiny which consumed a great deal of ammunition but thankfully resulted in very few casualties.

Mostly the work was establishing trig points for the 1:50,000 scale mapping programme which, as in Sarawak, gave straightforward surveying but logistical challenges. In contrast to the mostly uninhabited and trackless jungles of Sarawak, Sierra Leone had a well distributed population with villages linked by dirt roads. There were, needless to say, no helicopters, but we could usually get our Land Rovers within striking distance of our hills. Then it was onto our feet. Figure 7 shows a typical scene with a string of porters—one with that Tellurometer battery

**Fig. 7**  Porters heading for a trig (© Roy Wood 2020. All Rights Reserved)

on his head—setting off for the hill in the background. The frontispiece of the 1913 edition of the *Text Book of Topographical and Geographical Surveying* was a very similar photograph so I felt this was very much in the spirit of colonial times.

However, two other tasks stand out. One was to verify part of the Sierra Leone/ Liberia border so that it could be shown accurately on the new maps. The line through the bush had been agreed by a boundary commission in 1903 and I invited the local chiefs from both countries to meet me at our start point. Yes, they agreed that the border went up various rivers and tributaries and, as we walked the line, it all matched my crib from the 1903 report. This was fine until we reached the source of the last stream. 'So, what happens now?' I asked. 'Ah' said the Liberian, 'my grandfather told me that men came and made a pile of stones'. Sierra Leone agreed. We cleared back the undergrowth and there were mossy stones. The crib said, 'and here we built a cairn'. The chiefs knew exactly where the border was but, in that remote area, with families spread on both sides, they paid no attention to it at all.

The second task was to provide the framework for a new map of one of the towns. It was a flat area so, with almost continuous bush about 10 m high, intervisibility was a problem and I had to find a way to get above the trees. I didn't have access to prefabricated towers but noticed that palm trees, which were taller than the canopy, were dotted around the area. A local tree climber fixed a red flag on a palm where I wanted a point then we repeated the process around the town checking that previous flags could be seen from each new one. This worked well but the recent military mutiny had made the local authorities very jumpy and, as I gathered afterwards, there were rumours of communist mercenaries coming over the border from nearby Guinea. Someone thought my red flags were a signal and I found myself arrested. It was sorted fairly quickly but those hours in a police cell were an experience. Then it was back to the towers. As illustrated in Fig. 8, bush timber made the outside frame for the surveyor and 15 m of water pipe gave an independent support for the instruments. As ever, the observations were the easy bit.

Moving around the country we learned to avoid the extensive areas declared to be sacred by tribal secret societies. However, one encounter was quite scary. We had set up camp by a village near the coast and I was surprised to find that my African team put up tents very close to ours instead of finding beds in the village. I could get no explanation from them and they insisted on moving on early the next morning. A possible answer came from some Catholic missionaries we met a couple of weeks later. 'Those people mostly live on fish' they said 'but we hear that from time to time they get a meat hunger. Then in the evening when the women and children go down to the river to wash, a crocodile will come in and one them will be taken'. But it was just a crocodile skin with a man swimming underneath it and, although this was all hearsay, it seemed that perhaps cannibalism was still practiced. No wonder my team wanted to get away. In another area we heard that a number of people had disappeared in the run up to an election. It was not a subject to follow too closely but we did hear it said that eating the right bits of opposition supporters would help your candidate.

**Fig. 8** Timber tower (© Roy Wood 2020. All Rights Reserved)

## 4 Sabah

And finally, in 1969 it was back to Malaysia and Sabah at the north-east tip of
Borneo for my second DOS tour. About two-thirds the area of England, the state,
like Sarawak, was almost wholly covered by jungle. Apart from one route being
built while I was there, there were no roads beyond the immediate limits of the few
towns around the coast and most of the interior was uninhabited. In contrast to
Sierra Leone, getting anywhere depended on very limited air routes on the Borneo
Airways Dakota, by sea around the coast and then up the few navigable rivers or by
walking. I was fortunate in having the use of a 50-foot (15 m) Malaysian

**Fig. 9** Clearing sight lines (© Roy Wood 2020. All Rights Reserved)

government boat with a Malay skipper, a Chinese engineer and two local deck-hands. The boat had space for my team, a dozen reliable Ibans recruited in Sarawak and our kit. We also used it to tow a couple of hollowed out tree trunk longboats to work up rivers or inshore.

As in my previous tours, the main task was completing the framework for 1:50,000 mapping. I started in the interior where the terrain was similar to Sarawak but, with no access to aircraft, finding suitable trigs was more difficult and, as the mountains were lower, the trees were taller and clearing sight lines was harder work (Fig. 9).

The next task was a traverse around the easternmost peninsula of Sabah working from headland to headland along the low-lying coast. This simplified the logistics but required twenty-seven rather short legs to get between existing trigs to give us the vital opening and closing lines. We lived on the launch and used the longboats (Fig. 10) to weave through the coral reefs to our new trigs as we progressed along the coast. For this job timing was important. We started on the southern side of the peninsula towards the end of the north-east monsoon and I reckoned we would reach the turn round to the northern side as the winds changed round to the south-west. The plan worked and, apart from one very uncomfortable week out on the end with winds blowing in all directions, we stayed in the lee of the land and avoided the full impact of the weather.

**Fig. 10** Heading ashore in the longboat (© Roy Wood 2020. All Rights Reserved)

Apart from having to wait until the tide went out to complete one particularly low line, observing angles and measuring distances was straightforward. However, I was concerned that those twenty-seven legs gave an opportunity for errors to build up so, at four trigs spread along the traverse, I observed astronomical azimuths to provide an independent check on the angles. And, when we did the sums, there was a misclosure which was just outside our tolerances. However, comparing the traverse directions against the astro showed the error to be directly correlated to distance. We realized that the temperature difference between the jungle on the left and the sea on the right of each line caused the light to bend—a very unusual case of lateral refraction.

One final story about Sabah and that traverse. The captain of a Malaysian navy patrol boat had agreed to meet us on the end of the peninsula and bring me a case of beer. On the due day there was the ship but, as I took my longboat out to him, I realized that he was closed down and his gun was pointing at me. After a tannoy order to stop and raise my hands I eventually found that my friendly skipper had been moved elsewhere and that the new man knew nothing about me or the beer. More importantly, there had been an incident a few miles up the coast involving a smuggling ring which was said to involve Hong Kong, the Philippines, Sabah and Singapore. We had just avoided a delivery by Filipino pirates which had turned nasty and resulted in the deaths of several local people.

## 5  In Summary

Almost by definition, surveying in those newly independent countries meant working in the less developed and more difficult to reach areas that had not been tackled before. Measuring angles and distances and the resulting computations were much the same wherever we were, but the great challenge at that time was logistics. The real task was getting into the areas, finding suitable mountains for trigs in often demanding and uninhabited terrain, reaching the summits, and then sustaining the field parties, all of which required considerable initiative. Getting surveyors from helicopters onto jungle covered mountain tops, using helicopters for altimetry and designing and building bush towers were just some of the novel methods we developed as the situations demanded. However, it was extremely satisfying to have the independence to tackle jobs, often over periods of months, and to deliver the results with no in-country supervision. Working with such diverse people in remote areas was another of the pleasures of those days. There were some difficulties but the dedication and jungle craft of the Ibans, the shy simplicity of nomadic Punans, and the surprising sense of fun shown by my Sierra Leonean team, who would race me down from a summit at the end of a hard day on a hill, were all very special; the headhunters, the cannibals and the pirates along the way were a bonus. GPS now removes the need for trigs on intervisible mountaintops and also removes much of the adventure of those just post-colonial days. However, our work of fifty years ago is the basis of current maps of those areas and I feel very privileged to have been part of it.

**Major General Roy Wood M.A. M.Sc. FRICS FRGS** was a field surveyor in Sarawak, Sabah and Sierra Leone in the 1960s and, after military mapping and charting appointments in the UK, Germany and the USA, retired in 1993 as Director General of Military Survey. Previously an Honorary Secretary and Council member of the Royal Geographical Society, Chairman of the Association for Geographic Information, President of the Photogrammetric Society and Trustee of the Mount Everest Foundation, he is currently a Trustee (and past Chairman) of the disaster relief charity MapAction and President of the Defence Surveyors Association.

Printed by Printforce, the Netherlands